MINERAL GEOLOGY

MINERAL GEOLOGY

Editor

Claude Corda

DPH

DISCOVERY PUBLISHING HOUSE
INDIA

Published by:

DISCOVERY PUBLISHING HOUSE
4383/4B, Ansari Road, Darya Ganj
New Delhi-110 002 (India)
Phone : +91-11-23279245; 23253475; 43596065
Mobile : +91 9811179893 / +91 9871656464
E-mail : discoverybooksindia@gmail.com
orderdphbooks@gmail.com
namitwasan9@gmail.com
web : www.discoverypublishinggroup.com

Mineral Geology
Edited by: **Claude Corda**

International Standard Book Number: **978-81-19523-30-6** (Hardback)

PREFACE

Mineral Geology is a comprehensive book that deals with the minerals and their geological context. Written by a diverse array of esteemed authors and field experts from around the globe, this book offers a wealth of research, reviews, and technical papers, making it a valuable resource for professionals and enthusiasts alike.

At its core, mineral geology explores the formation, properties, distribution, and utilization of minerals within the Earth's crust. The book provides an in-depth overview of the fundamental principles governing mineral formation, including the processes of crystallization, metamorphism, and mineralization. It also delves into the various factors that influence the occurrence and abundance of minerals, such as geological settings, mineral associations, and tectonic processes.

As our understanding of Earth's geology continues to evolve, so does the study of minerals. The book covers cutting-edge research and emerging concepts, shedding light on new techniques for mineral identification, characterization, and analysis. It also explores the dynamic interplay between minerals and other scientific disciplines, such as geochemistry, petrology, and mineral physics, fostering a multidisciplinary approach to mineral geology.

Moreover, the book addresses the growing importance of minerals in various industries and their impact on society. As the demand for natural resources escalates, understanding the geological processes governing the formation and extraction of minerals becomes crucial. It provides insights into the economic significance of minerals, their role in the development of technologies, and their potential as sustainable resources. The book also delves into environmental considerations, exploring the environmental impact of mineral extraction and the development of environmentally friendly mining practices.

The book will serve as valuable guide for professionals, researchers, and students in the fields of geology, earth sciences, and mining engineering. It serves as an invaluable reference for geologists and mineralogists, providing them with a comprehensive source of information on mineral formation, classification, and occurrence. The book also appeals to researchers and academics seeking to stay abreast of the latest advancements in mineral geology, as it includes a collection of technical papers highlighting ongoing research and novel methodologies.

Furthermore, students pursuing degrees in geology or related disciplines will find this book to be an indispensable resource. It offers a comprehensive introduction to the study of minerals and their geological context, providing a solid foundation for further exploration in the field. The book's accessible language and extensive illustrations make complex concepts more understandable, making it suitable for both undergraduate and graduate students.

The editor of the book has gathered information from trustworthy and scientifically backed sources to give readers the most up-to-date and accurate information possible. The book is a collection of articles written by experts from all over the world, and the editor is grateful for their valuable input. The knowledge and expertise shared by these experts are insightful and will be beneficial for the readers.

Editor

CONTENTS

1

AN EXPLORATION INTO THE RELATIONSHIP BETWEEN MINERAL ELEMENTS (NITROGEN AND PHOSPHORUS) AND NUTRITIONAL QUALITY IN SOIL-WATERMELON (CITRULLUS LANATUS) SYSTEM

Shan He[1], Yuyan Zhao[1,3], Chuanpeng Liu[2], Zhenghe Li[2], Zeyu Zhang[2], Bing Li[1,3] and Xiaodan Tang[1,3]

[1]College of Geo-Exploration Science and Technology, Jilin University, Changchun, 130026, China

[2]The 7th Institute of Geology and Mineral Exploration of Shandong Province, Linyi, 276006, China

[3]Institute for Black Soils, Jilin University, Changchun, 130026, China

ABSTRACT

To fundamentally explore the correlation between crop nutritional quality and mineral elements, specific tests and accurate analysis were carried out on all organs and root soil of watermelon (Citrullus lanatus). The results showed that the distribution patterns of nitrogen and phosphorus at watermelon maturity were similar, and the average nitrogen and phosphorus contents were in the orders of leaf and seed > stem, peel, root > pulp > root soil, and peel > seed > root, stem, leaf > root soil > pulp, respectively. From the perspective of element geochemistry, biophile and lithophile elements had the strongest correlation with nitrogen and phosphorus, and watermelon did not antagonize soil nitrogen and phosphorus uptake. The prediction model of nitrogen translocation factor in watermelon organs to total acid was established by partial least squares with $R^2 = 0.81$. Significantly, when the isometric log-ratio of nitrogen to phosphorus in watermelon leaves was 1.97 to 2.19, the watermelon pulp showed better quality with total acid > 0.5%, total sugar > 5% and soluble solids > 10%. Therefore, the characterization of nitrogen and phosphorus in watermelon leaves can serve as a non-destructive analysis to predict watermelon fruit quality.

Keywords: *Nitrogen, Phosphorus, Soil, Watermelon, Migration and accumulation, Geochemical classification of elements, Fruit quality*

INTRODUCTION

In recent decades, nitrogen (N) emissions have increased sharply with the continuous intensification of human activities such as fossil fuel combustion, chemical fertilizer application and automobile exhaust emission, which has made more and more active N compounds accumulate in the atmosphere and sink into terrestrial and aquatic ecosystems (Bai et al. 2010). Human activity has also altered the global phosphorus (P) cycle. This anthropogenic eutrophication not only increases the primary productivity of ecosystems, but also alters the relative importance of nutrients in limiting productivity (Fourqurean and Zieman 2002). The research on mineral nutrition has become the main theme in ecology, agronomy and genetics (Tsige et al. 2022). N and P are the two most needed mineral nutrients in terrestrial ecosystem, which are closely related to the growth and physiological metabolism of plants (Nie et al. 2018; Dong et al. 2020). N is a component element of protein and nucleic acid in plant cells and participates in the synthesis of chlorophyll in chloroplast (Grzyb et al. 2021). P is an important component of nucleic acids and enzymes in plant cells and plays a key role in cell division, proliferation, genetic variation and other life activities (Richardson et al. 2009).

Plant growth and distribution patterns are determined by both external and internal resources. Water and nutrients in the soil and sunlight constitute external resources, while the preferential distribution, migration and transformation of elements in different organs of plants constitute internal resources. In the process of exploring external resources, there is no doubt that soil and its components have become the key factors affecting the life process of plants. The bioconcentration factor (BCF) can be applied to reflect the ability of plants to absorb elements from the soil. Most modern soils were formed in the Quaternary period, and the Quaternary geological processes also controlled the distribution of elements in the earth's surface system (Pan 2000). Soil-forming parent material is one of the five natural factors (parent material, climate, biology, topography and time) for soil formation. Its inherent mineralogical properties and physicochemical properties have great influence on soil characteristics and fertility. As early as the mid-nineteenth century, the concept of agricultural geology was put forward to explain the relationship between rock weathering and soil formation. In natural soil, soil fertility comes from natural fertility, which depends on the chemical composition of rocks and minerals to some extent, and is released gradually through weathering. Therefore, the content of mineral elements in soil affects the abundance and deficiency of elements in the parent material-cultivated soil–plant system and improves or limits the quality of crops (Ludvik and Schernthaner 2012). In order to improve the quality and yield of crops, scholars have done a lot of research on N and P fertilizer, but rarely explored the source of N and P in natural soil.

Previous studies have mostly focused on the source, migration and transformation of heavy metals rather than mineral nutrients in rock–soil–plant system (Rastegari Mehr et al. 2020; Zhao et al. 2021; Sun et al. 2022). A comprehensive understanding of mineral nutrients is helpful for agronomy, botany and ecology to set up plant varieties and manage fertilization measures according to local conditions. The migration and transformation of mineral nutrients in soil–

plant system are also an indispensable part of the internal resources of plant life activities.

With the gradual improvement of agricultural cultivation techniques and daily living standards, people's requirements for fruit quality are also constantly improving. Therefore, fruits are studied not only in quantity, but also in quality. Fruit nutritional quality is one of the most concerned problems in life, and its detection and evaluation have become a meaningful research topic. Meanwhile, watermelon, as a cash crop, is more and more favored by people. Because of its social importance, a variety of research on its fruit quality have been carried out (Ali et al. 2017). Shuanghou watermelon, well-known in China, has sweet flesh with little seeds and thin peel. It is a typical representative of watermelon varieties. During the watermelon quality analysis, the translocation factor (TF), one of the commonly used evaluation indexes, represents the mineral element enrichment ability of aboveground organs and reflects their internal vitality (Colle et al. 2009). The isometric log-ratio (ILR) method based on ion balance can not only explain nutrient concentration and balance, but also well correct the concentration deviation caused by plant physiological interaction (Geikloue et al. 2019).

To deeply investigate the relationship between nutritional quality and mineral elements in soil-watermelon system, this article analyzed the N and P concentrations in different organs and root soil of watermelon, explored the migration and enrichment modes of N and P during watermelon maturity, and elaborated the influence of rock-forming elements on N and P contents from the perspective of geochemistry.

MATERIALS AND METHODS

Plant and Soil Sample Preparation

Plant and soil samples were collected from plantation of watermelon, located in Yinan, Shandong, China (35°28′ N, 118°14′ E). The watermelon variety was "Shuanghou Watermelon" also known as "Sesame Grain Tight-skinned Watermelon". The area of Shuanghou watermelon plantation was about 10 km^2, and about one watermelon was planted per m^2 at creeping cultivation. In 2021, seedling was planted in mid to late April, and fruit was harvested in late June. Based on the principles of representativeness, typicality and timeliness, typical samples were collected from representative fields. A total of 30 watermelon samples and 30 soil samples were collected, and these samples were collected by five-point sampling method from 30 representative fields almost evenly distributed in Shuanghou watermelon plantation. Marginal effects and special individuals are avoided in sampling. Plant samples were collected from whole plants at maturity, and soil samples were collected from the top layer of soil 0–20 cm near the roots of the plants. The plant was fresh and intact, free from pests, diseases and mechanical damage. The soil was all sandy loam.

The watermelon samples were washed, and their organs were separated. The organs (except pulps) were dried at 105 °C for 30 min and then at 60 °C for 24 h, cooled to room temperature (25 °C), crushed with plant crushing machine (200 T, Huangdai, China), filtered through a 74-μm sieve, and sealed for use. Watermelon pulps were pressed into juice after removing seeds. The root–soil samples were air dried, ground with an agate mortar, screened through a 74-μm sieve, and sealed for use.

Main Reagent and Equipment

Sulfuric acid (H_2SO_4), nitric acid (HNO_3), perchloric acid ($HClO_4$), hydrochloric acid (HCl) and hydrofluoric acid (HF) were in guaranteed reagent grade. Copper sulfate, sodium hydroxide (NaOH), boric acid, methyl red, bromocresol green, 95% ethanol, hydrogen peroxide (H_2O_2), phenolphthalein and phenol were of analytical grade. They were all purchased from Tianjin Kermel Chemical Reagent Co., Ltd., China. The water needed for the experiment is ultrapure deionized water with a resistivity of 18.2 MΩ·cm at 25 °C. The national standard solution of P (GSB04-1741-2004) was 1000 μg/mL and obtained from National Center of Analysis and Testing for Nonferrous Metals and Electronic Materials. Its required concentration gradient was obtained by stepwise dilution. The certified reference materials (CRMs) (GBW07403 and GBW10012) were acquired from the Institute of Geophysical and Geochemical Exploration, Chinese Academy of Geological Sciences (Langfang, China) and applied to verify the accuracy and precision of the experimental methods.

Determination of Watermelon Pulp pH

The pH of watermelon juice was measured by a calibrated pH meter (PHBJ-260, Leici, China).

Determination of N Content in Plant Organs and Soil

1 g of sample, 5 g of copper sulfate and 10 mL of H_2SO_4 were accurately added into the digestion tube, digested in a graphite furnace (KDNX-20, Leici, China) at 480 °C for 1 h, cooled to room temperature and then distilled with 400 g/L NaOH in Kjeldahl analyzer (KDN-08C, lichen, China). The distilled liquid was absorbed with 20.00 mL 20 g/L boric acid and 3 drops of mixed indicator (0.1 g methyl red and 0.5 g bromocresol green dissolved in 100 mL 95% ethanol) and titrated with 0.1 mol/L HCl to obtain the N content.

Determination of P Content in Plant Organs and Soil

0.5 g of plant (except pulp) and soil samples, a few drops of ultrapure deionized water, 5 mL H_2SO_4 and 2 mL H_2O_2 were added into digestion tubes in sequence, digested at 340 °C for 1 h, cooled, added with 10 drops of H_2O_2, simmered for 5 min, and repeated until the solution was colorless or clear.

0.5 g of pulp samples, 10 mL HNO_3, 1 mL $HClO_4$ and 2 mL H_2SO_4 were accurately added into conical flasks and digested on the electric heating plate (SB-1.8–4, Shanghai Shiyan, China) at medium temperature until the digestion solution was colorless or yellowish. After cooling, the digestion solution was added with 20 mL ultrapure deionized water and driven out acid at high temperature until its volume was about 10 mL. The resulting solutions were used for the P content test in flow injection analyzer (FIA-6000+, Jitian, China).

Determination of Total Acid in Watermelon Pulp

25 g of watermelon pulp was accurately weighed into the conical flask, added with 40 mL ultrapure deionized water and 0.2 mL 1% phenolphthalein indicator, and titrated with 0.1 mol/L NaOH to obtain the total acid content.

Determination of Total Sugar in Watermelon Pulp

1 g of watermelon juice, 25 mL water and 10 mL HCl were added into a conical flask, shaken well, hydrolyzed in boiling water bath for 1 h, cooled to room temperature and filtered through a 0.45 μm membrane. The resulting solution was added with 1 mL phenol solution and 5 mL H_2SO_4, left for 10 min, shaken well, and cooled to determine the total sugar content by ultraviolet–visible spectrometer (T6, Persee, China).

Determination of Soluble Solid in Watermelon Pulp

The soluble solid of watermelon juice was measured by a calibrated Abbe refractometer (SW-32D, Suwei, China).

Determination of Major and Trace Element Content in Soil

4 g of soil sample was made into round cake under the pressure of 30 MPa for the determination of Al/Ca/Fe/K/Mg/Na/Si/Sr/Ti content by X-ray fluorescence spectrometer (EDX6000B, Skyray, China).

0.25 g of soil sample, 10 mL aqua regia (V_{HCl}: V_{HNO3} = 3: 1) and 2.5 mL HF were accurately added into the Teflon digestion tube, shaken well, heated to 125 °C for 2 h, cooled, fixed volume to 50 mL with ultrapure deionized water, and left overnight for the B content test by inductively coupled plasma mass spectrometry (NexION350D, PerkinElmer, America).

0.2 g of soil sample and 2 g of NaOH were accurately weighed into a nickel crucible, covered, heated to 570 °C for 20 min, cooled, dissolved with boiling hot water, transferred into a 100 mL volumetric flask, slowly added 5 mL 50% HCl, diluted with ultrapure deionized water to the mark, shaken well, and left for the F content test by the ion meter (PXSJ-216F, Leici, China).

Bioconcentration Factor (BCF)

BCF (Kelsey et al. 2006) refers to the ratio of element content in plants to the corresponding content in growing soil, and it is one of the specific manifestations of the element enrichment ability in plants. The formula is as follows:

$$\mathrm{BCF} = C_R/C_S \tag{1}$$

where C_R and C_S are the contents of an element in root and root soil, respectively.

Translocation Factor (TF)

TF (Kumar et al. 2018) refers to the ability of the above-ground parts of plants to accumulate elements, which can be served for evaluating the capacity of plant organs such as stems and leaves to absorb and transfer elements through roots. The formula is as follows:

$$\mathrm{TF} = C_{AP}/C_R \tag{2}$$

where C_{AP} and C_R are separately the contents of an element in the above-ground part (stem, leaf, peel, seed or pulp) and root of the plants.

Isometric Log-Ratio (ILR)

ILR (Parent 2011), an analysis method based on ionic equilibrium, fully considers the interaction between mineral ions since the interactions among plant ions are projected into a Euclidean space (Parent et al. 2013). It is accepted to be the most suitable description way for plant ionomics data at present. The ILR was computed as follows:

$$ILR_{[N|P]} = \left(\frac{1}{2}\right)^{\frac{1}{2}} \mathrm{Ln}\frac{g(C_N)}{g(C_P)} \qquad (3)$$

where $g(C_N)$ and $g(C_P)$, respectively, represent the content of N and P.

Statistical Analysis

The experimental data were analyzed and plotted by analysis of variance and regression using Excel and Origin. The clustering and correlation analysis and the partial least squares regression prediction were achieved by MATLAB. The coefficient of variation was mostly less than 1%. If the variation of element content exceeded 5%, an identical run would be undertaken, and closer data points were used. All mean values are the averages of three independent samples. It should be noted that both N and P contents obtained from the tests in this article were total N and total P contents, respectively. Blank and parallel controls were set in experiment.

RESULTS AND DISCUSSION

Validation of Analytical Methods

According to the requirements of the geological and mineral industry standard DZ/T0011-2015 of the People's Republic of China, the CRMs GBW07403 and GBW10012 were analyzed five times, and the logarithmic deviation (ΔlgC) and relative standard deviation (RSD) values between the measured mean value and the standard value of each element were computed and the results. The ΔlgC and RSD values of the two CRMs were all smaller than the standard monitoring limits (ΔlgC and RSD $\leq$7%), and the detection limits (DLs), estimated as three times standard deviation of the blank solution, were, respectively, 0.067 and 0.006 mg/g for N and P. These indicated that the accuracy, precision and detection abilities of above experimental methods could satisfy the requirement of analysis test.

Accumulation of N and P in Crop Organs

The N and P contents in root soils and various organs of watermelon were measured. The N and P contents were not significantly different between the 30 samples, indicating that there were no exceptional samples.

In different growth stages of crop, the accumulation of N varied greatly (Zhao et al. 2012). The N enrichment in different crop organs at maturity was also significantly different (Ookawa et al. 2003). The comparison result of N content in various organs of watermelon at mature stage is shown in Fig. 1a. Among watermelon organs, the N enrichment in seed was the highest, which was 30.35 ± 2.64 mg/g (mean ± standard deviation, the same below) and the lowest in pulp was 13.70 ± 4.54 mg/g. The order of average N accumulation was leaf and seed > stem, peel and root > pulp > root soil.

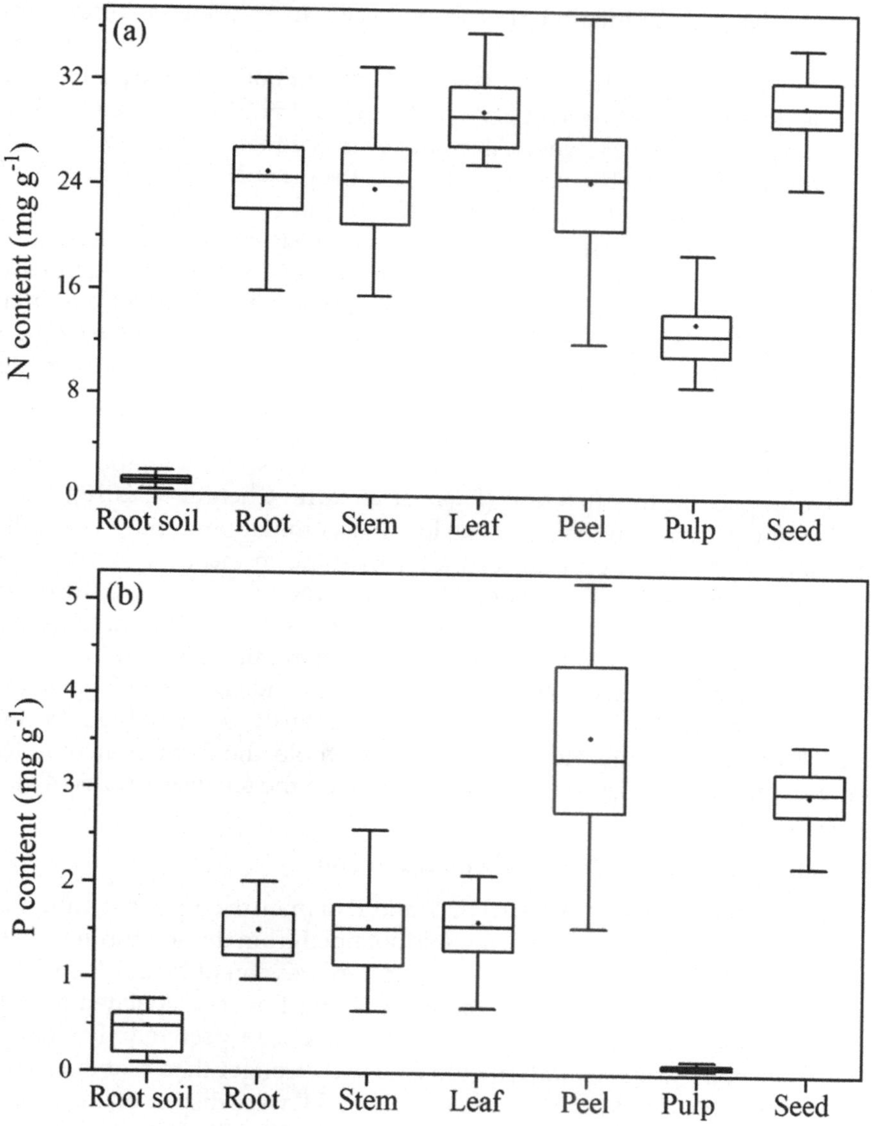

Figure 1. Comparison of N a and P b accumulation in root soil and various organs of watermelon.

In general, the N accumulation in the watermelon leaf and seed was almost twice as much as that in the pulp. N is the main constituent of protein and chlorophyll (Hikosaka 2016). In order to meet the needs of photosynthesis, leaves contain a large amount of chlorophyll and protein, and therefore, the N content in leaves is relatively high (Evans and Clarke 2019). During the different developmental stages of watermelon, the growth of organs is uneven. The growth center is the organ that develops first and has the greatest demand for organic material at different stages of watermelon development. Watermelon seeds are reproductive organs with high protein content and the growth center after fruit maturity, thus leading to a large accumulation of N in the seeds. In addition, watermelon peel contains high levels of protein amino acid, citrulline and minerals, which makes its N content relatively high as well. The low N accumulation in the pulp is mainly due to the degradation of protein after maturity, and N is transferred to the seed for storage (Bancal 2009). However, it should be noted that N redistribution occurs in crops at any time, which is usually related to the growth center of crop (Millard and Grelet 2010; Sanchez-Bragado et al. 2017).

Similarly, at mature stage, in different organs, the P enrichment varied greatly. The comparison result of P accumulation in various organs of watermelon at mature stage is presented in Figure 1b. In watermelon, the highest accumulation of P was 3.56 ± 1.17 mg/g in the peel, and the lowest one was 0.09 ± 0.02 mg/g in the pulp. The order of average P accumulation was peel > seed > root, stem, leaf > root soil > pulp.

The P enrichment in the watermelon pulp was the lowest, but that in peel was the highest. Peel plays an important role in plant stress tolerance, and P has a positive impact on this function (Rajsz et al. 2016; Chen et al. 2018); hence, P was relatively enriched in the peel. The low P accumulation in pulp was still due to the transfer of P from pulp to seed after ripening to meet the seed's demand for mineral elements. The P enrichment of each organ varies due to the different growth centers of plants at each developmental stage. When mineral nutrients migrate in organs, they follow the law of preferential distribution (Ding et al. 2020). Generally, before maturity, stems, leaves and the other organs are the growth center with high P accumulation. After maturity, some of the P in various vegetative organs will migrate to seeds. On the whole, the P enrichment in leaves is higher than that in pulp, which is consistent with the analysis results of various litters (Zhang et al. 2018).

Geochemical Analysis of N and P Elements in Soil

In plants, the migration patterns of N and P in mature crops are similar, while the demand for N and P in plant growth is obtained from the soil through the root of plants. The BCF of watermelon is the direct expression of N and P enrichment ability of the root to soil. The BCF values of N and P were calculated by Eq. (1). The means of BCF for N and P were 24.92% and 4.82%, respectively. Therefore, the enrichment capacity of N in watermelon root was stronger than that of P.

In Figure 2, the trends of BCF values for N and P were displayed together with those of N and P contents in roots. It was found that the BCF trend of N and P in samples was similar, in spite of the evident distinction of N and P contents in root. This suggested that the ability of each plant to obtain various mineral elements was similar and would not be significantly different due to the diversity of mineral element contents in soil or root. The similar enrichment degree of N and P in the plants revealed that the absorption of N and P in watermelon plants did not repel each other. In other words, there was no antagonistic in roots on the assimilation of N and P. It is an indisputable fact that the elemental composition of soil is inherited from parent material and is largely influenced by it (Chaplot 2013). In addition to human activities, N and P in soil are also affected by rocks and minerals. Thus, the contents of rock-forming elements in root–soil samples were tested. Correlation test was conducted on N, P and rock-forming elements in watermelon root soil and is listed in Table 1.

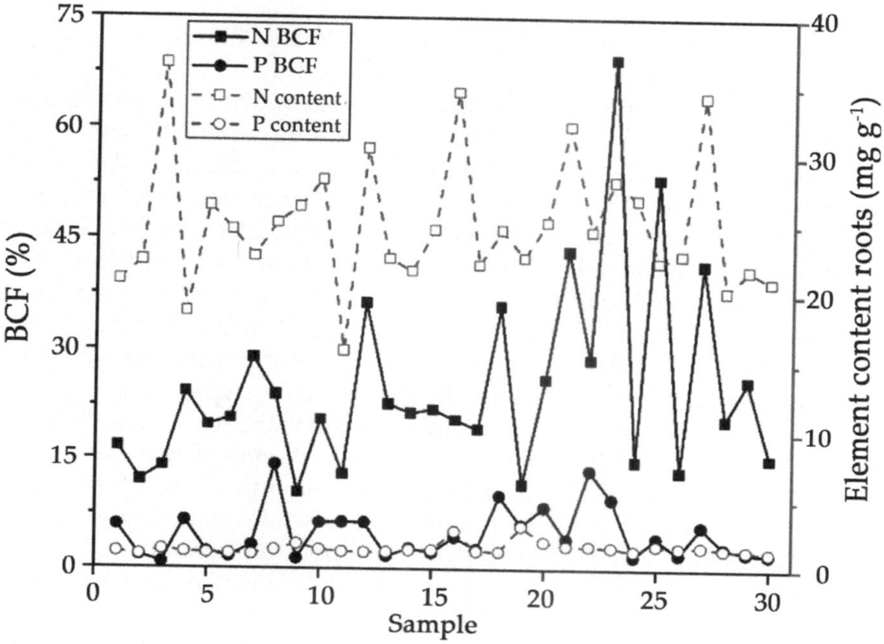

Figure 2. Trends of bioconcentration factor (BCF) and element content in roots.

Table 1. Pearson correlation between N, P and rock-forming elements in root soil (n = 30)

Element	Al	Ca	Fe	K	Mg	Na	Si	Sr	Ti	B	F
N	0.26	0.16	0.06	0.24	0.03	0.36*	0.25	0.26	0.17	0.74**	0.34
P	0.47**	0.48**	0.45*	0.45*	0.30	0.58**	0.35	0.35	0.48**	0.86**	0.53**

*Significant correlation at 0.05 level (both sides)

**Significant correlation at 0.01 level (both sides)

Table 2. Geochemical classification results of elements

Classification	N	P	Al	Ca	Fe	K	Na	Ti	B	F
Siderophile		√			√					
Lithophile		√	√	√		√	√	√	√	√
Sulfophilic				√	√					
Atmophile	√									
Biophile	√	√	√	√	√	√	√		√	√

Pearson correlation result showed that in root soil, there was correlation between Na, B and N, and between Al, Ca, Fe, K, Na, Ti, B, F and P. In the geochemical system, the elemental affinities are subject to the minimum energy law of the system. Under endogenous conditions, the elements with correlation are classified in Table 2, according to Goldschmidt's element geochemical classification.

N is an atmophile element. P is a siderophile and lithophile element. The elemental geochemical classification can reflect the distribution pattern of chemical elements in nature, and the characteristics of symbiotic combination of elements with each other. Ideally, elements with strong correlation with N should also belong to atmophile and biophile elements, and similarly, elements with strong correlation with P belong to siderophile, lithophile and biophile elements. In Table 2, most of the elements belonged to biophile elements, while no other elements except N belonged to atmophile; other elements related to P, such as Al, Ca, and K, belonged to lithophile elements, thus theoretically explaining that there were fewer elements related to N and more elements related to P. N and P in soil were partially derived from mineral decomposition. The soil-forming minerals were mainly gneiss, granite, diorite, quartz, feldspar, biotite, pyroxene, hornblende, kaolinite, limonite, calcite and dolomite in the study area. The mineral sources of N and P in root soil were remarkably different. There was almost no nitrogen in the above minerals, except for a small amount of fixed N (ammonium ions entering silicate minerals in isomorphism), while granite, gneiss, diorite, kaolinite, limonite, etc., all contained P, hence supporting again the correlation of N and P with other elements from a practical point. This may also provide some explanation for the correlation between elements under supergene condition. Since the mineral (biotite, kaolinite, quartz, feldspar, etc.) decomposition was also one of the main sources of B in soil, N in root soil was significantly correlated with B, and P was also highly correlated with B. In the endogenic products, the existing states of B, N and P have obvious specificity, they belong to the elements of salt acid radical, and this affinity is easy to appear in the state of salt mineral. Adding N (Kaundal et al. 2014) and P (Muhlbachova et al. 2017) fertilizer to the soil will reduce its B content, but the B content of soil itself largely depended on the physical and chemical properties of soil (Tkaczyk et al. 2017). Soil physical and chemical properties can be divided into two categories: nutrient indexes (related to N, P and K) and environmental indexes (related to pH, bulk density and cation exchange capacity). Therefore, there was a certain degree of correlation between N, P and B in soil-watermelon system. It was worth noting that P content also had a significant correlation with Ti content in the Pearson correlation of Table 1. Although Ti is not an essential element for crop growth, it is a beneficial element and plays an important role in growth and metabolism. Zahra et al. (2017) proved through experiments that the application of nano-TiO_2 in soil had a significant effect on P, that is, it could qualitatively increase P contents in crop roots, stems, and seeds.

The pedigree diagrams of N, P and rock-forming elements in watermelon root soil by cluster analysis (within groups, Euclidean distance) are shown in Figure 3. When d = 20, the clustering analysis was divided into three categories, ① K, F, Sr, Al, Na, Si; ② N, P, B, Ti; ③ Ca, Fe, Mg. With reference to the basic geological survey results, the bedrock types in the study area were mainly acidic rocks such as Cenozoic granitic gneiss, Mesozoic granite and diorite, whose rock-forming minerals are mainly quartz SiO_2, feldspar $(K,Ca,Na)[AlSi_3O_8]$, biotite $K(Mg,Fe)_3[Si_3AlO_{10}](OH,F)_2$, pyroxene $(Ca,Na)(Mg,Fe,Al,Ti)_2[(Si,Al)_2O_6]$ and hornblende $(Ca,Na)_{2-3}(Mg,Fe,Al)_5[Si_6(Si,Al)_2O_{22}](OH,F)_2$. After the weathering of these primary minerals, secondary minerals such as kaolinite $Al_4(Si_4O_{10})(OH)_8$, limonite $FeO(OH)\cdot nH_2O$, calcite $CaCO_3$ and dolomite $CaMg(CO_3)_2$ were formed. All of the above primary and secondary minerals were retained in the soil and formed soil minerals. Therefore, the three types representing the light mineral rock-forming elements, N and P nutrient elements, and dark mineral rock-forming elements obtained by cluster analysis corresponded well with the geological conditions of the study area. When d = 25, watermelon soils were divided into two groups. N, P and light-colored mineral elements were consolidated into one group. According to the chemical formula of these minerals, primary mineral orthoclase, albite, muscovite and quartz do not contain dark mineral-forming elements. During the magmatic activity under the endogenous conditions, these four minerals can exist in the granite at the same time. While in supergene environment, Na, K, Si and Al in the light-colored mineral-forming elements are alkali metal elements, which are the most abundant in shale and argillaceous rock and the least abundant in carbonate rock. Ca, Mg, and Sr in the dark mineral-forming elements are alkaline earth metal elements, with the highest content in carbonate and the lowest content in sandstone.

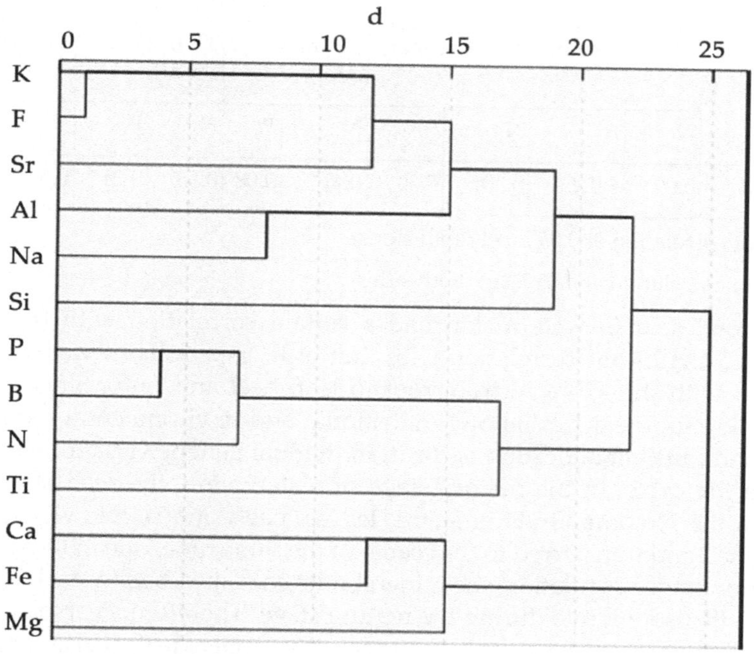

Figure 3. Cluster analysis pedigree diagram of N, P and rock-forming elements in watermelon root soil.

In the cluster analysis, N and P were classified into the same category with the light mineral rock-forming elements, which suggested that there was a certain relationship between N, P and rock-forming elements. Since the soils in different areas had experienced their own soil-forming process, the category of N and P might also be caused by the soil-forming process in different areas. Therefore, the classification of mineral nutrient elements could explain the source of soil parent material.

Influence of Mineral Elements on Nutritional Quality

N and P are essential mineral nutrients in plants. They can increase crop yield, improve the quality of agricultural products, and increase sugar content in crops. Thus, the effects of N and P on fruit quality parameters should be further explored after finding out the accumulation of N and P in various organs of crops at maturity. Two methods including correlation and regression analysis were used to study the connection of them.

Fruit quality parameters include external quality (such as color and size) and internal quality (such as soluble solid, total sugar and total acid) (Ali et al. 2017). The internal quality parameters concerned in watermelon. In order to explore the internal relationship between nutritional quality and N and P accumulation in each organ of watermelon, the correlation test was conducted on, and no significant correlation was found. However, there was a certain correlation between TF and total acid, and their Pearson correlation results are listed in Table 3.

Table 3. Pearson correlation between total acid and watermelon organs' translocation factor (n = 30)

Watermelon	TF (root-stem)		TF (root-leaf)		TF (root-peel)		TF (root-seed)		TF (root-pulp)	
	N	P	N	P	N	P	N	P	N	P
Total acid	0.09	– 0.25	0.39*	0.20	0.41*	0.08	0.35	0.01	0.51**	0.06

*Significant correlation at 0.05 level (both sides)

**Significant correlation at 0.01 level (both sides)

The total acid in watermelon had a certain correlation with the TF of N, but had no significant correlation with that of P. Especially, it was significantly correlated with the TF of N from root to leaf, peel and pulp. Since each stage of plant development has its own nutritional and environmental requirements, growth and nutrient allocation occur in individual genetic variations throughout the plant life cycle. In the mature stage of watermelon, the seed is the growth center, so the N compounds in stems, leaves, peels and pulps will be partially decomposed and transferred to the seeds. At the same time, due to the preferential distribution under regulation, the mineral nutrients absorbed by each organ from the root will also migrate during the mature stage. Therefore, Pearson correlation results in Table 3 suggested that N migration from root to leaf, peel and pulp might cause the change of total acid quality in watermelon.

In order to gain a clearer relationship between total acid and the TF of N, partial least squares method was applied to predict the total acid. The basic principle of this method was to extract the principal components (several comprehensive variables with the best explanatory power to both the independent variables TF values, pH and the dependent variable total acid) from the independent variable system consisting of pH and TF values of N from root to leaf, peel and pulp, establish the regression equation of total acid to the principal components and then restore the above equation to the partial least squares regression equation of total acid to pH and TF values of N from root to leaf, peel and pulp. In this article, the measured values of pH and TF values of N from root to leaf, peel and pulp of watermelon were set as predictive variables, the measured values of total acid at corresponding points were set as response values, the calculation of partial least squares regression equation was realized by invoking "plsregress" function in MATLAB software, and the prediction results are shown in Figure 4. The partial least squares regression equation of total acid and the TF of N from root to leaf, peel and pulp was as follows:

$$\text{Total acid} = 43.14 - 0.32 \times TF_{N(\text{root - leaf})} + 2.49 \times TF_{N(\text{root - peel})} + 2.64 \times TF_{N(\text{root - pulp})} - 7.36 \times \text{pH} \quad (4)$$

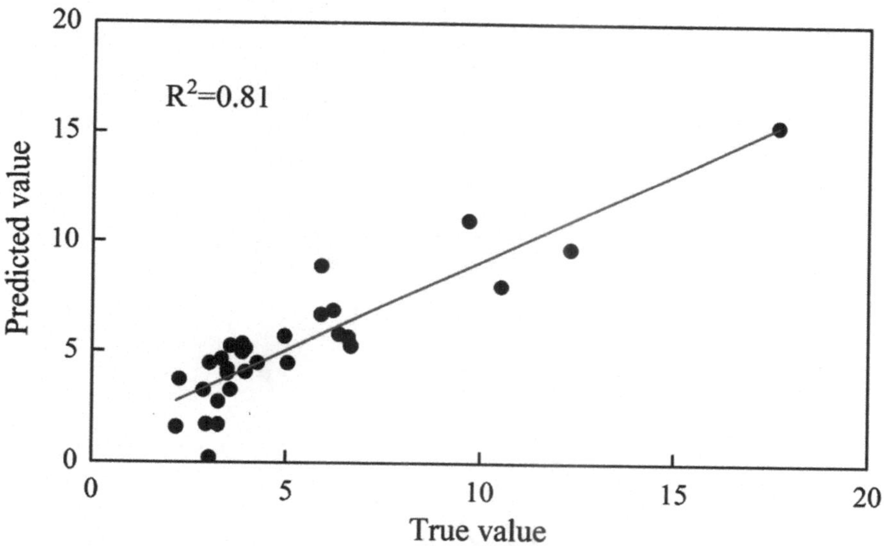

Figure 4. Partial least squares regression prediction results for total acid.

The coefficient of regression equation indicated the specific degree of each component's influence on the total acid. The order was as follows: pulp > peel > leaf. This was consistent with Pearson correlation results of Table 3.

In the diagnosis of crop nutrition problems, element ratio was a good calculation method and has been widely employed. The Ca to Mg ratio ([Ca | Mg]) reflects geographical location and soil minerals in agroecosystem, and the N to P ratio ([N | P]) reflects the balance between two basic life processes (synthesis of protein and r-RNA) (Loladze and Elser 2011). In other words, [Ca | Mg] is related to the external factors acting on plant growth, such as soil composition and geological characteristics, while [N | P] reflects the internal cause of nutrient balance in plants. In view of the internal reason [N | P], many research directions have been carried out. The characteristics of N and P content in leaves are particularly important when exploring crop quality. Determination of the foliar [N | P] has been suggested as a method to assess the nutrient requirements for plant growth. Marchand et al.

(2013) utilized the ILR to analyze the nutritional characteristics of crops in the study of cranberry and achieved remarkable effects; hence, the ILR was also used here to explore the crop nutritional quality. The $ILR_{[N|P]}$ of watermelon leaves computed by Eq. (3) was 2.08 ± 0.21 (ranged from 1.64 to 2.86), and its correlation with crop nutritional quality (total sugar, soluble solids and total acid) is presented in Figure 5.

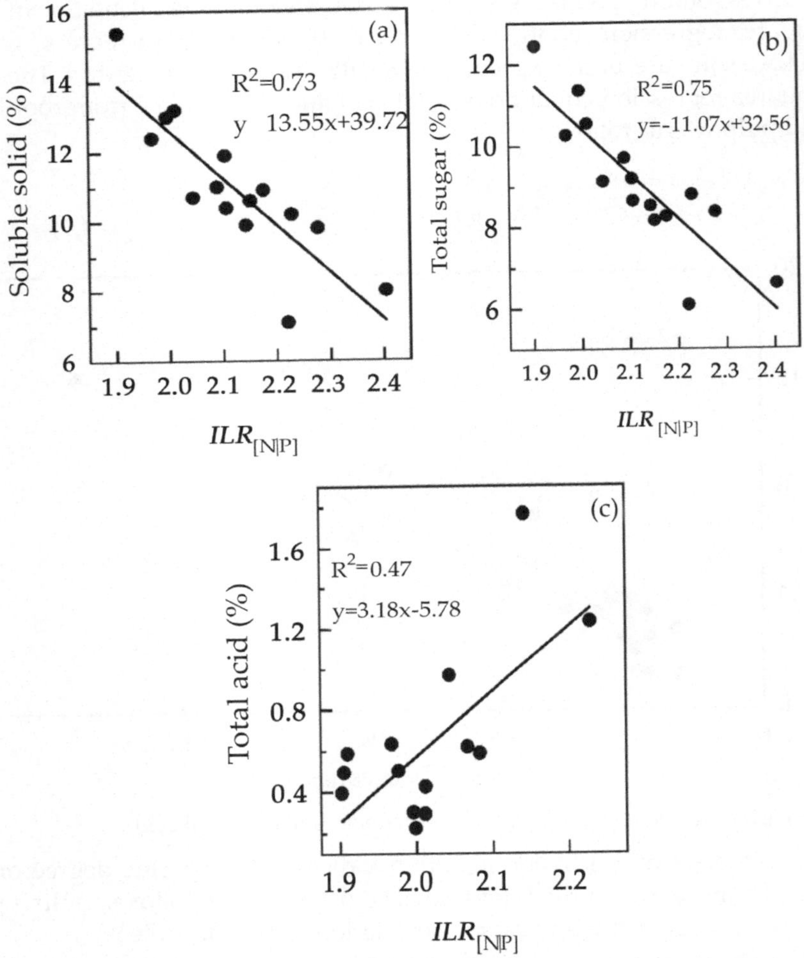

Figure 5. Correlation between leaf isometric log-ratio of N to P ($ILR_{[N|P]}$) and soluble solid (a), total sugar (b), total acid (c) ($P < 0.05$).

The correlation trends of $ILR_{[N|P]}$ in watermelon leaves to soluble solids and total sugars were the same because the soluble solids were related to sugar content. Particularly, 85% of the soluble solids in watermelon were total sugar (Macgillivray 1947). The soluble solid content was also related to the fruit viscosity. With the increase in soluble solid content, the fruit viscosity would also increase (Fukai and Matsuzawa 1999). Moreover, the soluble solid and total sugar contents of watermelon with pulp browning and softening disorder were lower than those of normal watermelon. The soluble solid content of most delicious fruits is more than 10% (Choudhary et al. 2012). In Figure 5a, when the leaf $ILR_{[N|P]}$ was less than 2.19, the soluble solid content of watermelon was more than 10%. The higher the soluble solid content, the better the taste. If the total sugar was more than 5%, the taste will be more delicious. Based on Figure 5b, when leaf $ILR_{[N|P]}$ was less than 2.48, the total sugar of watermelon was more than 5%. Total acids include inorganic acids and organic acids, organic acids have important nutritional value for the human body, malic acid with natural flavor is beneficial to the absorption of amino acids, while citric acid can increase appetite and promote the absorption of Ca and P. Usually, the total acid content above 0.5% makes the fruit taste better (Gao et al. 2018). In light of Figure 5c, when leaf $ILR_{[N|P]}$ was greater than 1.97, the total acid of watermelon was more than 0.5%. In general, watermelons with a leaf $ILR_{[N|P]}$ of 1.97 to 2.19 have suitable acid and sweet flavor, just as 50% of Shuanghou watermelons in this experiment. The method of predicting fruit quality by leaf ILR characteristics was non-destructive to fruit. There are also other non-destructive techniques for watermelon quality assessment, such as dielectric spectroscopy, laser Doppler vibrometry and magnetic resonance imaging, but they have the disadvantages of complex mathematical modeling, cumbersome instrument adjustment and expensive equipment costs (Ali et al. 2017). While leaf ILR analysis is technically simple and low cost. Therefore, the internal quality prediction by leaf N and P data is believed to be another idea for future non-destructive analysis of watermelon quality.

CONCLUSIONS

In this study, the migration and enrichment characteristics of N and P in various organs of watermelon were clarified by testing the collected watermelon and its root–soil samples. The influence of rock-forming elements on N and P content was discussed by means of geochemistry. The internal correlation between fruit quality parameters and N, P of watermelon was quantified on the basis of N and P distribution.

The migration and accumulation patterns of N and P in mature watermelon were similar. The seed was the growth center of this period, so part of the mineral elements in each organ was transferred to seeds for storage. Since photosynthetic leaves required protein and chlorophyll which were synthesized by much N, there was abundant N accumulation in leaves. The special physiological function of P made it rich in the peel. The order of average N accumulation was leaf and seed > stem, peel and root > pulp > root soil, while that of average P accumulation was peel > seed > root, stem, leaf > root soil > pulp.

The BCF results demonstrated that there was no antagonistic but synergetic effect on the uptake of N and P by watermelon plants. From the geochemical point of view, biophile elements, lithophile elements and mineral elements were the most correlated.

N migration from roots to organs in watermelon would limit the total acid quality of pulp, and the TF of N from root to leaf, peel and pulp could predict the total acid content of pulp accurately together with the pulp's pH. Moreover, the contents of total acid, total sugar and soluble solid were the best when the $ILR_{[N|P]}$ of watermelon leaves was 1.97–2.19.

In conclusion, there was indeed an internal correlation between mineral elements and fruit nutritional quality. This study only took watermelon as an example to confirm the correlation between some nutritional quality parameters and the distribution of N and P accumulation in watermelon, and the indicating effect of mineral elements on the comprehensive nutritional quality of crops should be further investigated in future.

Abbreviations

- N: Nitrogen
- P: Phosphorus
- BCF: Bioconcentration factor
- TF: Translocation factor
- ILR: Isometric log-ratio
- H_2SO_4 : Sulfuric acid
- HNO_3 : Nitric acid
- $HClO_4$: Perchloric acid
- HCl: Hydrochloric acid
- CRMs: Certified reference materials
- ΔlgC: Logarithmic deviation
- RSD: Relative standard deviation
- DLs: Detection limits

REFERENCES

1. Ali MM, Hashim N, Bejo SK, Shamsudin R. Rapid and nondestructive techniques for internal and external quality evaluation of watermelons: a review. Sci Hortic-Amster. 2017;225:689–99.
2. Bai YF, Wu JG, Clark CM, Naeem S, Pan QM, Huang JH, Zhang LX, Han XG. Tradeoffs and thresholds in the effects of nitrogen addition on biodiversity and ecosystem functioning: evidence from inner Mongolia Grasslands. Global Change Biol. 2010;16(1):358–72.
3. Bancal P. Decorrelating source and sink determinism of nitrogen remobilization during grain filling in wheat. Ann Bot-London. 2009;103(8):1315–24.
4. Chaplot V. Impact of terrain attributes, parent material and soil types on gully erosion. Geomorphology. 2013;186:1–11.
5. Chen N, Song B, Tang S, He J, Zhou Y, Feng J, Shi S, Xu X. Overexpression of the ABC transporter gene TsABCG11 increases cuticle lipids and abiotic stress tolerance in arabidopsis. Plant Biotechnol Rep. 2018;12(5):303–13.
6. Choudhary BR, Pandey S, Singh PK, Singh R, Rai M. VRW-3′: a yellow rinded watermelon with high total soluble solids. Acta Hort. 2012;936:471–5.
7. Colle C, Madoz-Escande C, Leclerc E. Foliar transfer into the biosphere: review of translocation factors to cereal grains. J Environ Radioact. 2009;100(9):683–9.
8. Ding G, Lei GJ, Yamaji N, Yokosho K, Mitani-Ueno N, Huang S, Ma JF. Vascular cambium-localized AtSPDT mediates xylem-to-phloem transfer of phosphorus for Its preferential distribution in arabidopsis. Mol Plant. 2020;13(1):99–111.
9. Dong S, Zhang J, Zha T, Dai X, He M. Increased plant density with reduced nitrogen input can improve nitrogen use efficiency in winter wheat while maintaining grain yield. Arch Agron Soil Sci. 2020;66(12):1707–20.

10. Evans JR, Clarke VC. The nitrogen cost of photosynthesis. J Exp Bot. 2019;70(1):7–15.
11. Fourqurean JW, Zieman JC. Nutrient content of the seagrass Thalassia testudinum reveals regional patterns of relative availability of nitrogen and phosphorus in the Florida Keys USA. Biogeochemistry. 2002;61(3):229–45.
12. Fukai Y, Matsuzawa T. Physicochemical properties of flesh of watermelon with physiological disorder. J Japn Soc Food Sci Technol-Nippon Shokuhin Kagaku Kogaku Kaishi. 1999;46(12):839–41.
13. Gao L, Zhao S, Lu X, He N, Liu W. 'SW', a new watermelon cultivar with a sweet and sour flavor. HortScience. 2018;53(6):895–6.
14. Geikloue A, Reyhanitabar A, Najafi N. Investigating the balance status of nutrients in wheat plant using isometric log ratio and field validation of balances. Agron J. 2019;111(5):2404–10.
15. Grzyb A, Wolna-Maruwka A, Niewiadomska A. The significance of microbial transformation of nitrogen compounds in the light of integrated crop management. Agronomy. 2021;11(7):1415.
16. Hikosaka K. Optimality of nitrogen distribution among leaves in plant canopies. J Plant Res. 2016;129(3):299–311.
17. Kaundal A, Sharma SK, Kumar P, Sankhyan N, Dutta J. Distribution of boron forms in relation to soil characteristics, chemical fertilizers, and amendments in an acid alfisol of northwestern Himalayas. Commun Soil Sci Plan. 2014;45(13):1772–83.
18. Kelsey JW, Colino A, Koberle M, White JC. Growth conditions impact 2,2-bis (p-chlorophenyl)-1,1-dichloroethylene (p, p'-DDE) accumulation by Cucurbita pepo. Int J Phytoremediation. 2006;8(3):261–71.
19. Kumar V, Singh J, Chopra AK. Assessment of plant growth attributes, bioaccumulation, enrichment, and translocation of heavy metals in water lettuce (Pistia stratiotes L.) grown in sugar mill effluent. Int J Phytoremediation. 2018;20(5):507–21.
20. Loladze I, Elser JJ. The origins of the Redfield nitrogen-to-phosphorus ratio are in a homoeostatic protein-to-rRNA ratio. Ecol Lett. 2011;14(3):244–50.
21. Ludvik B, Schernthaner G. Diabetes Care Austria 2009: registry for type 2 diabetic patients in general practitioners' ordinations in Austria. Wien Klin Wochenschr. 2012;124(3–4):69–77.
22. Macgillivray JH. Soluble solids content of different regions of watermelons. Plant Physiol. 1947;22(4):637–40.
23. Marchand S, Parent S-E, Deland J-P, Parent L-E. Nutrient signature of Quebec (Canada) cranberry (Vaccinium macrocarpon Ait.). Rev Bras Frutic. 2013;35(1):292–304.
24. Millard P, Grelet G-A. Nitrogen storage and remobilization by trees: ecophysiological relevance in a changing world. Tree Physiol. 2010;30(9):1083–95.
25. Muhlbachova G, Cermak P, Vavera R, Kas M, Pechova M, Markova K, Kusa H, Ruzek P, Hlusek J, Losak T. Boron availability and uptake under increasing phosphorus rates in a pot experiment. Plant Soil Environ. 2017;63(11):483–90.
26. Nie P, Dong T, He Y, Xiao S, Qu F, Lin L. The effects of drying temperature on nitrogen concentration detection in calcium soil studied by NIR spectroscopy. Appl Sci. 2018;8(2):269.
27. Ookawa T, Naruoka Y, Yamazaki T, Suga J, Hirasawa T. A comparison of the accumulation and partitioning of nitrogen in plants between two rice cultivars, Akenohoshi and Nipponbare, at the ripening stage. Plant Prod Sci. 2003;6(3):172–8.

28. Pan G. Soil science of earth surface system. Beijing: Geology Press; 2000.

29. Parent L-E. Diagnosis of the nutrient compositional space of fruit crops. Rev Bras Frutic. 2011;33(1):321–34.

30. Parent S-E, Parent LE, Jose Egozcue J, Rozane D-E, Hernandes A, Lapointe L, Hebert-Gentile V, Naess K, Marchand S, Lafond J, Mattos D Jr, Barlow P, Natale W. The plant ionome revisited by the nutrient balance concept. Front Plant Sci. 2013;4:39.

31. Rajsz A, Warzybok A, Migocka M. Genes encoding cucumber full-size ABCG proteins show different responses to plant growth regulators and sclareolide. Plant Mol Biol Rep. 2016;34(4):720–36.

32. Rastegari Mehr M, Keshavarzi B, Moore F, Hooda PS, Busquets R, Ghorbani Z. Arsenic in the rock–soil–plant system and related health risk in a magmatic–metamorphic belt, West of Iran. Environ Geochem Health. 2020;42(11):3659–73.

33. Richardson AE, Hocking PJ, Simpson RJ, George TS. Plant mechanisms to optimise access to soil phosphorus. Crop Pasture Sci. 2009;60(2):124–43.

34. Sanchez-Bragado R, Serret MD, Araus JL. The nitrogen contribution of different plant parts to wheat grains: exploring genotype, water, and nitrogen effects. Front Plant Sci. 2017;7:1986.

35. Sun W, Zhao Y, Zhang Z, Li B, Li Z, Tang X. Greenhouse characterization of inorganic mercury, methyl mercury and ethyl mercury migration and transformation in Indian mustard and Chinese pakchoi. Anal Lett. 2022;55(9):1425–39.

36. Tkaczyk P, Bednarek W, Dresler S, Krzyszczak J, Baranowski P, Slawinski C. Relationship between assimilable-nutrient content and physicochemical properties of topsoil. Int Agrophys. 2017;31(4):551–62.

37. Tsige BA, Dechassa N, Tana T, Laekemariam F, Alemayehu Y. Effect of mineral nitrogen, phosphorus, and potassium fertilizers on the productivity of faba bean (Vicia faba L.) in acidic soils of Wolaita Zone, Southern Ethiopia. Int J Agron. 2022;2022:2232961.

38. Zahra Z, Waseem N, Zahra R, Lee H, Badshah MA, Mehmood A, Choi H-K, Arshad M. Growth and metabolic responses of rice (Oryza sativa L.) cultivated in phosphorus-deficient soil amended with TiO_2 nanoparticles. J Agr Food Chem. 2017;65(28):5598–606.

39. Zhang Y, Yang J, Yang W, Tan B, Fu C, Wu F. Climate, plant organs and species control dissolved nitrogen and phosphorus in fresh litter in a subalpine forest on the eastern Tibetan Plateau. Ann Forest Sci. 2018;75(2):51.

40. Zhao Y, Zhang Z, Li B, Zhao Y, Lu J, Tang X. Accurate determination and comprehensive evaluation of heavy metals in different soils from Jilin Province in Northeast China. Anal Lett. 2021;54(12):1901–28.

41. Zhao B, Ren P, Liu J, Zhao M: Nitrogen accumulation, retranslocation and partitioning in oats (Avena L.) as affected by different water supply and nitrogen fertilization. In: 2nd International conference on energy, environment and sustainable development (EESD 2012), Jilin, People's Republic China.

2

A NEW METHOD FOR THE EMPIRICAL CONVERSION OF LOGGING DATA TO CLAY MINERAL FRACTION IN THE NANKAI ACCRETIONARY PRISM

Suguru Yabe[1], Rina Fukuchi[2], Yohei Hamada[3] and Gaku Kimura[4]

[1]Geological Survey of Japan, National Institute of Advanced Industrial Science and Technology (AIST), Tsukuba Central 7, 1-1-1 Higashi, Tsukuba, Ibaraki, 305-8567, Japan
[2]Geoengineering Headquarters, Dia Consultants Co., Ltd., 2-272-3 Yoshino-cho, Kita-ku, Saitama, 331-0811, Japan
[3]Institute for Extra-cutting-edge Science and Technology Avant-garde Research, Kochi Institute for Core Sample Research, Japan Agency for Marine-Earth Science and Technology, 200 Monobe Otsu, Nankoku-shi, Kochi, 783-8502, Japan
[4]Department of Marine Environment and Resources, Tokyo University of Marine Science and Technology, No. 2 Building 311, 4-5-7 Konan, Minato-ku, Tokyo, 108-8477, Japan

ABSTRACT

Sediment lithology, especially the clay mineral fraction, is basic geologic information, which is important for understanding faulting and evolutionary process of the accretionary prism. During scientific drilling, the clay mineral fraction is measured using X-ray diffraction analyses of samples from the acquired cores. However, coring in an accretionary prism is sometimes difficult, which hinders the acquisition of clay mineral fraction data. This study details the development of a new method used to empirically estimate the clay mineral fraction from the electrical resistivity and natural gamma ray logs obtained from the Kumano section of the Nankai Tough, where multiple scientific drilling expeditions have been conducted. This method is composed of two steps. First, porosity is estimated using electrical resistivity logs. In order to compensate for the effect of porosity, the natural gamma ray logs are then normalized using (1 – porosity). Second, the normalized natural gamma ray logs are converted into clay mineral fractions using an empirical conversion function. This empirical function was determined by comparing all available normalized natural gamma ray logs and clay mineral fractions data collected from the Kumano section of the Nankai Trough. As a byproduct of the porosity estimations, thermal conductivity and temperature depth profiles were also estimated for all of the logging sites. As electrical resistivity logs and natural gamma ray logs are essential measurements made during scientific drilling, this new method can be applied to future drilling expeditions in the Nankai Trough. Although the empirical conversion function established in this study is valid only at the Kumano section of the Nankai trough, the same procedure can be applicable to other subduction zones to establish local empirical conversion functions if enough data are available. The clay mineral fraction depth profile estimated using this method can provide useful information for various geological studies, including sedimentology and structural geology of the shallow accretionary prism.

Keywords: *Clay mineral fraction, Logging while drilling, Natural gamma ray log, Accretionary prism, Nankai Trough*

INTRODUCTION

Understanding the lithology of sedimentary materials in accretionary prisms is important for a variety of reasons. It is basic sedimentological information and is useful for understanding the provenance of the material (e.g., Moore et al. 2015a, b; Pickering et al. 2013). The mechanical properties of the sediment also change according to the lithology. The clay mineral fraction of the sediment is especially important, as sediment with a high clay mineral fraction has a weaker frictional strength, which influences the formation process of the shallow plate interface in an accretionary prism (Moore et al. 1986; Moore and Shipley 1988; Moore et al. 1990; Vrolijk 1990; Ujiie et al. 2003; Sawai et al. 2014; Oohashi et al. 2015; Hüpers et al. 2017). To determine the lithology of sediments in shallow accretionary prisms, scientific drilling is a useful tool. This study focuses on the Nankai Trough, where the International Ocean Discovery Program (IODP) has conducted multiple expeditions as part of the Nankai Trough Seismogenic Zone Experiment (NanTroSEIZE) using the drilling vessel (D/V) Chikyu (Tobin and Kinoshita 2006), which aimed to understand seismogenesis and evolutionary process of the Nankai trough.

Core samples and logging data are acquired during scientific drilling. The clay mineral fraction is measured from core samples using X-ray diffraction (XRD) analyses (Expedition 315 Scientists 2009a). Due to much effort (Expedition 319 Scientists 2010a; Guo and Underwood 2012; Saffer et al. 2017; Strasser et al. 2014a; Underwood 2017a, b, c; Underwood and Guo 2013, 2017; Underwood 2017a, b, c; Underwood and Song 2016; Yamaguchi et al. 2020), clay mineral assemblages have been determined and used to interpret the evolution of the accretionary prism (Underwood 2018; Underwood and Guo 2018). However, XRD data is limited by the fact that data cannot be acquired when core samples are not recovered. This is especially true in the accretionary prism, which is composed of easily collapsible sediment, and coring has sometimes been difficult due to poor hole conditions throughout the NanTroSEIZE project (Expedition 315 Scientists 2009b, Expedition 316 Scientists 2009a; Kitajima et al. 2020). In difficult drilling conditions, logging while drilling (LWD) techniques are more promising than coring. LWD can acquire data continuously without pauses in drilling, which is advantageous for maintaining satisfactory hole conditions. Among the acquired logging data, natural gamma ray (NGR) log data are used to infer the lithology of the drilled sediment. The NGR log measures the number of gamma rays encountered by the detector in the logging tool (e.g., Expedition 314 Scientists 2009a). If spectral NGR logs are available, which can distinguish the source elements of the gamma radiation (Expedition 319 Scientists 2010b), mineralogical modeling can be used to quantitatively estimate the downhole lithology (e.g., Hampson et al. 2005). However, this is not possible when NGR logs that count only the total amount of NGR radiation are available. In such a case, the values from the NGR logs are usually used as a qualitative index of shale content. Because NGR logs are highly dependent on the lithology and chemical composition of geological formations, which can vary widely, it is difficult to establish a general conversion method from NGR log data to the clay mineral fraction of the formation. However, if specific regions with a relatively common lithology are analyzed in detail, the NGR logs could be quantitatively related to the clay mineral fraction. Because logging data are acquired continuously with depth, a depth profile of the downhole clay mineral fraction could be estimated in more detail using logging data than from XRD analyses of core samples.

This study developed a new method to empirically estimate the sediment clay mineral fraction from logging data obtained from the Kumano section of the Nankai Trough. During NanTroSEIZE project, the spectral NGR logs are acquired only at one site, which hinders mineralogical modeling from logging data at other sites. Therefore, using logging data and core measurements acquired throughout the NanTroSEIZE project, we established an empirical conversion function from the NGR log data to the clay mineral fraction. Because the number of radioactive elements around the detector in the logging tool is affected by the porosity of the formation, we compensated for this effect by estimating the porosity structure from electrical resistivity logs using the method of Yabe et al. (2019).

In this paper, we provide a brief description of the tectonic setting in the Nankai Trough, compile all the available NGR logs from the NanTroSEIZE project, compare the porosity-normalized NGR logs with the clay mineral fractions measured from core samples to establish an empirical conversion function between the two, and show that the estimated clay mineral fraction depth profiles can be used for discussions of various aspects, including sedimentology and structural geology.

TECTONIC SETTING AND DATA

The Nankai trough is located off the coast of southwest Japan, where the Philippine Sea Plate (PSP) subducts at a rate of ~ 6 cm/year (Miyazaki and Heki 2001; DeMets et al. 2010; Kimura et al. 2018). During subduction, thick sediments on the incoming PSP have accreted onto the overriding plate, forming an accretionary prism. The Kumano transect, which is the main target of the NanTroSEIZE project (Tobin and Kinoshita 2006), is located off the Kii Peninsula. Seismic surveys and a series of IODP expeditions have been conducted along this transect to understand the geologic structure and history of this subduction zone (Figure 1). The forearc prism is composed of inner and outer wedges separated by the outer ridge, where the megasplay fault intersects the seafloor (Kimura et al. 2007). The outer wedge formed rapidly after 2 Ma, when the megasplay fault became active (Strasser et al. 2009; Kimura et al. 2011; Moore et al. 2015a, b). In contrast, the inner wedge is composed of older sediments, suggesting that the inner wedge formed before 6 Ma (Kimura et al. 2018). Uplift caused by motion along the megasplay fault forms a forearc basin on the inner wedge prism, which is called the Kumano Basin.

The sediments that make up the accretionary prism have hemipelagic and terrigenous sources. In the Nankai Trough, the incoming PSP is covered by thick hemipelagic sediment (Shikoku Basin sediment), whereas terrigenous sediments are transported into the margin as trench wedge fill and slope apron facies. The geologic background of the Shikoku Basin sediments has been reviewed in detail by Underwood (2018). The Shikoku Basin formed in response to backarc rifting along the Izu-Bonin Arc at ~ 30 Ma, and spreading continued until ~ 15 Ma (Okino et al. 1994, 1999; Sdrolias et al. 2004). Several bathymetric highs around the Shikoku Basin, which are related to basin formation, control the sediment supply into the basin. The Kinan seamount chain is the spreading axis of the Shikoku Basin and was volcanically active until approximately 7–8 Ma (Ishii et al. 2000). The western side of the Shikoku Basin is defined by the Kyushu-Palau ridge, which is the remnant portion of the Izu-Bonin arc (Ishizuka et al. 2011). Although pelagic clays were deposited on the basaltic basement when the Shikoku Basin was located far offshore from the Japan arc, terrigenous turbidite sediments mixed with the pelagic sediments as the PSP moved closer to the Japan arc. Two major turbidite facies are observed along the Kumano transect. The Kyushu Fan is a mixed turbidite facies observed above the pelagic clay layer along the entire Nankai Trough, with an age of ~ 12–15 Ma (Pickering et al. 2013). The Kyushu Fan deposits are thought to have been transported from the East China Sea, although the current route was cut at ~ 7 Ma by the opening of the Okinawa Trough (Underwood and Guo 2018). The Zenisu Fan is a volcanic turbidite facies observed only along the Kumano transect, with an age of ~ 8–9 Ma (Pickering et al. 2013).

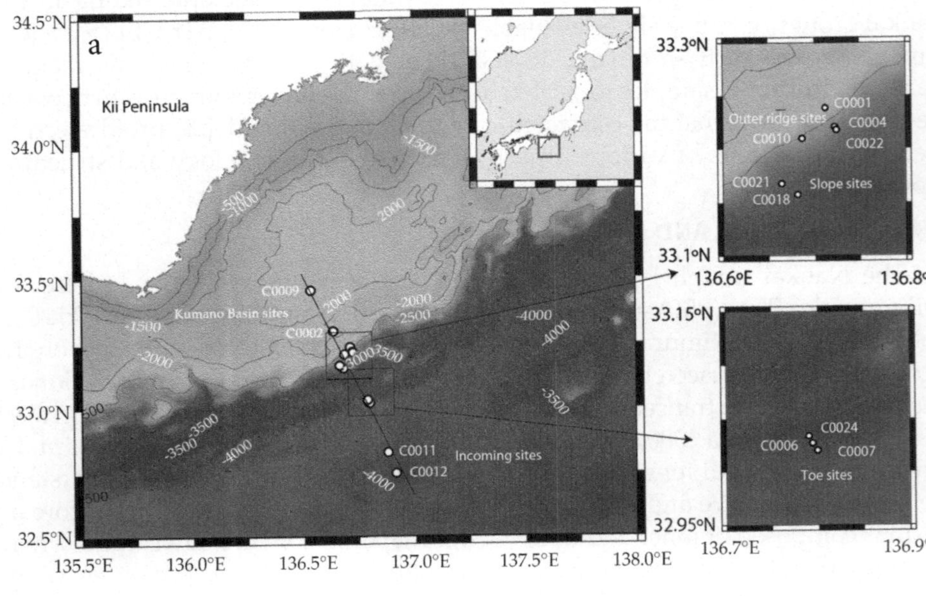

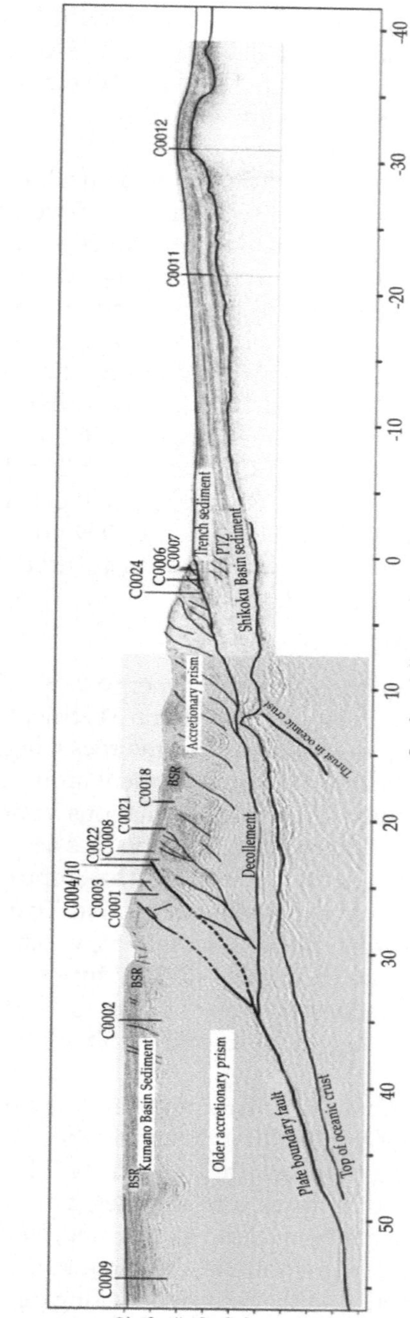

Figure 1. Regional map of the Kumano transect in the Nankai Trough. a Map view of drilled sites in NanTroSEIZE project. Background color and contours show the bathymetry. White dots are the NanTroSEIZE drilling sites. A seismic line shown in b is presented by a black line. b Seismic profile along the Kumano transect after Figure F2A of Tobin et al. (2020).

The Zenisu Fan is thought to be a tuffaceous gravity flow produced by the Izu-Honshu collision zone (Underwood and Guo 2018). Along the Kumano transect, the Zenisu Fan sediments are called the Middle Shikoku Basin (MSB) sediments, and sediments above and below this unit are called the Upper and Lower Shikoku Basin (USB and LSB) sediments, respectively.

The NanTroSEIZE project has conducted ocean drilling in the various tectonic environments along the Kumano transect (Figure 1). Two sites (C0011 and C0012) were drilled on the oceanic plate outside the subduction zone. Three sites (C0006, C0007, and C0024) were drilled at the prism toe, where subduction initiates. These sites were drilled through the plate boundary décollement. Sites C0018 and C0021 were drilled through the slope basin on the trench slope to determine past megathrust earthquake seismicity by analyzing the turbidites in this section. Four sites (C0001, C0004, C0010, and C0022) were drilled around the outer ridge near the seafloor intersection of the megasplay fault. Site C0001 is located on the hanging wall, while sites C0004 and C0010 were drilled through the megasplay fault. Site C0022 is located in the slope basin, where the megasplay fault extension is expected to pass through. Sites C0002 and C0009 are located in the Kumano Basin. Site C0002 is the deepest hole of all the sites drilled during the NanTroSEIZE project, with a depth of ~ 3265 mbsf.

GAMMA RAY LOG NORMALIZATION

The NGR logs measure the gamma ray energy that radiates from radioactive elements in geologic formations (e.g., Expedition 314 Scientists 2009a). Because the radioactive elements are in the solid parts of the rocks, higher porosities result in decreased amounts of radioactive elements present around the detector, resulting in smaller NGR log values. Therefore, formations containing the same clay mineral fraction can exhibit different NGR log values according to their porosity, which makes it difficult for scientists to establish an empirical conversion function between NGR log values and the clay mineral fraction. To exclude this effect due to variations in porosity (ϕ) from the NGR variations, we calculated the normalized NGR, which is defined as the NGR value divided by $1-\phi$ (Hüpers et al. 2015). In this study, we applied a spatial smoothing of 10 m to the datasets used to smooth out fine-scale variations in logging data and depth shifts between logging holes and core holes.

To calculate the normalized NGR, porosity distributions were prepared using three different datasets. The first dataset is the porosity values measured onboard D/V Chikyu using the moisture and density (MAD) method (Expedition 315 2009a). As this method requires core samples, porosity was measured only at depths, where cores were recovered. In addition, because LWD and coring cannot be conducted at the same time in the same hole, there could be a depth shift between the core depth and the recorded logging depth.

The second dataset is porosities converted from electrical resistivity log data. Electrical resistivity is known to correlate empirically with porosity (Archie 1942). Although the thermal structure of the hole is required to convert the electrical resistivity into porosity, Yabe et al. (2019) developed a method to estimate the porosity and thermal structure simultaneously from electrical resistivity log data. In their method, porosity and thermal structure are estimated iteratively until they converge to values that satisfy correlations between porosity, thermal conductivity, and electrical resistivity. Their correlations between physical properties were also developed using measurements made onboard D/V Chikyu. Porosity is converted from the electrical resistivity log data using the assumed thermal structure and empirical relationships between electrical resistivity and porosity. The calculated porosity is further converted to thermal conductivity using the established relationship. Then, the thermal structure is updated based on the thermal conductivity structure and observed heat flow. Although a thermal steady state is assumed when calculating the thermal structure, corrections are applied to some sites according to the depositional history of each site using a correction factor α. A detailed description of this process is described in Yabe et al. (2019). The iterative estimation process continues until the results converge. Yabe et al. (2019) applied this method to Site C0002 and showed that the estimated structures are consistent with the porosities and thermal conductivities measured in core samples, as well as with porosities calculated from density log data. In this study, we applied the method developed by Yabe et al. (2019) to all NanTroSEIZE sites, where electrical resistivity logs are available. The bit resistivity was used for the conversions in this study, because it usually has the greatest investigation depth among the acquired electrical resistivity logs and is less influenced by the drilling mud invasion effect and/or borehole conditions.

The third dataset is porosities converted from density log data. Because grain density and fluid density are estimated from shipboard MAD measurements as 2.7 and 1.0 g/cm^3, respectively (Yabe et al. 2019), density logs can be converted to porosities. Although it is the most reliable dataset measured in situ, density log data are only available for sites C0001, C0002, and C0009.

Incoming Sites

Two sites (C0011 and C0012) are located on the PSP outside of the subduction zone (Figure 1). Site C0011 is located at the foot of the Kashinosaki knoll, which represents the normal sedimentary stratigraphy of the PSP. Site C0012 is located on the Kashinosaki knoll, which is a topographic high related to volcanic activity. Hence, the sediments observed at site C0012 have been more influenced by heating than those at site C0011.

At site C0011, LWD was performed during IODP Expedition 322 (Expedition 322 Scientists 2010a), and coring was performed during both Expeditions 322 (Expedition 322 Scientists 2010a) and 333 (Expedition 333 Scientists 2012a). At site C0012, LWD was performed during IODP Expedition 338 (Strasser et al. 2014b) and coring was performed during Expeditions 322 (Expedition 322 Scientists 2010b) and 333 (Expedition 333 Scientists 2012b). According to the site reports, the stratigraphy can be divided into Upper, Middle, and LSB sediments, in addition to a thin pelagic claystone layer and the Kashinosaki Knoll basement below it (Figure 2). USB sediments (Unit 1) are divided into two or three sub-units according to the abundance of ash layers. MSB sediments (Unit 2) are dominated by volcaniclastic sandstones. The Lower Shikoku Basin sediments are divided into three sub-units (Units 3–5). The upper unit is dominated by hemipelagic sediments, whereas the middle and lower units are rich in terrigenous components (Kyushu Fan sediments) and volcaniclastic-rich facies, respectively.

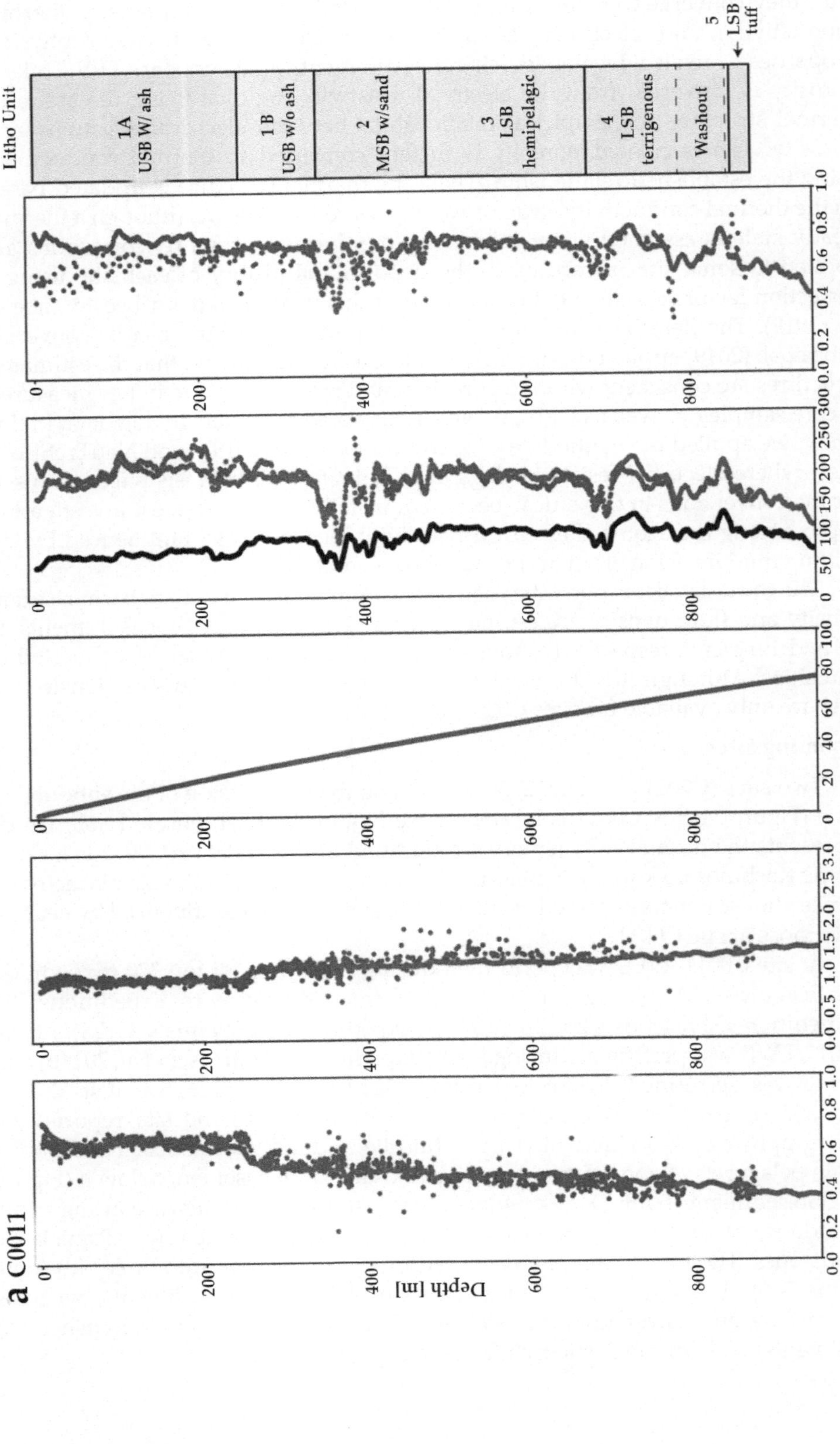
a C0011
Depth [m]
Porosity
Thermal conductivity [W/mK]
Temperature [°C]
Natural Gamma Rat [gAPI]
Clay mineral fraction
Litho Unit
1A USB W/ ash
1 B USB w/o ash
2 MSB w/sand
3 LSB hemipelagic
4 LSB terrigenous
Washout
5 LSB tuff

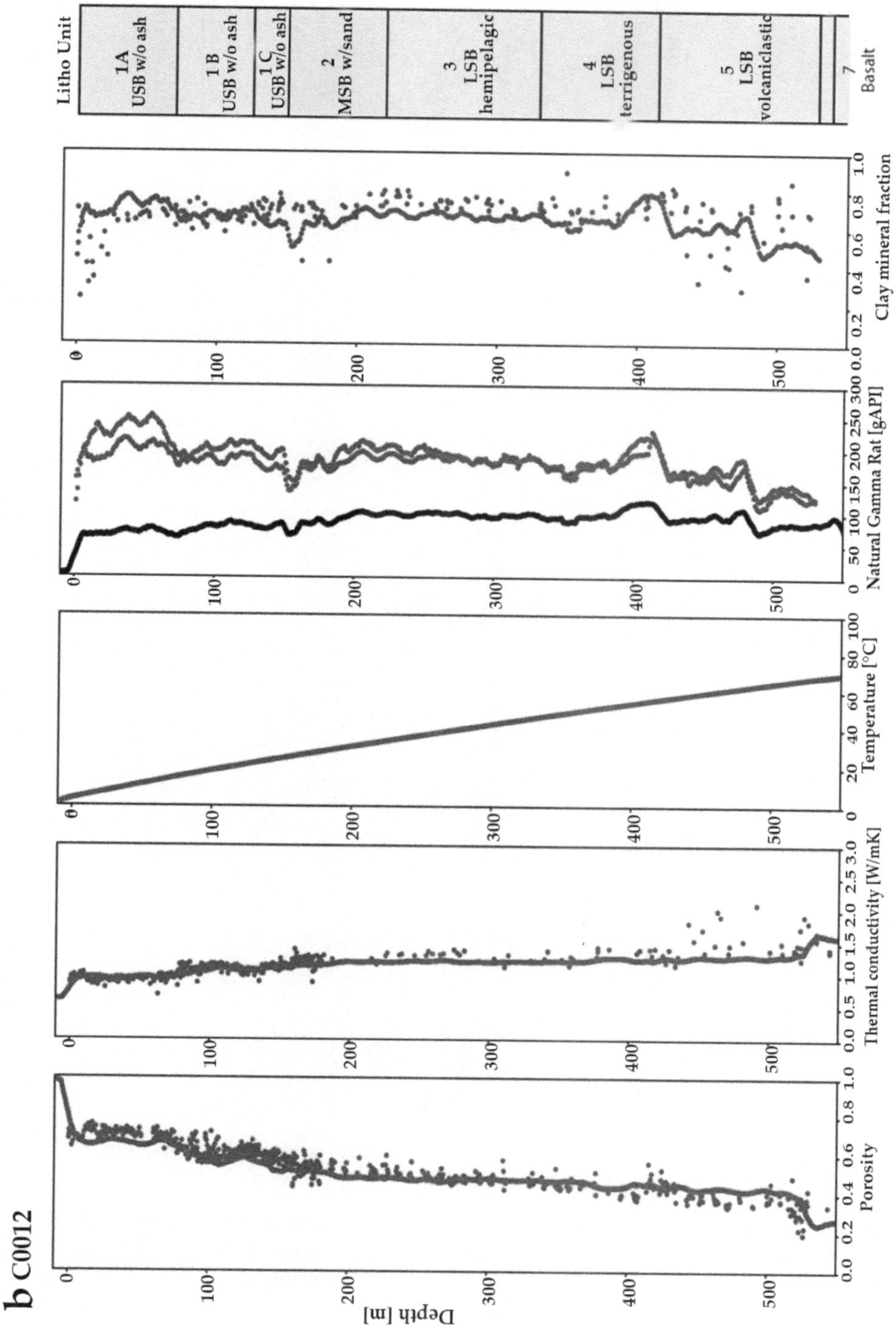
b C0012
Litho Unit
1A USB w/o ash
1 B USB w/o ash
1 C USB w/o ash
2 MSB w/sand
3 LSB hemipelagic
4 LSB terrigenous
5 LSB volcaniclastic
7 Basalt
Depth [m]
Porosity
Thermal conductivity [W/mK]
Temperature [°C]
Natural Gamma Rat [gAPI]
Clay mineral fraction

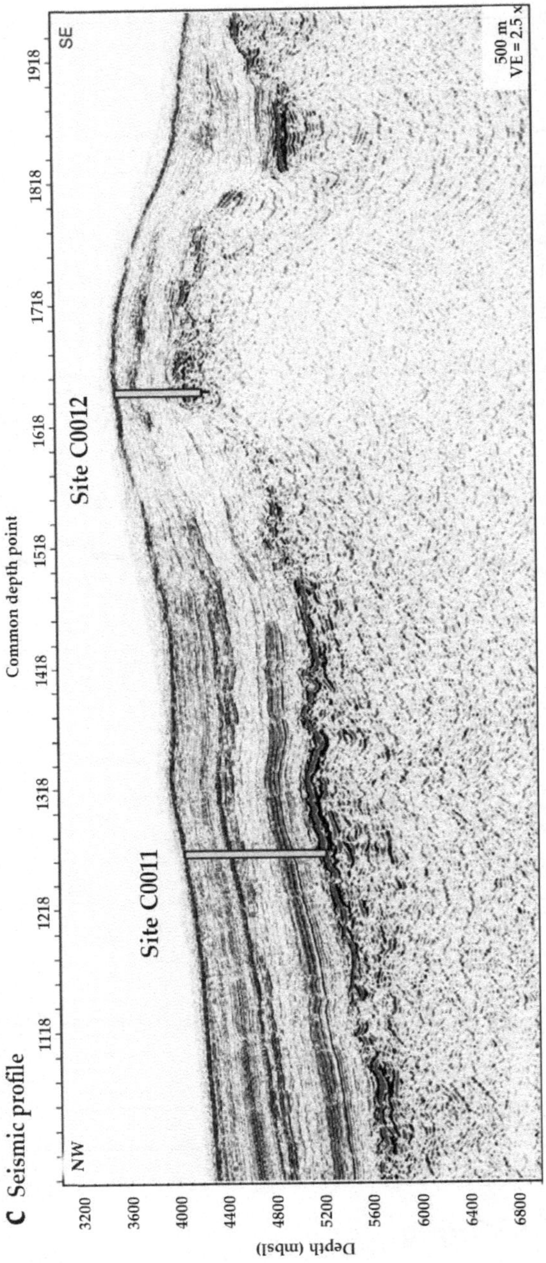

Figure 2. Data from the incoming sites. a Depth distributions of porosity, thermal conductivity, temperature, NGR, clay mineral fraction, and lithologic units at site C0011. Green symbols in the porosity, thermal conductivity, and clay mineral fraction plots represent data from core sample measurements. Red symbols in the porosity, thermal conductivity, and temperature plots indicate data estimated using the electrical resistivity logs. Black symbols in the NGR plot are the original (not normalized) NGR values, and green and red symbols are the NGR normalized using the MAD porosities and porosities estimated

from the electrical resistivity logs, respectively. Red symbols in the clay mineral fraction plot are results estimated using the new method developed in this study. In the rightmost panel, USB, MSB, and LSB represent the upper, middle, and lower Shikoku Basin sediments, respectively. Lithological units colored by red represent the standard sediments, for which the empirical conversion function is established. Lithological units colored by green represent the upper part of USB sediments or the slope apron facies younger than 1.5 Ma, which shows higher normalized NGR log values than the standard sediments possibly due to volcanic ashes included in sediments. b Same as a, but for site C0012. Lithological units colored by gray are excluded from comparison in Figure 7 due to poor data condition. c A seismic profile around incoming sites after Figure F2 of Expedition 333 Scientist (2012b).

Porosity was estimated from the resistivity logs using heat flow values of 89.5 mW/m^2 at site C0011 (Expedition 333 Scientists 2012a) and 141 mW/m^2 (Expedition 333 Scientists 2012b) at C0012. Site C0012 exhibits a higher heat flow value, because it is located on the Kashinosaki Knoll, which was produced by local volcanic activity. As these sites are located far offshore and on the incoming plate, a sedimentation rate is slow enough to keep a thermal steady state. Therefore, no correction for thermal structure was applied. Figure 2 shows the porosities, thermal conductivities, and thermal structures at sites C0011 and C0012. The porosities and thermal conductivities measured from the core samples are consistent with those estimated from the electrical resistivity logs. As the method developed by Yabe et al. (2019) only considers sedimentary material, the estimated values for the basaltic formations are not reliable. Although porosity and thermal conductivity exhibit a gradual decrease and increase with depth, respectively, they also exhibit a relatively sharp decrease and increase, respectively, at ~ 250 mbsf at site C0011 and at ~ 70–125 mbsf at site C0012, which correspond to lithologic Unit 1B.

Figure 2 also shows the NGR log data and the clay mineral fractions. Clay mineral abundances at these sites have been reported previously by Underwood and Guo (2013, Underwood 2017a, b, c). The original NGR logs show an increasing trend with depth, although the normalized NGR logs show an almost constant or a slightly decreasing trend with depth, with several negative spikes. The negative spikes are especially prominent at ~ 370–470 mbsf at site C0011 and at ~ 150–190 mbsf at site C0012, which shipboard scientists interpreted as sand-rich layers of MSB sediment (Zenisu Fan sediments). Two large negative spikes around ~ 700 mbsf at site C0011 were also interpreted as sandy interbeds (Kyushu Fan sediments). The clay mineral fraction exhibits an increasing trend from ~ 45% to ~ 65% in the shallowest part of the sections (0–100 mbsf at site C0011 and ~ 70 mbsf at site C0012). Below this layer, the clay mineral fraction is almost constant (around 60–70%), except for the lower part of LSB sediments (Unit 4 and 5) showing scattering clay mineral fraction with interbedding sandy layers (Expedition 322 Scientists 2010b).

Toe Sites

At site C0006, LWD was performed during IODP Expedition 314 (Expedition 314 Scientists 2009b) and coring was performed during Expedition 316 (Expedition 316 Scientists 2009a). At site C0024, LWD and coring were both performed during Expedition 358 (Yamaguchi et al. 2020). According to the site reports, the stratigraphy at these sites is divided into three lithologic units (Figure 3). The upper unit is a thin slope apron facies. The middle unit is the accreted trench wedge sediment. The shallow part of this unit is composed of sandy coarse sediments, whereas the deeper part is composed of muddy sediments. The lower unit is the accreted USB sediment. Although coring was successful only down to this unit, LWD successfully penetrated the potential décollement at ~ 710 mbsf at site C0006 and at ~ 850 mbsf at site C0024.

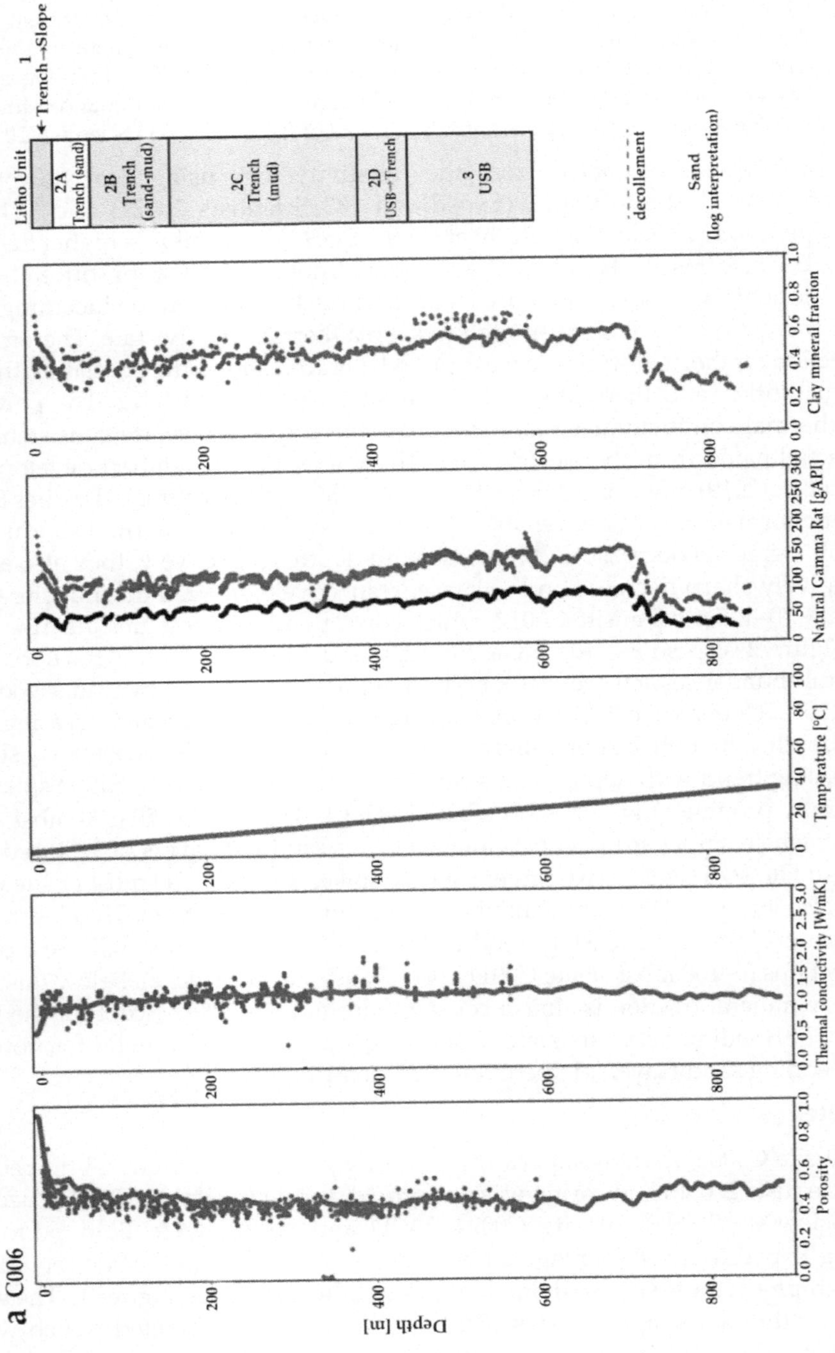

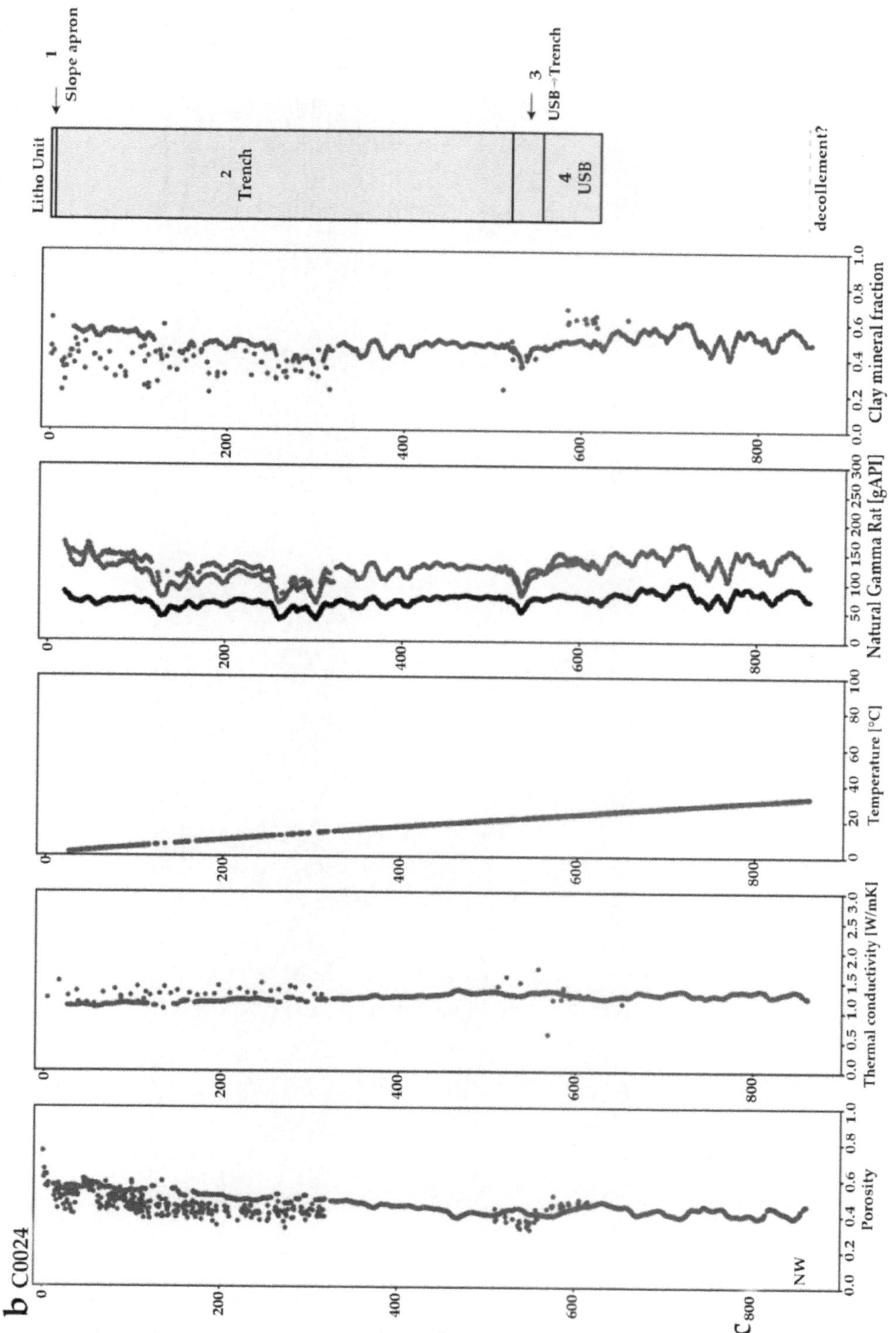
b C0024
c
Depth [m]
NW
Porosity
Thermal conductivity [W/mK]
Temperature [°C]
Natural Gamma Rat [gAPI]
Clay mineral fraction
Litho Unit
1 Slope apron
2 Trench
3 USB→Trench
4 USB
decollement?

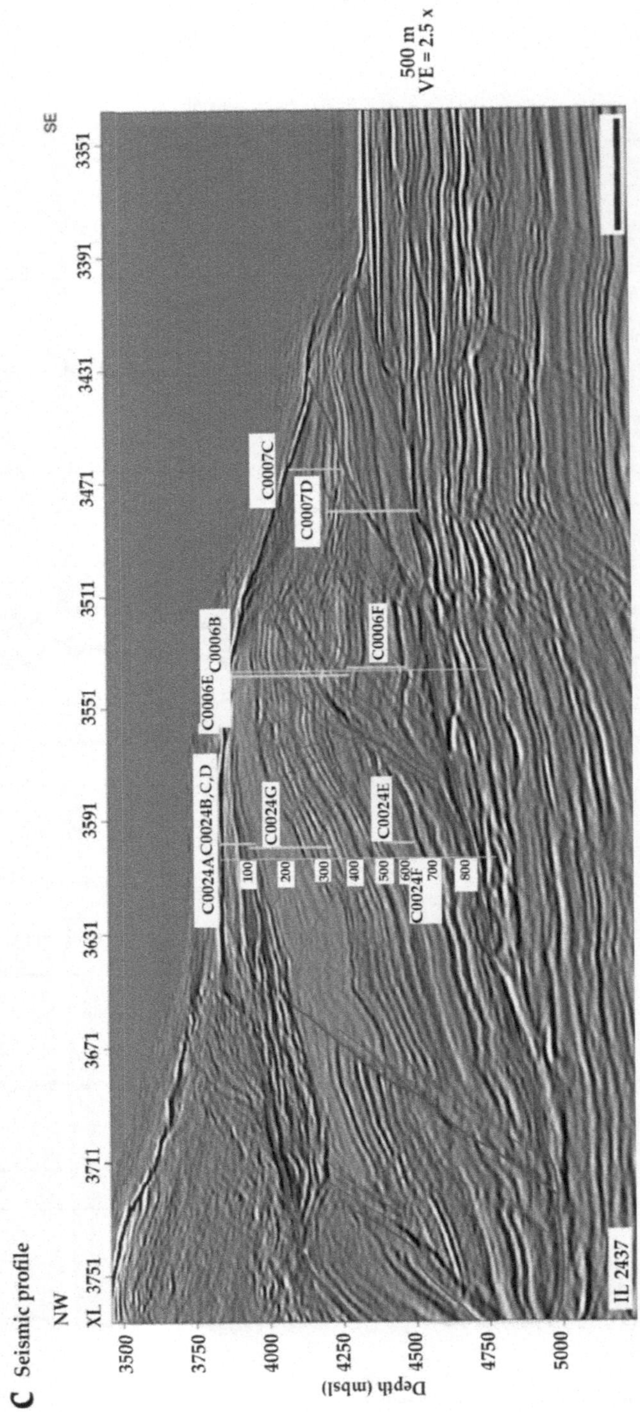

Figure 3. Data from the toe sites. a Site C0006, b site C0024, and c a seismic profile around the prism toe after Figure F7 of Tobin et al. (2020). Symbols are the same as in Figure 2.

The observed heat flow at site C0006 is very low (~ 35 mW/m^2) (Expedition 316 Scientists 2009a; Harris et al. 2011). However, according to Harris et al. (2011), the heat flow value corrected for the effects of bathymetry and rapid sedimentation was 45 mW/m^2. Hence, we used this value to estimate the porosity using the resistivity log data. We set the thermal structure correction factor α to 0.8 at the seafloor, and it was linearly increased to 1.0 at 25 mbsf, where the boundary between Units 1 and 2 is estimated. This was done, because Unit 1 had a rapid sedimentation rate as it accumulated at the slope apron (Expedition 316 Scientists2009a). In contrast, the observed heat flow at site C0024 is high (~ 98 mW/m^2) (Yamaguchi et al. 2020). No correction for thermal structure was applied, because the slope apron facies unit at site C0024 is very thin.

Figure 3 shows the porosities, thermal conductivities, and thermal structures of sites C0006 and C0024. The porosities and thermal conductivities measured from the core samples are in good agreement with the values estimated at site C0006, whereas they show systematic differences at site C0024, which could be due to insufficient information on the heat flow and thermal structure of this site. Porosity exhibits a slightly decreasing trend in the accreted trench sediments. The porosity then remains almost constant, with some fluctuations in the accreted USB sediments. Porosity increases from 0.4 to 0.5 across the potential décollement at site C0006, whereas this increase is not observed at site C0024.

Figure 3 also shows the NGR log data and the clay mineral fraction. Clay mineral abundances at these sites have been previously reported by Guo and Underwood (2012) and Yamaguchi et al. (2020). Both the original and normalized NGR logs at C0006 show constant values in the accreted trench wedge sediment, which increase in the transitional unit from the trench wedge sediment to the USB sediment. The clay mineral fraction also increases rapid from 40 to 50% to 60–70% at the boundary between these two units. The clay mineral fraction of the accreted USB sediment is consistent with those of the incoming sites. The original and normalized NGR logs show a dramatic drop at the possible décollement at site C0006, where the estimated porosity also increases. The low NGR values observed below this depth have been interpreted as underthrust sand layers (Expedition 314 Scientists 2009b).

Slope Sites

At site C0018, LWD was performed during IODP Expedition 338 (Strasser et al. 2014c), and coring was performed during Expedition 333 (Expedition 333 Scientists 2012c). At site C0021, LWD and coring were performed during Expedition 338 (Strasser et al. 2014d). According to the site reports, the stratigraphy at these sites consists of a single lithologic unit (Figure 4). These sites are located on the slope basin, and the entire drilled section is slope basin sediments. The single unit is further divided into two sub-units according to the sand content.

Porosity was estimated from the resistivity log using heat flow values of 62 mW/m^2 at site C0018 (Expedition 333 Scientists 2012c) and 55 mW/m^2 at site C0021. We used a lower heat flow at site C0021, because it is located ~ 2 km landward of site C0018. This value is consistent with the heat flow estimated using bottom-simulating reflector (BSR) in this region (Ohde et al. 2018). As no detailed information on the bathymetric and sedimentation rate corrections for the heat flow value are available for these sites, no thermal structure corrections were applied.

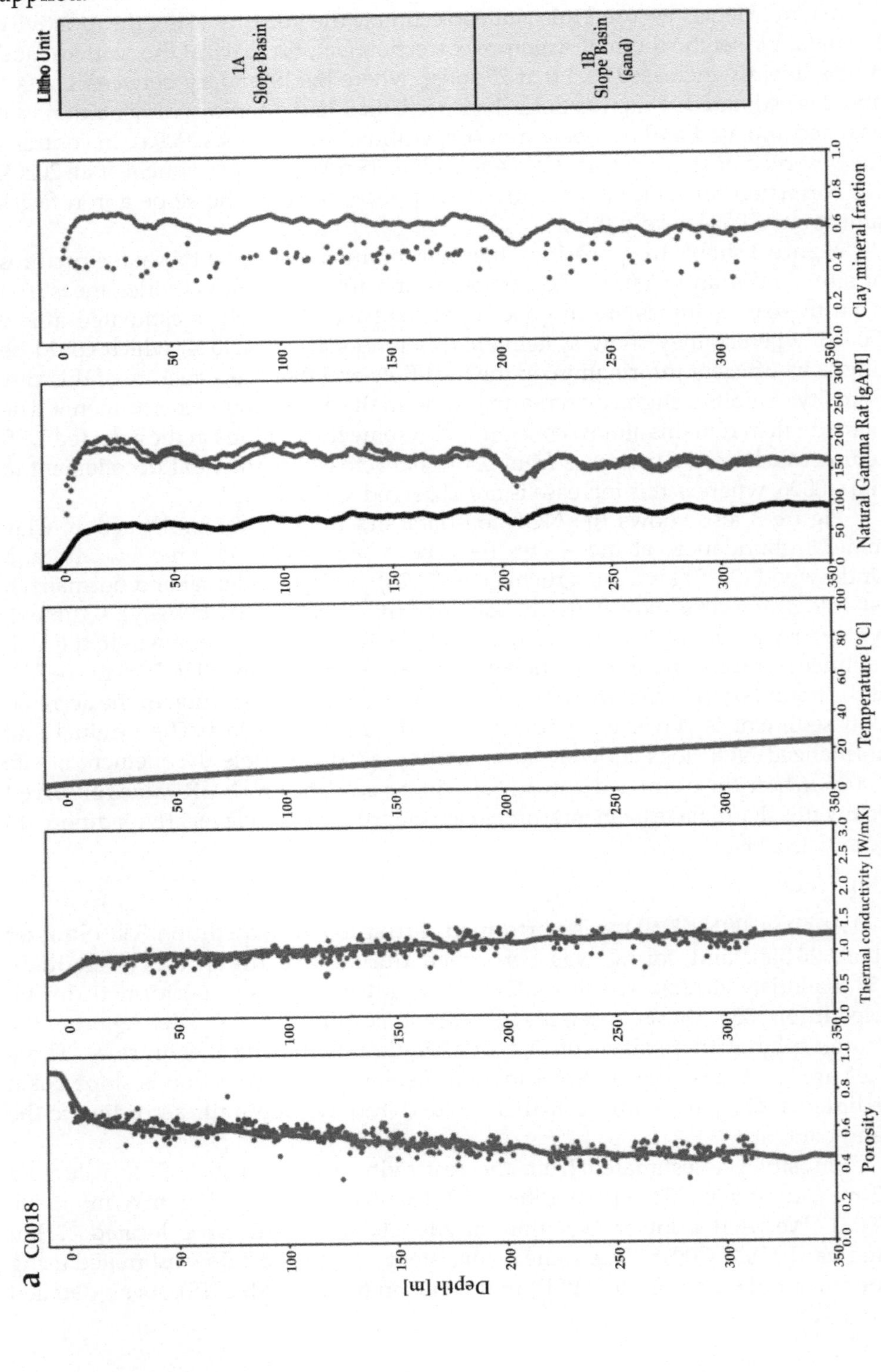

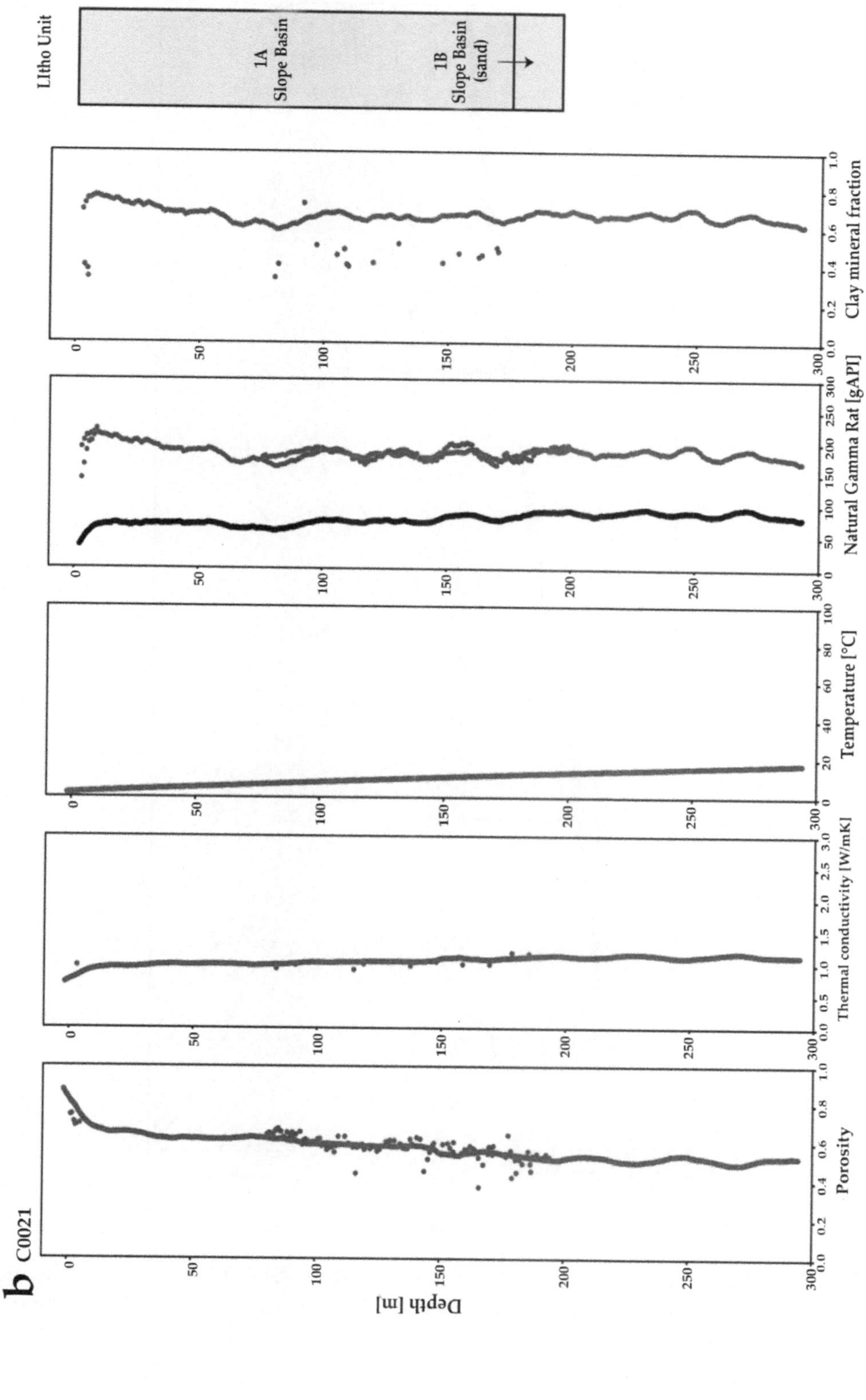
b C0021
Depth [m]
Porosity
Thermal conductivity [W/mK]
Temperature [°C]
Natural Gamma Rat [gAPI]
Clay mineral fraction
Litho Unit
1A
Slope Basin
1B
Slope Basin
(sand)

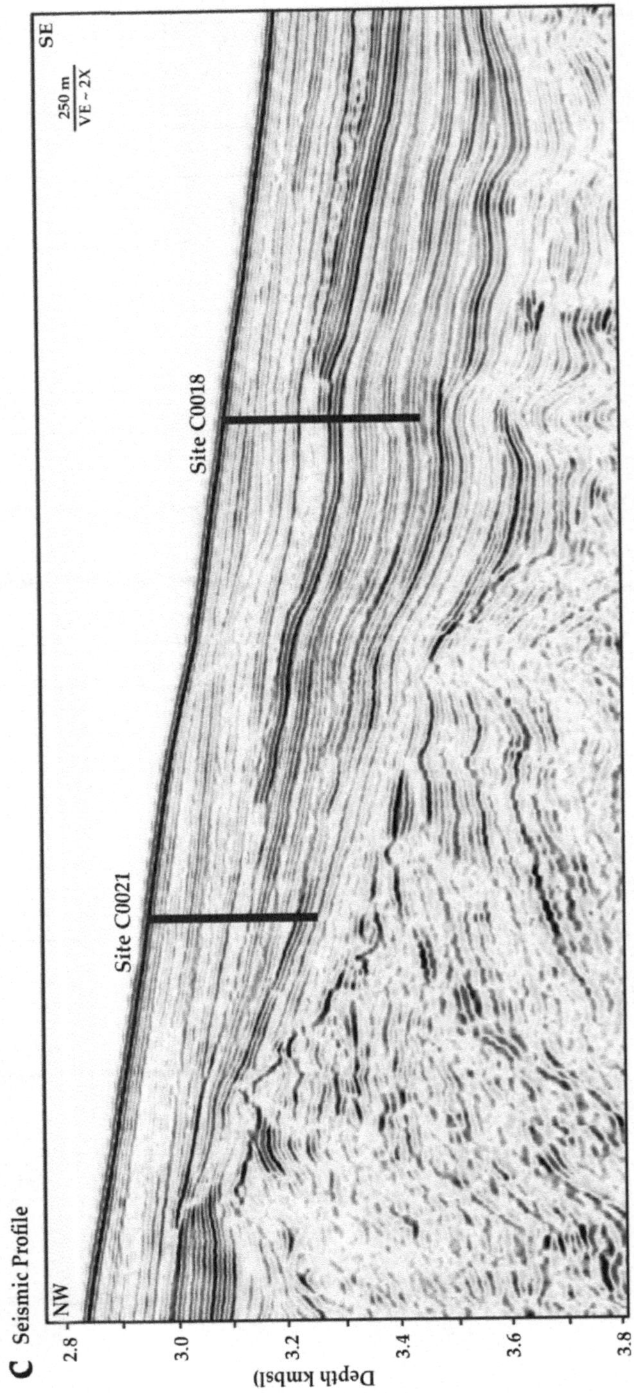

Figure 4. Data from the slope sites. a Site C0018, b site C0021, and c a seismic profile around the slope apron after Figure F3 of Strasser et al. (2014d). Symbols are the same as in Figure 2.

Figure 4 shows the porosities, thermal conductivities, and thermal structures at sites C0018 and C0021. The porosities and thermal conductivities measured from core samples are consistent with those estimated from the electrical resistivity logs. The porosities and thermal conductivities show simple decreasing and increasing trends with depth, respectively, with some fluctuations. Figure 4 also shows the NGR logs and clay mineral fractions. Clay mineral abundances at these sites have been reported previously by Underwood (2017a). The original NGR logs from site C0018 have almost constant values with several fluctuations. On the other hand, site C0021 exhibits an increasing trend with depth. A negative spike in the original NGR logs around 210 mbsf is interpreted as interbedding sandy layer (Strasser et al. 2014c). The normalized logs from site C0018 decrease across the sub-unit boundary, which corresponds to a higher sand content in the lower sub-unit. The clay fraction in the lower sub-unit exhibits lower values than those of the upper sub-unit, though they are scattered and include some higher values. In contrast, the normalized NGR logs from site C0021 exhibit almost constant values, with some fluctuations. Clay mineral abundances were measured in samples from the upper sub-unit only.

Outer Ridge Sites

At sites C0001 and C0004, LWD was performed during IODP Expedition 314 (Expedition 314 Scientists 2009c, d) and coring was performed during Expedition 315 (Expedition 315 Scientists 2009b; Expedition 316 Scientists 2009b). At site C0010, LWD was performed during Expedition 319 (Expedition 319 Scientists 2010c) and coring was performed during Expedition 365 (Saffer et al. 2017). At site C0022, LWD and coring were performed during Expedition 338 (Strasser et al. 2014a). Sites C0004, C0010, and C0022 were drilled through the splay fault, whereas site C0001 was drilled only through the hanging wall. According to the site reports, the stratigraphy of the hanging wall consists of two lithologic units (Figure 5): the slope apron facies and the accreted inner prism sediments. As ash layers are rare in the inner prism sediments, which are frequently included in USB sediments, shipboard scientists interpreted this unit as the trench fill deposited during a period of low sand–silt influx, such as Unit 2C or 2D at site C0006. The footwall of the splay fault is the underthrust slope apron facies observed at sites C0004 and C0010. A fault-bounded section is also observed around the splay fault at site C0004.

The observed heat flow at these sites is low: ~ 47 mW/m^2 at site C0001 (Expedition 315 Scientists 2009b; Harris et al. 2011) and 54–60 mW/m^2 at site C0004 (Expedition 314 Scientists 2009d; Harris et al. 2011). However, according to Harris et al. (2011), the heat flow value corrected for the bathymetry and rapid sedimentation effects was 60 mW/m^2 at both sites. Therefore, we used this value to estimate the porosities from the resistivity log data at sites C0001 and C0004. We also used this value at site C0022, as it is consistent with the heat flow estimated using BSR in this region (Ohde et al. 2018). We used a heat flow of 54 mW/m^2 at site C0010 (Expedition 319 Scientists 2010c). Although this value is slightly lower than the value used for the other sites, we used this value, because site C0010 is located westward of the other sites. This value is also consistent with the heat flow estimated using BSR in the region (Ohde et al. 2018). We only set the thermal structure correction factor α at site C0001 to 0.77 at the seafloor and it was linearly increased to 1.0 at 205 mbsf, where the boundary between Units 1 and 2 is estimated to occur, because Unit 1 has a rapid sedimentation rate and accumulated in the trench (Expedition 315 Scientists 2009b).

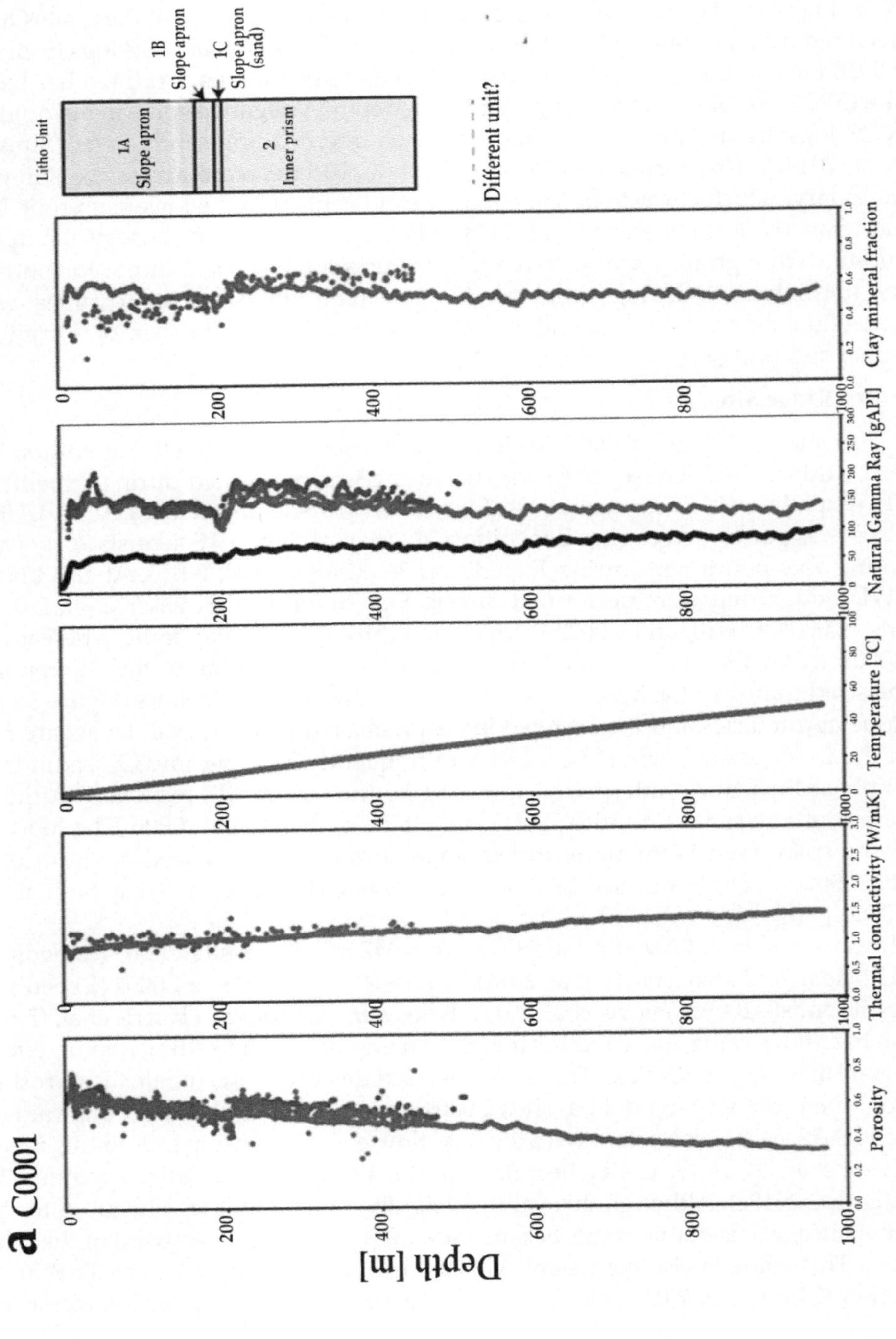

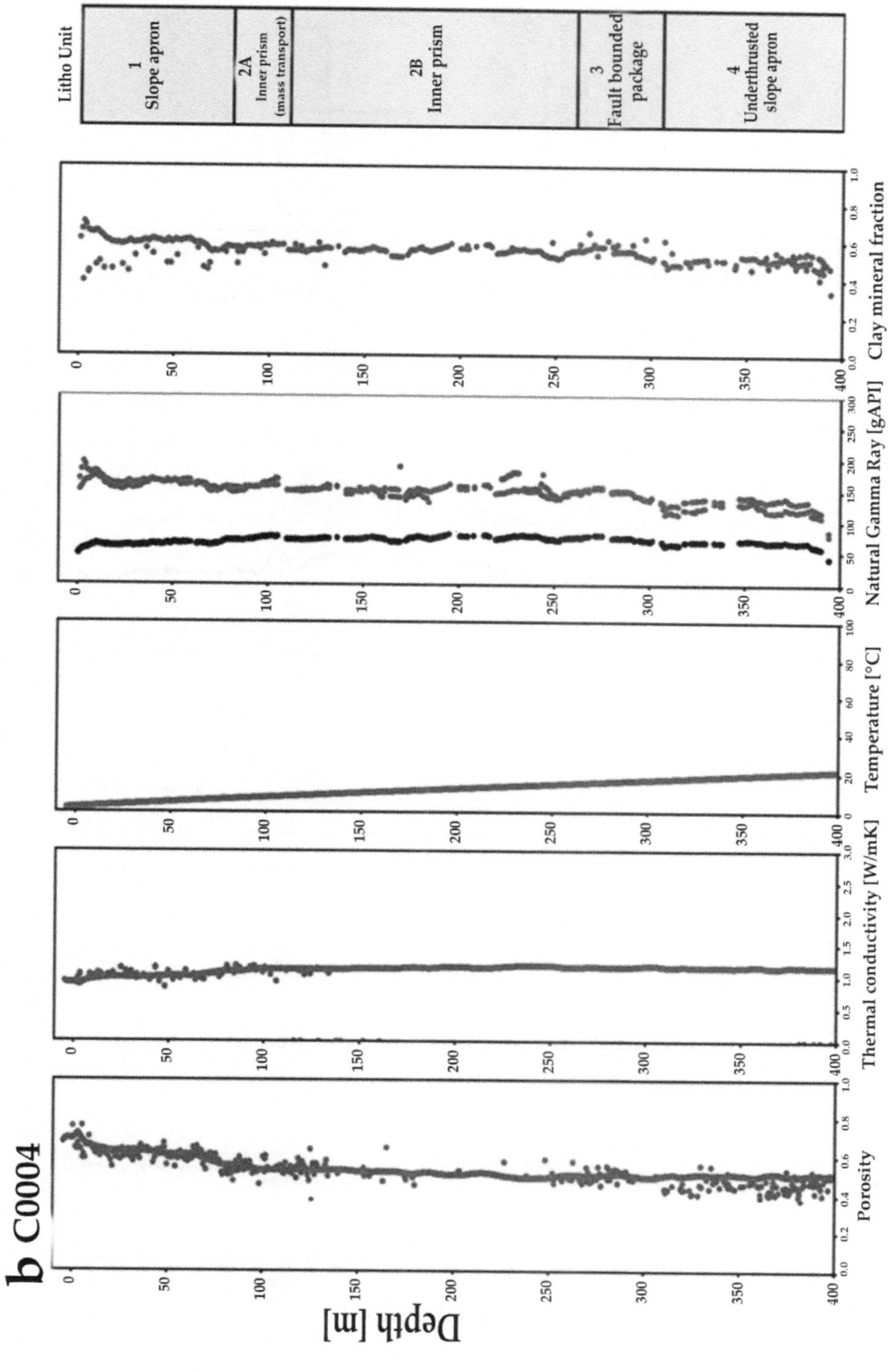
b C0004
Depth [m]
Porosity
Thermal conductivity [W/mK]
Temperature [°C]
Natural Gamma Ray [gAPI]
Clay mineral fraction
Litho Unit
1
Slope apron
2A
Inner prism
(mass transport)
2B
Inner prism
3
Fault bounded
package
4
Underthrusted
slope apron

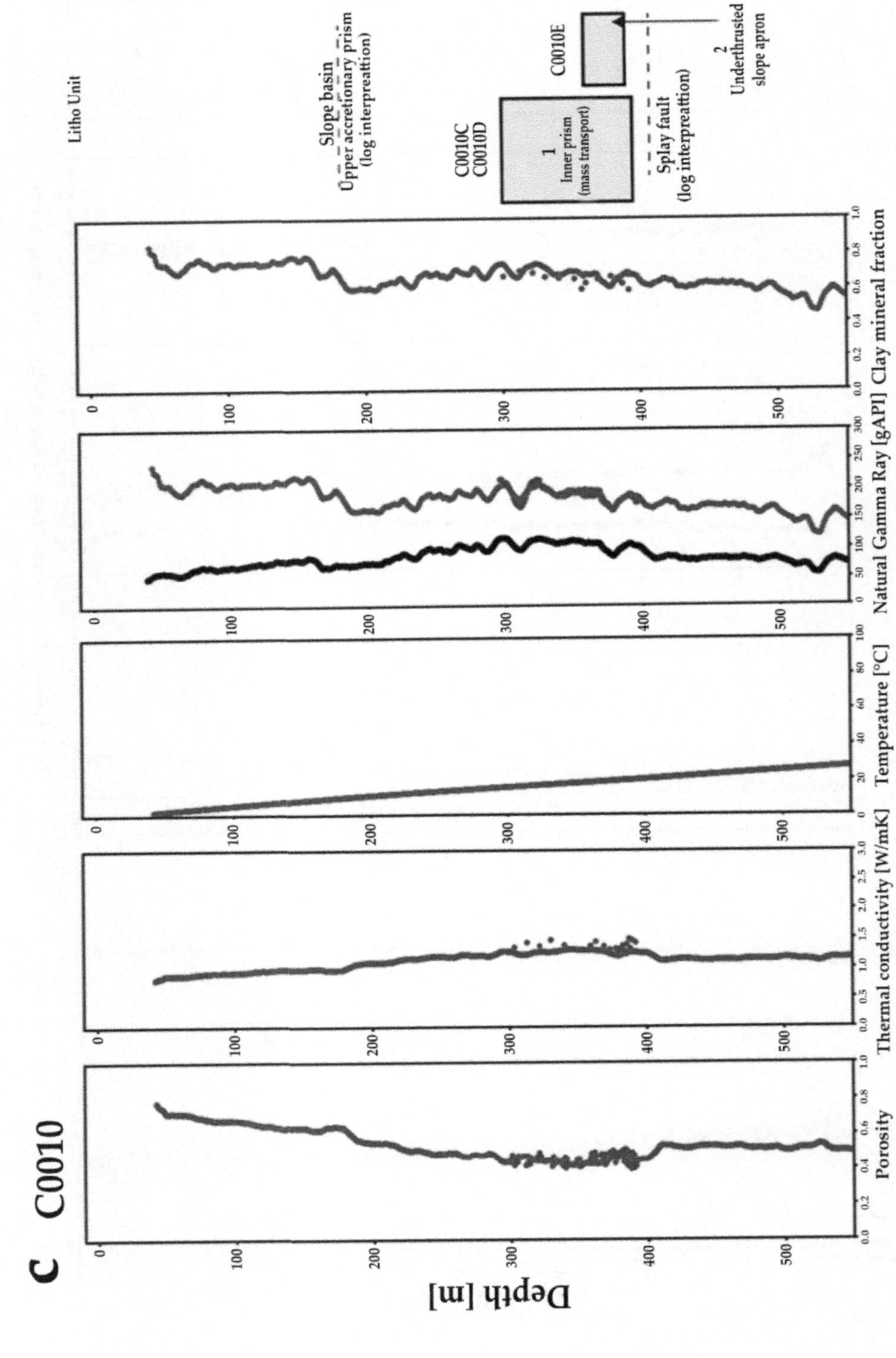
C
C0010
Depth [m]
Porosity
Thermal conductivity [W/mK]
Temperature [°C]
Natural Gamma Ray [gAPI]
Clay mineral fraction
Litho Unit
Slope basin
Upper accretionary prism
(log interpreattion)
C0010C
C0010D
1
Inner prism
(mass transport)
C0010E
Splay fault
(log interpreattion)
2
Underthrusted
slope apron

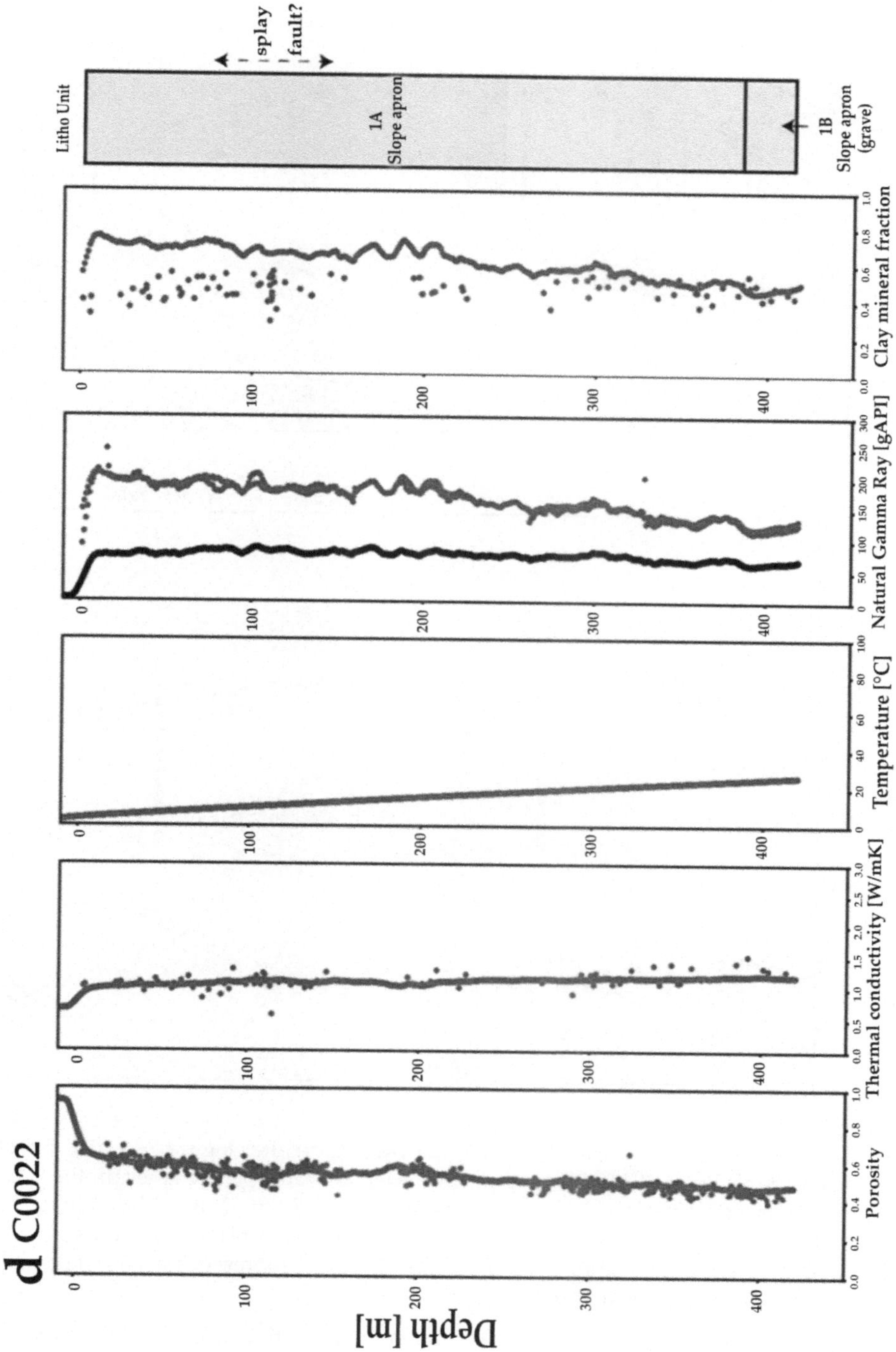
d C0022
Depth [m]
Porosity
Thermal conductivity [W/mK]
Temperature [°C]
Natural Gamma Ray [gAPI]
Clay mineral fraction
Litho Unit
1A
Slope apron
1B
Slope apron
(grave)
splay
fault?

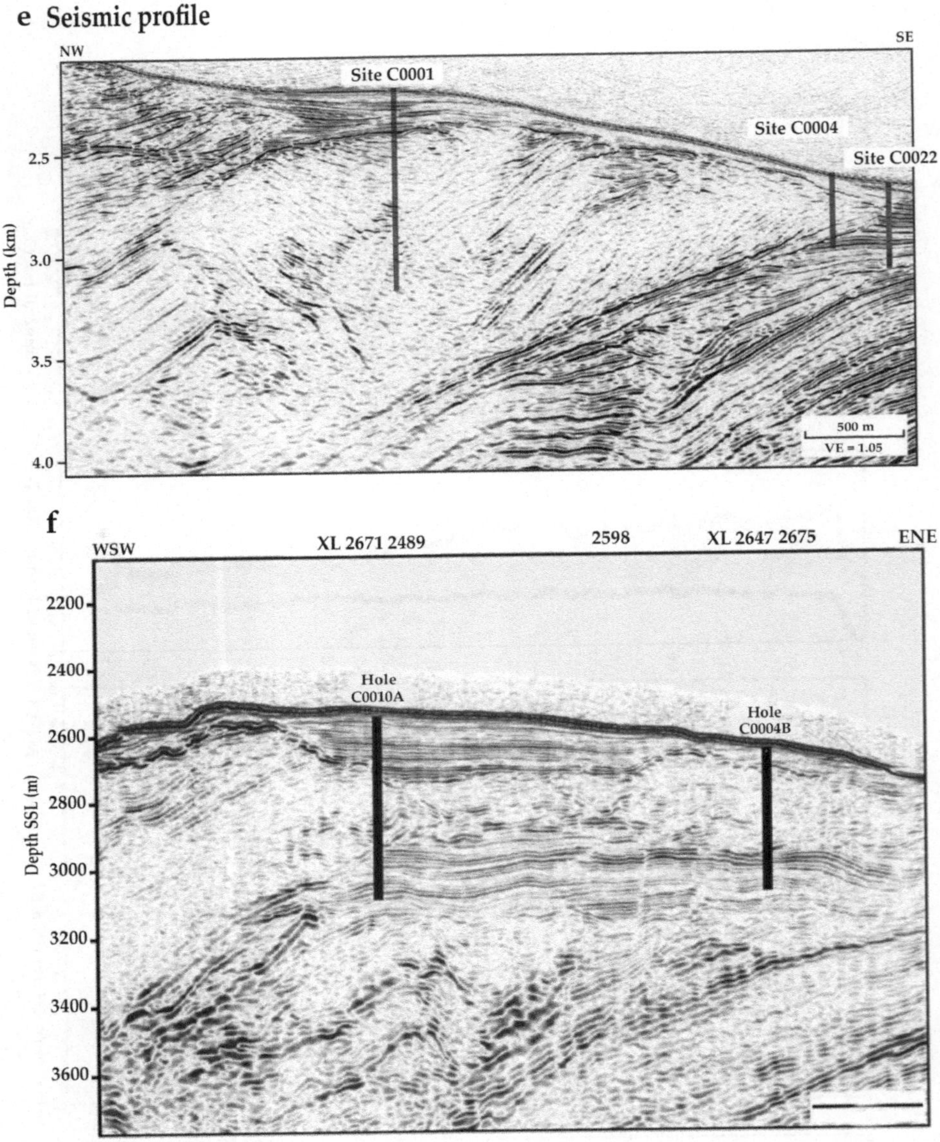

Figure 5. Data from the outer ridge sites. a Site C0001, b site C0004, c site C0010, d site C0022, and e a seismic profile around the outer ridge (left) after Figure F9A of Tobin et al. (2009) and (right) Figure F30 of Expedition 319 Scientists (2010c). The latter profile is taken in the along-trench direction. Symbols are the same as in Figure 2. In addition to the symbols explained in Figure 2, the blue symbols in the porosity plot show the porosities calculated from the density logs. Blue symbols in the NGR plot represent the NGR normalized using the density log-derived porosities.

Figure 5 shows the porosities, thermal conductivities, and thermal structures at sites C0001, C0004, C0010, and C0022. The porosities and thermal conductivities measured from the core samples are consistent with those estimated from the electrical resistivity logs, although they show slight differences in Unit 2 at site C0001, which could be due to an insufficient thermal structure correction. A step increase in porosity is observed across the unit boundary between the slope apron facies and the inner prism at site C0001, whereas it is not observed at site C0004.

Figure 5 also shows the NGR logs and clay mineral fractions. Clay mineral abundances at these sites have been previously reported by Guo and Underwood (2012), Saffer et al. (2017), and Strasser et al. (2014a). At site C0010, we only used the clay mineral fractions from Holes C0010C and C0010D, because Hole C0010E contains data from the underthrust sediments, even though the logged hole is within the inner prism unit at this depth interval. The NGR value could be lower than the true value due to poor hole conditions at several depth intervals (530–590, 625–650, 730–750, 780–860, and 900–925 mbsf) at site C0001 (Expedition 314 Scientists 2009c). In the slope apron facies, the original NGR logs show an increasing trend with depth, whereas the normalized NGR logs show almost constant values or a slightly decreasing trend with depth. The clay mineral fraction correspondingly increases from ~ 35 to ~ 45%. At site C0001, both the original and normalized NGR logs show a step increase across the unit boundary between the slope apron facies and the inner prism sediments. The clay mineral fraction increases correspondingly at this depth from 45 to 65%. However, such an increase was not observed at site C0004. In contrast, the normalized NGR values decrease in the underthrust slope apron facies at sites C0004 and C0010. The clay mineral fraction decreases correspondingly to ~ 45% at site C0004.

Kumano Basin Sites

At site C0002, LWD was performed during IODP Expeditions 314, 338, 348, and 358 (Expedition 314 Scientists 2009e; Strasser et al. 2014e; Tobin et al. 2015a; Kitajima et al. 2020) and coring was performed during Expeditions 315 and 338 (Expedition 315 Scientists 2009c; Strasser et al. 2014e). During Expeditions 348 and 358, short cores were also recovered. During Expeditions 338, 348, and 358, cutting samples were recovered during LWD (e.g., Tobin et al. 2015b). At site C0009, wireline logging was conducted during IODP Expedition 319 (Expedition 319 Scientists 2010a). According to the site reports, the stratigraphy of these sites is categorized into three lithologic units (Figure 6). The upper unit is the Kumano Basin sediments, which are terrigenous sediments that accumulated behind the uplifting outer ridge. The middle unit is the slope apron facies, which was deposited before the uplift of the outer ridge occurred. The lower unit is the inner prism sediments.

The observed heat flow at site C0002 is low at ~ 40 mW/m^2 (Expedition 315 Scientists 2009c; Harris et al. 2011). However, according to Harris et al. (2011), the heat flow value corrected for bathymetry and rapid sedimentation effects was 57 mW/m^2. This value is consistent with the value estimated by Sugihara et al. (2014). Therefore, we used this value to estimate the porosities using the resistivity logs. We set the thermal structure correction factor α to 0.78 at the seafloor, and it was linearly increased to 1.0 at 830 mbsf, where the unit boundary between Units 2 and 3 is estimated to occur, because the Kumano Basin sediments have a rapid sedimentation rate (Expedition 315 Scientists 2009c). As this site was investigated previously by Yabe et al. (2019), refer to their study for details. At site

C0009, wireline logging was conducted only in the deeper part of the hole, which prohibited the application of the method developed by Yabe et al. (2019) at this site. Instead, we refer to the porosities estimated from the density log data.

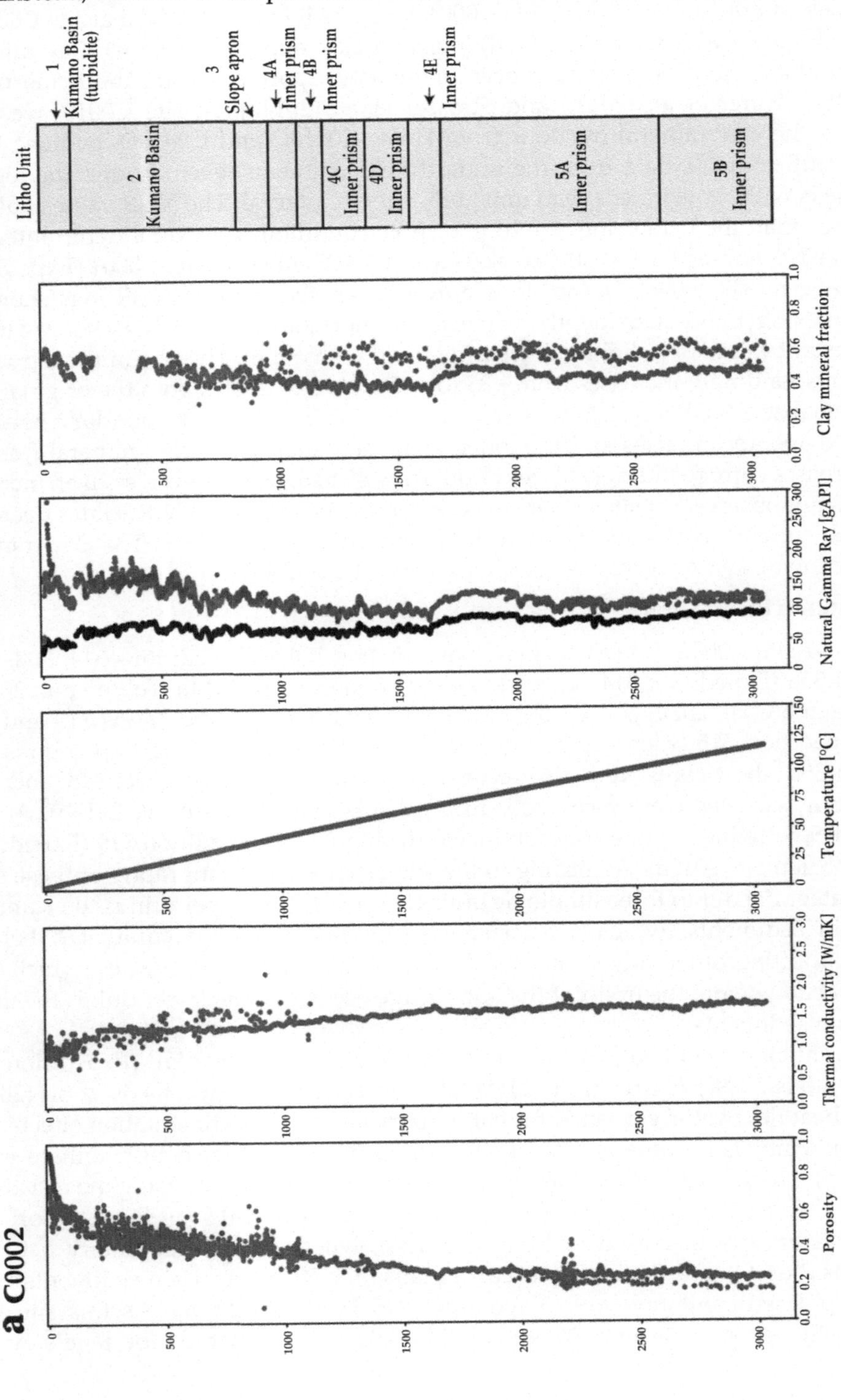

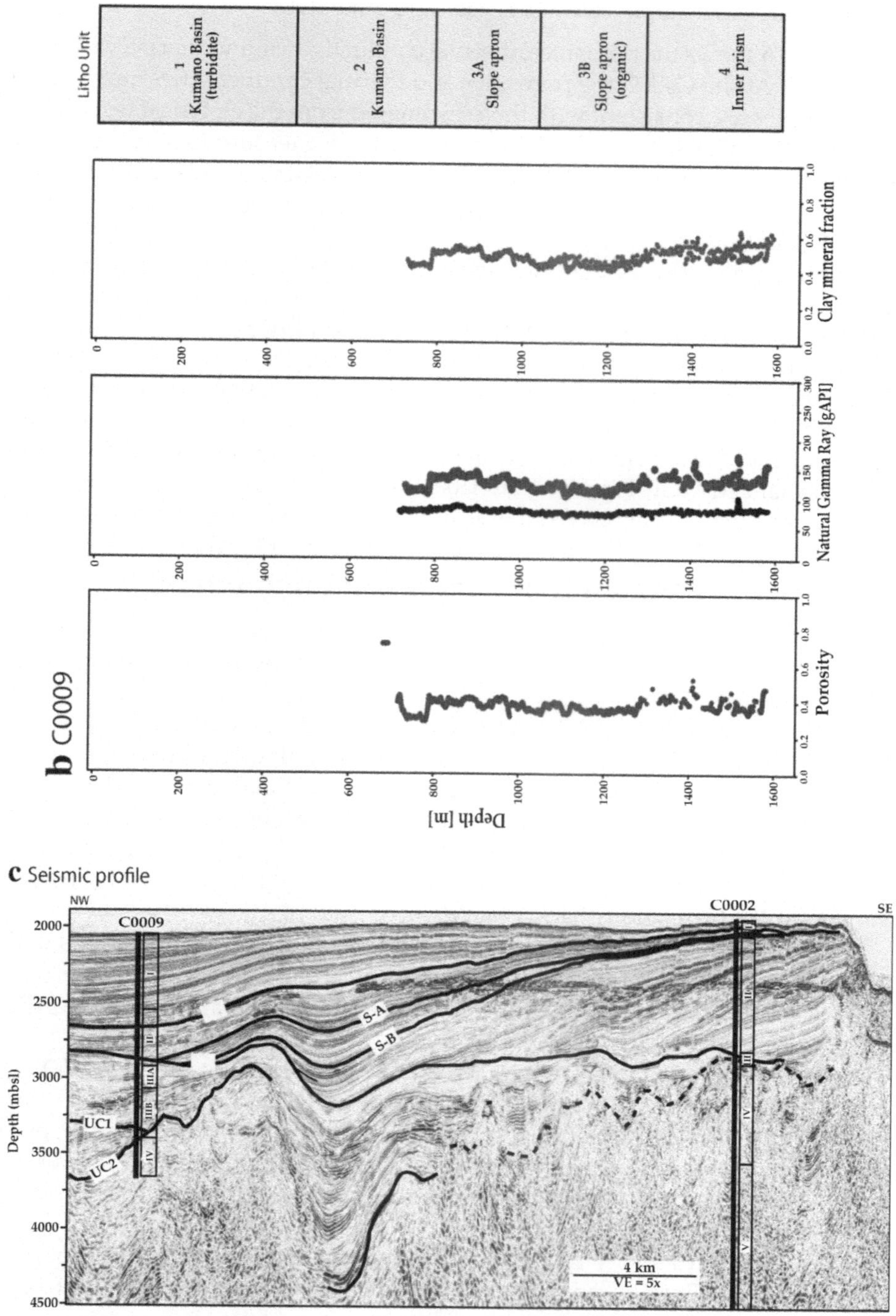

Figure 6. Data from the Kumano Basin sites. a Site C0002, b site C0009, and c a seismic profile around the Kumano Basin after Figure F116 of Expedition 319 Scientist (2010a). Symbols are the same as those in Figs. 2 and 5. Lsithological units colored by light blue represent the

inner prism sediment below the Kumano Basin, which shows lower NGR log values than the standard sediments possibly due to poor hole condition.

Figure 6 shows the porosities, thermal conductivities, and thermal structures at site C0002. At site C0002, the porosities and thermal conductivities measured from core samples are consistent with those estimated from the electrical resistivity logs above ~ 1100 mbsf. Below ~ 2000 mbsf, porosity was measured mainly from cutting samples, and is roughly consistent with the values estimated using resistivity logs, although there are some differences. Yabe et al. (2019) discussed this previously in further detail.

Figure 6 also shows the NGR logs and clay mineral fractions. Clay mineral abundances at these sites have been previously reported by Guo and Underwood (2012), Underwood and Song (2016), Underwood (2017b, c), and Expedition 319 Scientists (2010a).

In the Kumano Basin sediments at site C0002, the original and normalized NGR logs exhibit a decreasing trend with depth below ~ 500 mbsf. Correspondingly, the clay mineral fraction decreases from ~ 55% at ~ 500 mbsf to ~ 45% at ~ 830 mbsf. The normalized NGR log from site C0009 also exhibits a change in trend in the slope apron facies. The clay mineral fraction also decreases from 50% at ~ 1100 mbsf to 40% at ~ 1200 mbsf, then increases to 50% at ~ 1300 mbsf.

NORMALIZED GAMMA RAY LOGS VS. CLAY MINERAL FRACTIONS

In this section, the normalized gamma ray logs calculated in the previous section are compared with the clay mineral fractions measured from the core and cutting samples.

We estimated the porosity in three ways (MAD porosity, porosity estimated from electrical resistivity logs, and porosity estimated from density logs). The three types of estimated porosities generally exhibited good agreement with each other, as shown in the porosity depth profiles (Figs. 2, 3, 4, 5, 6). Therefore, we used the NGR logs normalized by the resistivity log derived porosity as representative data (Figure 7), because logging data represent the bulk rock values for the borehole.

Figure 7 shows the relationship between the normalized NGR logs and the clay mineral fraction in each geologic environment. At the incoming sites (Figure 7a), the normalized NGR values and clay mineral fractions are ~ 150–200 gAPI and 60–80%, respectively (red plots in Figure 7a).

An exception is Unit 1A of the USB sediments (green plots in Figure 7a), which has a lower clay mineral fraction (50–65%), although the normalized NGR values exceeded 200 gAPI. MSB sediments were excluded from this comparison, because the Zenisu Fan is composed of sandy sediments, as suggested by their low NGR values, although only a few measurements of the clay mineral fraction indicate low values.

This could be due to a sampling bias of the clay mineral abundance measurements. Units 4 and 5 of the LSB sediments were also excluded from this comparison, because the clay mineral fractions measured from core samples exhibited scattered values possibly due to interbedding sandy layers (Expedition 322 Scientists 2010b).

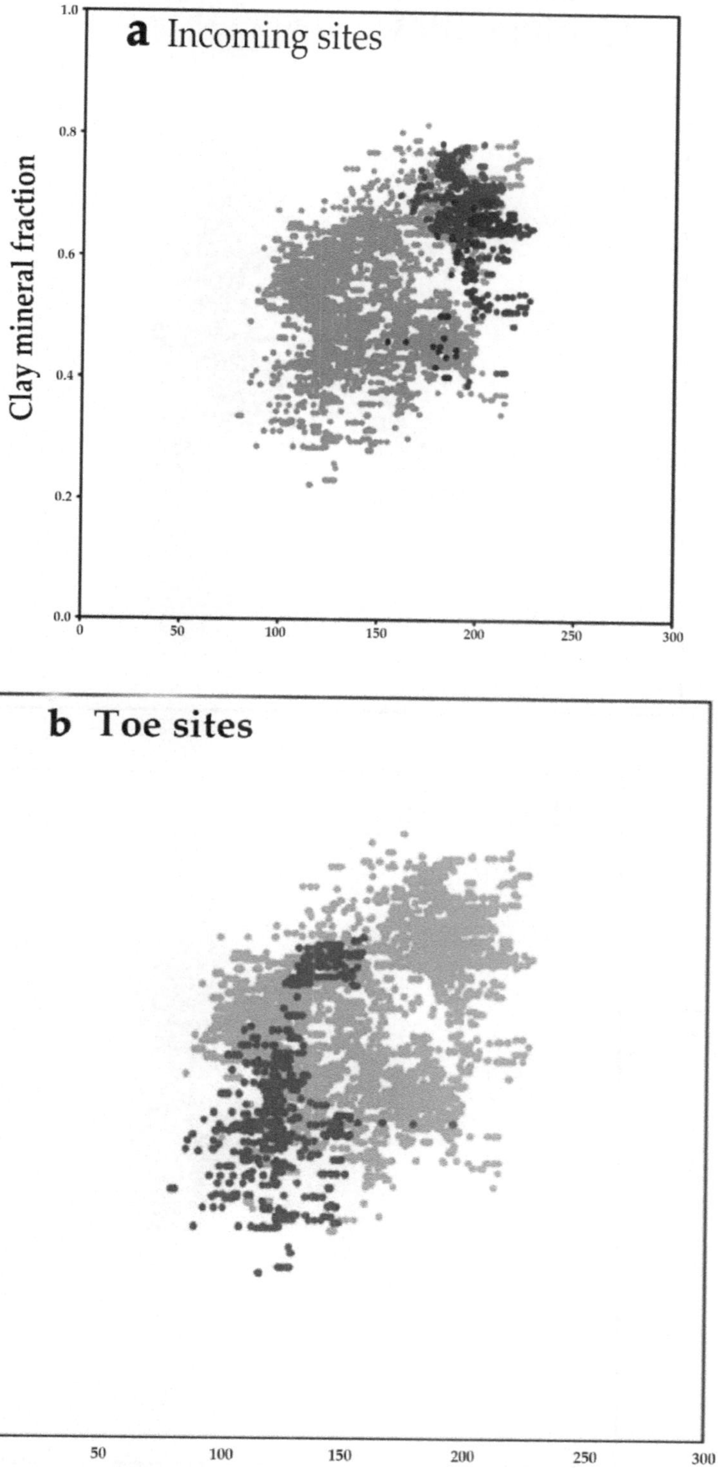
a Incoming sites
Clay mineral fraction
1.0
0.8
0.6
0.4
0.2
0.0
0
50
100
150
200
250
300
b Toe sites
1.0
0.8
0.6
0.4
0.2
0.0
0
50
100
150
200
250
300

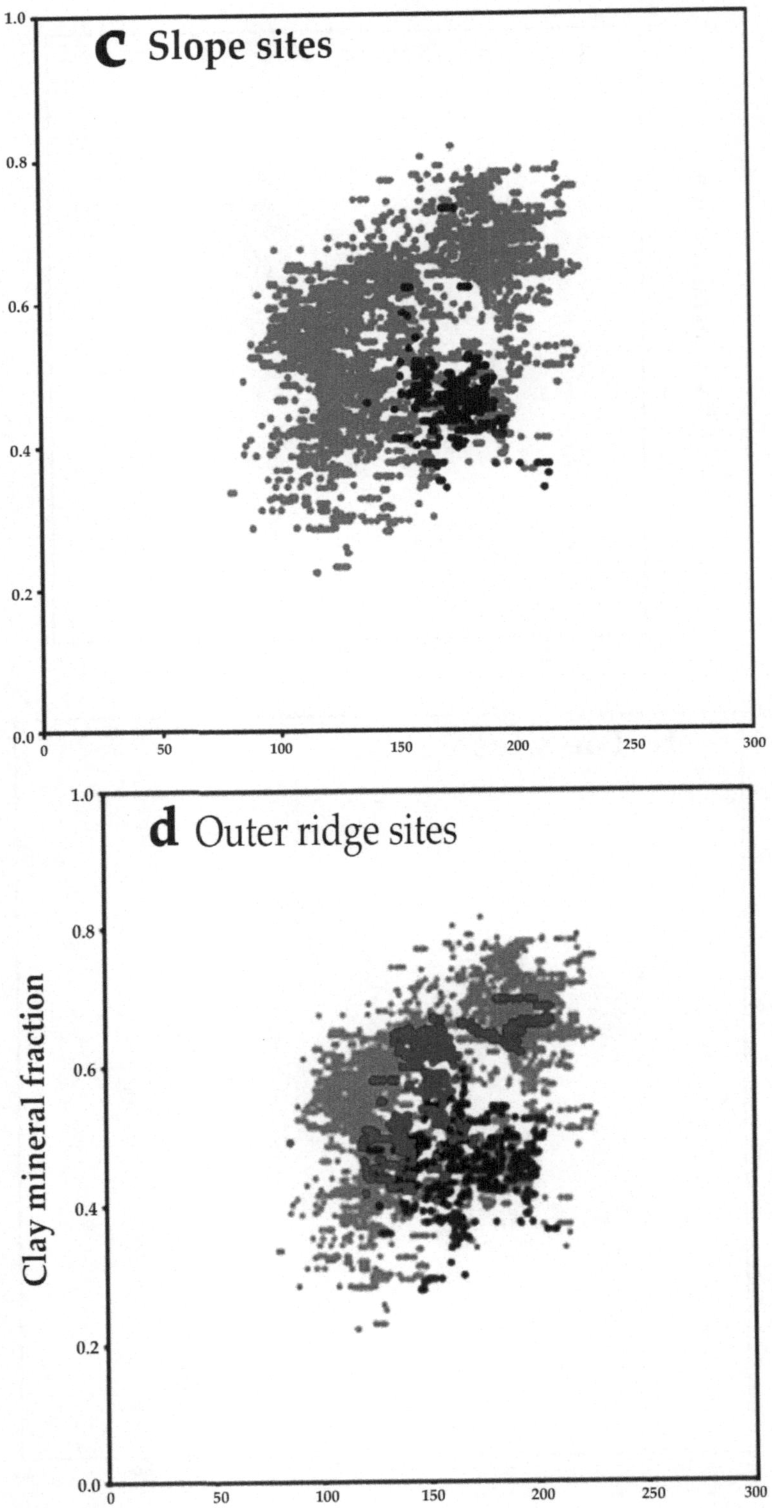
c Slope sites
1.0
0.8
0.6
0.4
0.2
0.0
0
50
100
150
200
250
300
d Outer ridge sites
Clay mineral fraction
1.0
0.8
0.6
0.4
0.2
0.0
0
50
100
150
200
250
300

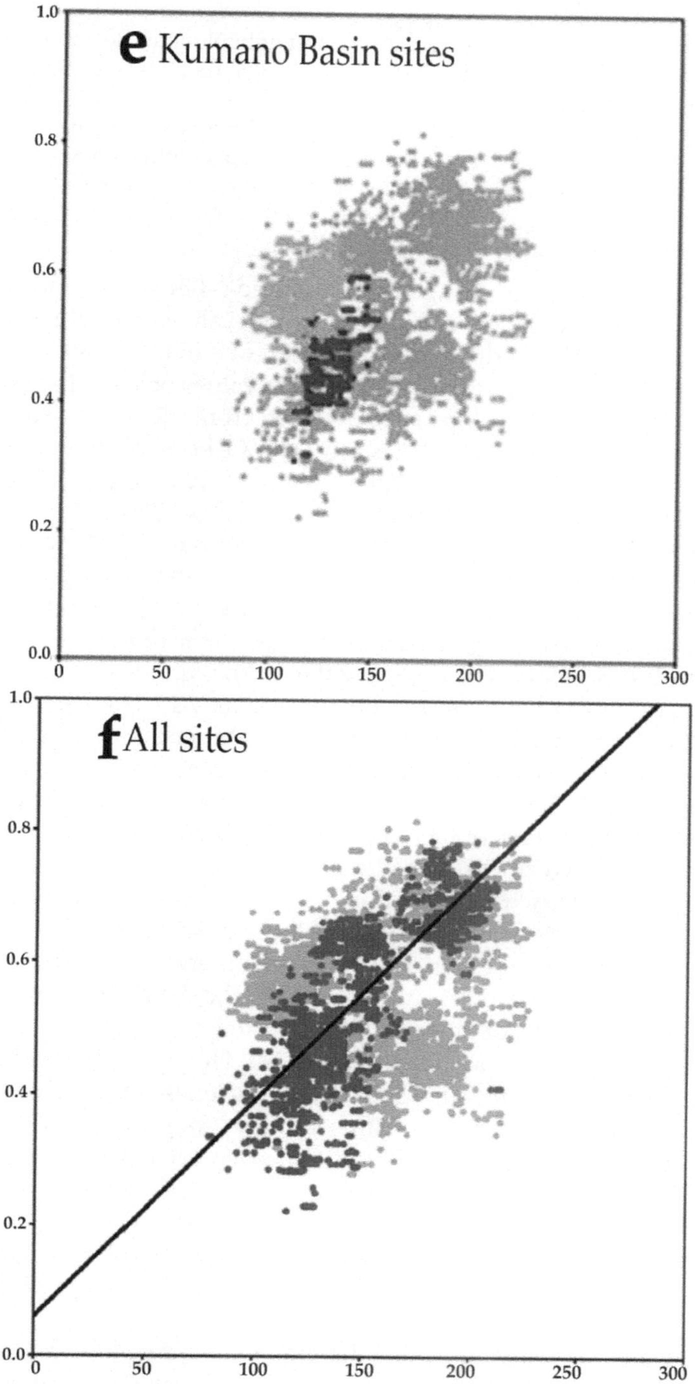

NGR logs normalized porisity estimated elecrtical resistivity logs [gAP]

Figure 7. Comparison between the NGR logs normalized using porosity estimated from the electrical resistivity logs and the clay mineral fractions. The panels show results from the a incoming sites, b toe sites, c slope sites, d outer ridge sites, e Kumano Basin sites, and f all sites combined. Color code (red, green, and light blue) for dots in Figure 7 is common to that used for lithological units in Figs. 2, 3, 4, 5, 6. Red dots represent the standard sediments, for which the empirical conversion function is established. Green dots represent the upper part of USB sediments or the slope apron facies younger than 1.5 Ma, which shows higher normalized NGR log values than the standard sediments possibly due to volcanic ashes included in sediments. Light blue dots are the inner prism sediment below the Kumano Basin, which shows lower NGR log values than the standard sediments possibly due to poor hole condition. Grey dots indicate all data.

The toe sites cover a wide range of normalized NGR values and clay mineral fractions (Figure 7b). The trench wedge facies exhibit values of ~ 100 gAPI and ~ 40%, whereas the accreted USB sediments have values of ~ 150 gAPI and ~ 65%. Trends observed at the toe sites can extend to the higher values observed at the incoming sites. On the other hand, the slope sites deviated from this trend (Figure 7c). The clay mineral fraction at the slope sites is low (~ 45%), because they consist of slope apron facies, although they exhibit high normalized NGR values of 150–200 gAPI. Similar trends were observed at the outer ridge sites (Figure 7d). Slope apron facies at the top of the formation that are younger than 1.5 Ma (Expedition 333 Scientists 2012c; Strasser et al. 2014d), have low clay mineral fractions, but exhibit high normalized NGR values (green dots in Figure 7d). On the other hand, the inner prism sediments exhibit normalized NGR values of ~ 150 gAPI and clay mineral fractions of ~ 65%, which are similar to the values of the accreted USB sediments found at the toe sites (red dots in Figure 7d). The slope apron facies at site C0022 older than 1.5 Ma (Strasser et al. 2014a) and underthrust sediments at site C0004, which are also older than 1.5 Ma (Expedition 316 Scientists 2009b), exhibit normalized NGR values of ~ 120 gAPI and clay mineral fractions of ~ 50%, which are close to the values of the accreted trench wedge sediments found at the toe sites (red dots in Figure 7d). The Kumano Basin sediments also exhibit similar values as the accreted trench wedge sediments found at the toe sites (Figure 7e). The Kumano Basin sediments examined here are located below 500 mbsf at site C0002, and have ages of ~ 1.5 Ma or older. The inner prism sediments below the Kumano Basin sediments deviate from the overall trend of the data. They have a relatively high clay mineral fraction of 50–65%, although the normalized NGR values are small, at ~ 100 gAPI (light blue dots in Figure 7e). However, this deviation in the inner prism below the Kumano Basin sediments is less evident at site C0009. The clay mineral fraction of the inner prism sediments found at site C0009 is ~ 55%, which is close to the value at site C0002. However, NGR values normalized by porosity estimated from the density logs are ~ 140 gAPI, which is much higher than those at site C0002 (Figure 6). Inner prism sediments at site C0001 also exhibit normal NGR values (Figure 5). Hence, the low NGR values observed at site C0002 could be due to poor hole conditions (Expedition 314 Scientists 2009e; Tobin et al. 2015a).

Although the datasets compiled herein are scattered and contain some exceptions (Unit 1A of the USB sediments and slope apron sediments younger than ~ 1.5 Ma, which are presented by green dots in Figure 7, and inner prism sediments below the Kumano Basin sediments, which are presented by light blue dots in Figure 7), they show a roughly linear relationship between the normalized

NGR log values and the clay mineral fraction (red dots in Figure 7). The scattering may be attributed to several factors, such as variable borehole conditions, minor differences in the lithology of a stratigraphic unit, insufficient calibration, and different spectral detectability of the logging tools. However, the overall trend shows a clear linear relationship between the normalized NGR values and the clay mineral fraction. We used this relationship to establish an empirical conversion function between the normalized NGR logs and the clay mineral fractions by applying least-square fitting of a linear function. We weighted the data using 1 − porosity when applying the fit, because higher porosities produce larger uncertainties in the normalized NGR logs. The conversion function was estimated to be:

$$\text{Claymineralfraction}=3.3(\pm1.0)\times10^{-3}\times\text{NGR} + 0.06(\pm0.15) \qquad (1)$$

when using porosity estimated from the electrical resistivity logs (Figure 7f). The coefficient of determination is 0.60. When using MAD porosity, the conversion function was estimated to be:

$$\text{Claymineralfraction}=2.9(\pm0.8)\times10^{-3}\times\text{NGR} + 0.12(\pm0.12) \qquad (2)$$

with the coefficient of determination of 0.68 and for porosity estimated from density logs the conversion function was estimated to be

$$\text{Claymineralfraction}=2.6(\pm2.2)\times10^{-3}\times\text{NGR} + 0.14(\pm0.30) \qquad (3)$$

with the coefficient of determination of 0.75. Those conversion functions are consistent with each other within the fitting uncertainties. Although the coefficient of determination is highest for the equation based on porosity estimated from density logs (Eq. 3), the uncertainty for the estimated parameters is largest. This is due to the lesser amount of data for density logs. The standard deviations for fitting residuals are about 5–8% in every case. With these conversion functions, we can estimate the clay mineral fraction at depth intervals, where only logging data are available.

The predicted clay mineral fraction with the established conversion function (Eq. 1) at all sites are shown in Figs. 2, 3, 4, 5, 6 and summarized in Table 1. At incoming sites (Figure 2), the predicted clay mineral fractions basically show good agreement with the measured ones except for Unit 1A of USB sediments. At the middle part of C0012 (Unit 2–4), the predicted clay mineral fraction is systematically ~ 5% lower than the measured ones, though its difference is within the standard deviation for fitting residuals of established conversion function. At toe sites (Figure 3), the predicted clay mineral fractions for trench wedge sediments at C0006 are in good agreements with the measured ones. At C0024, the predicted clay mineral fractions for trench wedge sediments are systematically higher than the measured ones. This could be due to porosity estimated systematically higher than the MAD porosity at this site. The predicted clay mineral fractions for the accreted USB sediments at toe sites tend to be lower than the measured ones. This deviation of ~ 10% is slightly larger than the standard deviation for fitting residuals of established conversion function, though its causes are uncertain. At slope sites (Figure 4), the predicted clay mineral fractions for slope basin sediments are higher than the measured ones. The same trend can be observed at outer ridge sites (Figure 5). The predicted clay mineral fractions for slope apron sediments are higher than the measured ones.

Table 1. A summary table for all sites studied in this study

Setting	Site	Unit	Lithological interpretation	Normalized NGR log value [gAPI]	Predicted clay mineral fraction	Clay mineral fraction	comment
Incoming sites	C0011	1A	USB	200–220	~ 0.7–0.8	0.55–0.65	High NGR log values due to volcanic ash
		1B	USB	~ 200	~ 0.7	~ 0.7	
		2	MSB	Scattered (90–210)	Scattered (0.4–0.7)	Scattered (0.4–0.7)	Scattered due to interbedding sandy layers
		3	LSB	~ 200	~ 0.7	~ 0.7	hemipelagic
		4	LSB	130–220	0.55–0.8	Scattered (0.3–0.75)	Terrigenous, Scattered due to interbedding sandy layers
		5	LSB	~ 160	~ 0.6	Scattered (0.5–0.9)	tuff, Scattered due to interbedding sandy layers
	C0012	1A	USB	190–210	0.65–0.75	0.25–0.65	High NGR log values due to volcanic ash
		1B	USB	~ 180	~ 0.65	~ 0.65	
		1C	USB	~ 180	~ 0.65	~ 0.65	Rich in volcanic ash
		2	MSB	140–190	0.5–0.7	0.45–0.8	Sandy
		3	LSB	~ 190	~ 0.7	~ 0.75	hemipelagic
		4	LSB	160–220	0.6–0.8	0.65–0.9	terrigenous
		5	LSB	130–190	0.45–0.7	Scattered (0.3–0.9)	Volcaniclastic, Scattered due to interbedding snady layer
		6	Pelagic clay	–	–	–	Too thin to be analyzed
		7	Basalt	–	–	–	N/A
Toe sites	C0006	1	Slope apron	90–150	0.35–0.55	~ 0.4	Transitional to trench wedge
		2A	Trench wedge	~ 110	~ 0.45	0.3–0.5	Sandy
		2B	Trench wedge	~ 110	~ 0.45	0.35–0.6	Sandy to muddy
		2C	Trench wedge	~ 110	~ 0.45	0.35–0.6	muddy
		2D	Trench wedge	~ 130	~ 0.5	0.5–0.6	Transitional to USB
		3	USB	~ 150	~ 0.55	~ 0.65	
	C0024	1	Slope apron	–	–	–	Too thin to be analyzed
		2	Trench	90–150	0.4–0.55	0.2–0.5	There exists bias in porosity estimation
		3	Trench	80–120	0.4–0.5	0.35–0.5	Transitional to USB
		4	USB	~ 130	~ 0.5	~ 0.6	
Slope sites	C0018	1A	Slope basin	~ 190	~ 0.65	0.35–0.55	High NGR log values due to volcanic ash
		1B	Slope basin	~ 160	~ 0.6	0.3–0.6	Sandy, High NGR log values due to volcanic ash
	C0021	1A	Slope basin	180–210	0.65–0.8	0.3–0.5	High NGR log values due to volcanic ash
		1B	Slope basin	~ 180	~ 0.65	–	Sandy, No clay mineral fraction from core samples
Outer ridge sites	C0001	1A	Slope apron	140–160	0.5–0.55	0.25–0.5	High NGR log values due to volcanic ash
		1B	Slope apron	~ 140	~ 0.5	~0.5	High NGR log values due to volcanic ash
		1C	Slope apron	~ 120	~ 0.45	~ 0.35	Sandy, High NGR log values due to volcanic ash
		2	Inner prism	~ 150	~ 0.55	~ 0.65	
	C0004	1	Slope apron	~ 160	~ 0.6	0.45–0.55	High NGR log values due to volcanic ash
		2A	Inner prism	~ 150	~ 0.55	0.55–0.6	Mass transport deposit
		2B	Inner prism	~ 150	~ 0.55	0.55–0.6	
		3	Fault bounded package	~ 150	~ 0.55	0.55–0.6	
		4	Underthrust slope apron	~ 140	~ 0.5	~ 0.5	
	C0010	1	Inner prism	~ 190	~ 0.7	~ 0.65	
		2	Underthrust slope apron	–	–	–	Removed from analysis due to too large depth shift
	C0022	1A	Slope apron	~ 190/120–150	~ 0.7/0.45–0.6	0.35–0.55/0.4–0.55	0–250 m (younger than ~ 1.5 Ma)/250 m – (older than ~ 1.5 Ma)
		1B	Slope apron	~ 120	~ 0.45	0.4–0.55	Rich in gravel
Kumano basin sites	C0002	1	Kumano Basin	~ 150	~ 0.55	–	Turbidite, No clay mineral fraction from core samples
		2	Kumano Basin	130–150	0.45–0.55	0.4–0.6	
		3	Ancient slope apron	~ 140	~ 0.5	0.4–0.6	
		4A	Inner prism	~ 100	~ 0.4	0.5–0.65	Poor hole condition?
		4B	Inner prism	~ 100	~ 0.4	0.5–0.65	Poor hole condition?
		4C	Inner prism	~ 100	~ 0.4	0.5–0.65	Poor hole condition?
		4D	Inner prism	~ 100	~ 0.4	0.5–0.65	Poor hole condition?
		4E	Inner prism	~ 100	~ 0.4	0.5–0.65	Poor hole condition?
		5A	Inner prism	110–140	0.4–0.55	0.5–0.65	Poor hole condition?
		5B	Inner prism	~ 120	0.45	0.5–0.65	Poor hole condition?
	C0009	1	Kumano Basin	–	–	–	Turbidite, No clay mineral fraction from core samples
		2	Kumano Basin	~ 110	~ 0.45	–	No clay mineral fraction from core samples
		3A	Ancient slope apron	~ 130	~ 0.5	–	No clay mineral fraction from core samples
		3B	Ancient slope apron	~ 110	~ 0.45	~ 0.45	Rich in organic materials
		4	Inner prism	~ 130	~ 0.5	0.5–0.6	Poor hole condition?

The predicted clay mineral fractions for other units are consistent with the measured ones except for the inner prism at C0001. This deviation could be due to porosity estimated systematically lower than MAD porosity in this unit. At Kumano Basin sites (Figure 6), the predicted clay mineral fractions at C0009 agrees with the measured ones, whereas the predicted clay mineral fractions for the inner prism at C0002 is systematically lower than the measured ones possibly due to poor hole condition.

Colors for each column represent quality of data. Red columns represent the standard sediments, for which the empirical conversion function is established. Green columns represent the upper part of USB sediments or the slope apron facies younger than 1.5 Ma, which shows higher normalized NGR log values than the standard sediments possibly due to volcanic ashes included in sediments. Light blue columns are the inner prism sediment below the Kumano Basin, which shows lower NGR log values than the standard sediments possibly due to poor hole condition. Gray columns are lithological units not used in the analysis

DISCUSSION

Spectral NGR Log at Site C0009

NGR logs count gamma ray radiation produced by a geological formation. There are several possible gamma ray radiation sources in geological formations (e.g., Faure 1986). The first is potassium, which is abundant in crustal rocks and exists in various rock-forming minerals such as micas, feldspars, and amphiboles. The other two are uranium and thorium, which are usually included in granites, are well absorbed by clay minerals, and can be used as a shale indicator. As we observed correlations between the normalized NGR logs and clay mineral fractions, thorium and uranium are expected to be sources of these NGR log variations. Gamma ray radiation produced by potassium, uranium, and thorium can be discriminated by observing the energy band of the gamma ray radiation. However, the usual NGR log count only accounts for the total amount of gamma ray radiation in the entire energy band. To discriminate between the gamma ray radiation sources, spectral NGR logging is required, although these observations were not commonly performed during the NanTroSEIZE project. Spectral NGR logging was conducted only at site C0009 (Expedition 319 Scientists 2010a).

Figure 8 shows the spectral NGR log data, as well as standard NGR log data. Thorium content fluctuated with depth, which visually correlate well with standard NGR data. Potassium content had peaks around ~ 900 and ~ 950 mbsf, though such peaks are not evident in standard NGR logging data. To quantitatively evaluate which component most influenced the standard NGR logging data, we applied multiple regression analyses and calculated the standard partial regression coefficients. The calculated coefficients were -0.69 for potassium, 1.76 for uranium, and 2.53 for thorium, which suggest that the variations in NGR are due to variations in the clay mineral fraction in the formation.

Hüpers et al. (2015) calculated normalized NGR logs for the incoming sites and used them to estimate the abundances of volcanic ash in the Shikoku Basin sediments. Their findings contrast with those of the study, which used normalized NGR logs to estimate the clay mineral fractions. As they focused only on the incoming sites, where the clay mineral fraction was almost constant (60–70%) (Figs. 2 and 3), variations in the normalized NGR logs are expected to be due to causes other than variations in the clay mineral fraction, such as the presence of volcanic ash. On the other hand, this study used all available data from the Nankai Trough, which spans larger variations in clay mineral fraction, from less than 40% to more than 70%. When the clay mineral fraction varies so widely, it can be expected that the normalized NGR logs represent variations in the clay mineral fraction, as NGR logs vary according to thorium and uranium contents (Figure 8).

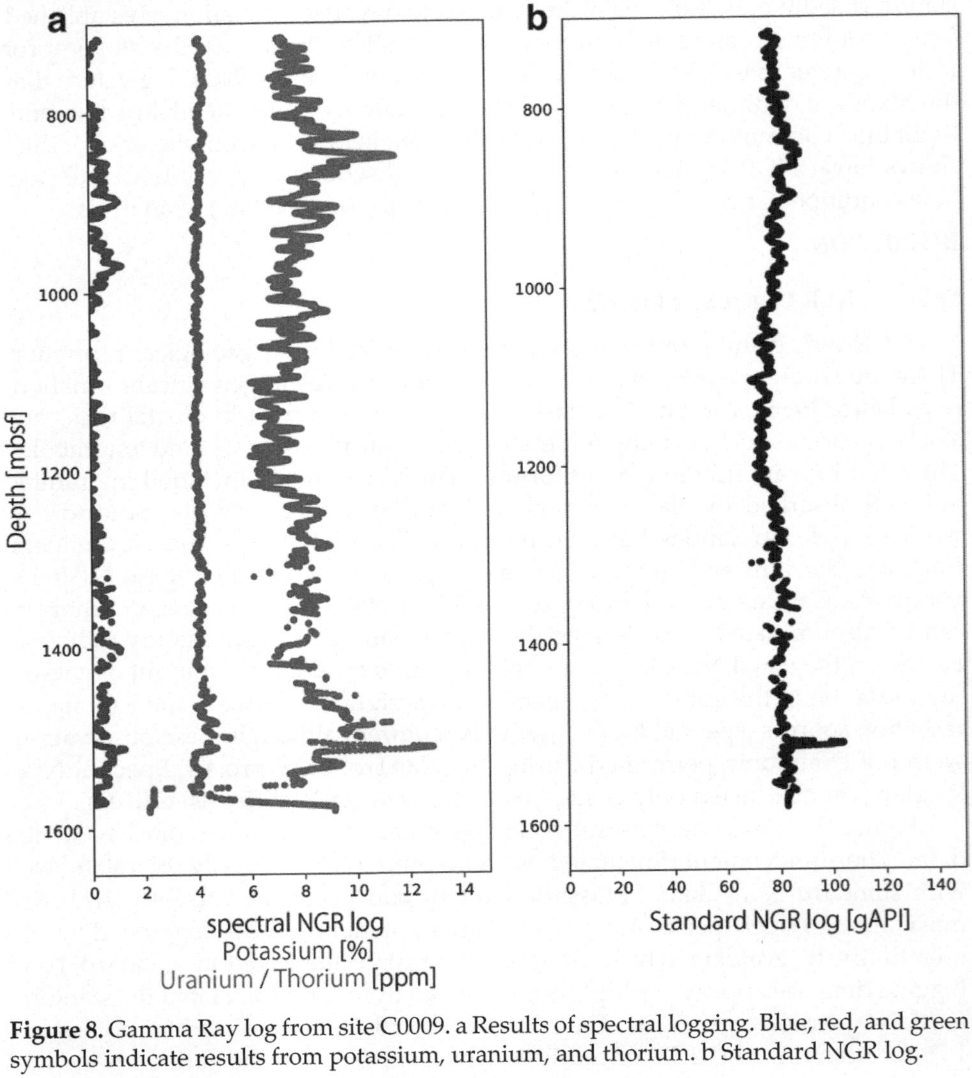

Figure 8. Gamma Ray log from site C0009. a Results of spectral logging. Blue, red, and green symbols indicate results from potassium, uranium, and thorium. b Standard NGR log.

Exceptions to the Empirical Conversion Function

We compiled normalized NGR logs and clay mineral fractions measured from core and cutting samples, and obtained an empirical conversion function between them. However, we excluded several facies that deviated from the main trend when constructing the conversion function. The reason for the deviations can be explained from a sedimentological perspective.

The first exception is Unit 1A of the USB sediments observed at the incoming sites, which exhibits higher normalized NGR values than expected from the other data. Volcanic ash is frequently observed within these strata (Expedition 333 Scientists 2012a, b). Hüpers et al. (2015) also estimated that the volcanic ash fraction of Unit 1A is high in the USB sediments. As volcanic ash layers include abundant radioactive elements, this could be a potential cause of the higher normalized NGR values.

The second exception is the slope apron facies, which also exhibit higher normalized NGR values than expected from the other data. Site C0022 (Figure 5d) clearly contains higher normalized NGR values in the slope apron sediments shallower than ~ 250 mbsf, which correspond to ages younger than ~ 1.5 Ma (Strasser et al. 2014a). This is consistent with the underthrust slope apron sediments observed at site C0004, which are older than 1.5 Ma (Expedition 316 Scientists 2009b) and do not exhibit deviating normalized NGR values (Figure 5b). According to Strasser et al. (2014a), ~ 250 mbsf at site C0022 does not correspond to a lithologic unit boundary, although a sub-unit boundary in log unit is documented at ~ 213 mbsf. This depth may correlate with temporal changes in the riverine sediment supply. According to Usman et al. (2014), the Kumano Basin sediments and slope apron sediments were originally supplied by the Kumano River on the Kii Peninsula and the Tenryu River in the Tokai region. At ~ 1 Ma, megasplay fault activity uplifted the outer ridge (Strasser et al. 2009; Moore et al. 2015a, b) and cut off the sediment supply from the Kumano River to the slope apron. At this time, the sediment supply from the Fuji River to the trench axial current became dominant. Sediments from the Izu-Honshu collision zone supplied by the Fuji River are rich in volcanic materials (Usman et al. 2014), resulting in higher normalized NGR values. However, this hypothesis contradicts the normal NGR values observed in the trench wedge sediments at the toe sites. In contrast, Clift et al. (2013) suggested that the influence of the Fuji River on the trench wedge sediments is weak, even along the Kumano section. Further studies are, therefore, needed to resolve these contradictory findings.

The last exception is the inner prism sediments below the Kumano Basin at site C0002, which exhibit lower normalized NGR values than expected. As discussed in "Normalized gamma ray logs vs. clay mineral fractions" section, this could be due to poor hole conditions at this site, because this deviation is less evident at sites C0001 and C0009. However, it might also be due to different sediment sources. Underwood (2018) suggested that the inner prism sediments are supplied not from the PSP, but from the Pacific Plate, based on analyses of the clay mineral assemblages. If the sediment origin is completely different from that of the Shikoku Basin sediments, the empirical conversion function established herein cannot be applied to the inner prism sediments below the Kumano Basin sediments.

When applying our new method in future drillings, it is possible to expect whether the predicted clay mineral fraction deviates from the actual values by monitoring borehole conditions and knowing which layer is drilled through using information from seismic surveys. Although comparisons with the clay mineral fraction measured for core samples are required to validate the predicted values, dense measurements, as has been done during NanTroSEIZE project, are not needed saving the effort of onboard scientists.

Lithological Interpretations without Core Samples

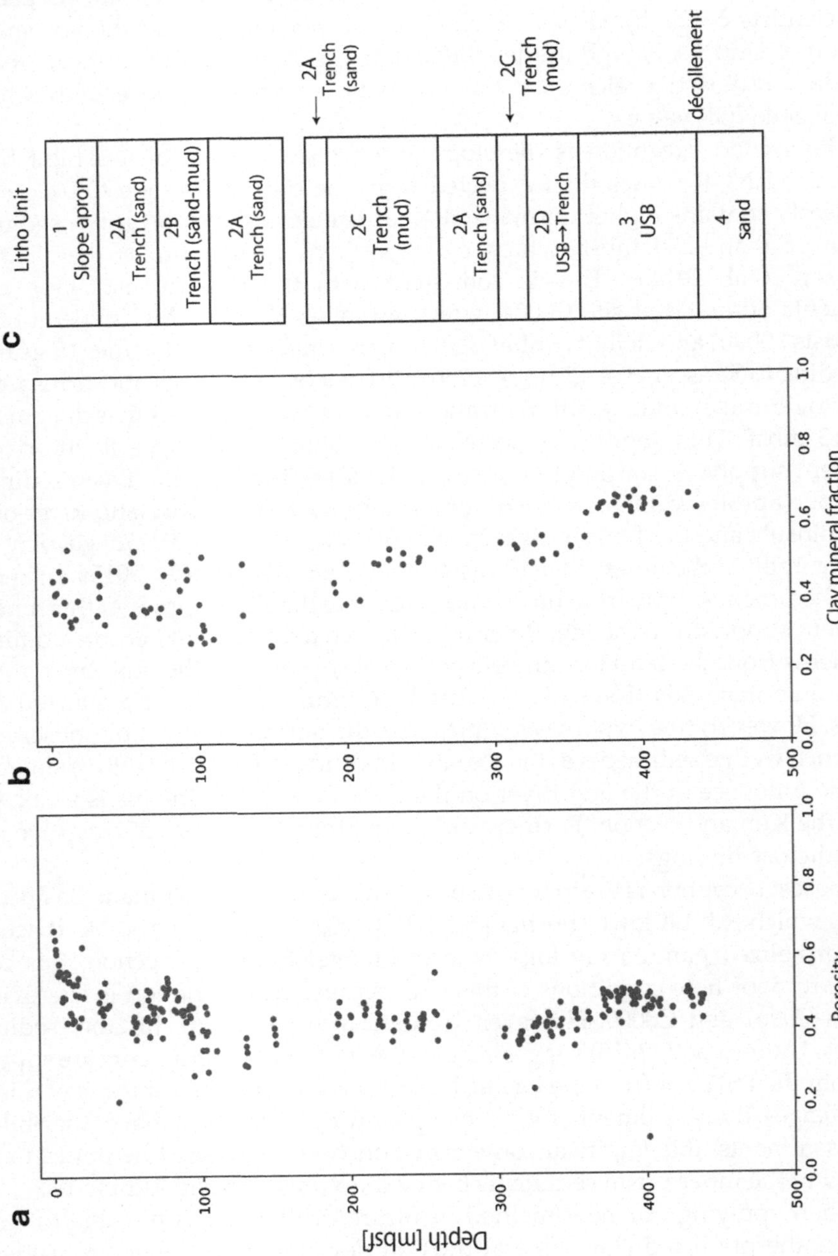

Figure 9. Data from site C0007. Depth profiles of a porosity and b clay mineral fraction measured from core samples are shown. Lithologic unit descriptions are also shown in c.

The empirical conversion function determined in this study provides estimations of clay mineral fractions at depths, where both electrical resistivity logs and NGR logs are available. Using this new conversion method, it is possible to interpret the subsurface lithology, even when core samples are not recovered or when the clay mineral fraction is not measured directly.

At the incoming sites, MSB or Zenisu Fan sediments are observed as sandy turbidite facies (Expedition 322 Scientists 2010a, b; Expedition 333 Scientists 2012a, b; Pickering et al. 2013). However, the clay mineral fractions measured from the core samples are around 70% (Figure 2), which does not correspond to the visual core descriptions. In contrast, the clay mineral fractions estimated from the logging data exhibit low values of ~ 40% at site C0011 and ~ 50% at site C0012. In addition, the clay mineral fraction is not uniform; rather, there are several sandy layers interbedded in the clayey sediments, which is consistent with the visual core descriptions (Expedition 322 Scientists 2010a, b; Expedition 333 Scientists 2012a, b; Pickering et al. 2013). Therefore, the clay mineral fractions estimated from the logging data more accurately reflect the characteristics of the bulk rock in the borehole.

At the toe sites, the shallow décollement was drilled at sites C0006, C0007, and C0024 (Expedition 314 Scientists 2009b; Expedition 316 Scientists 2009a, c; Yamaguchi et al. 2020). Unfortunately, site C0007 was only cored, whereas the shallow décollement at sites C0006 and C0024 were only logged with LWD (Figure 2). At site C0007, the underthrust layer below the décollement is sandy sediments, based on visual core descriptions (Expedition 316 Scientists 2009a), although the clay mineral fraction was not measured in this layer due to very low core recovery (Figure 9).

The clay mineral fraction of the underthrust layer is estimated to be ~ 30% from the logging data collected at site C0006 (Figure 2). This value is consistent with the lowest values of clay mineral fraction measured in the trench wedge sediment (Unit 2 of C0006 and C0024). In contrast, this low clay mineral fraction is not estimated at site C0024 (Figure 2), which suggests that the hole did not penetrate the underthrust sandy layer. The clay mineral fraction estimated from the logging data also provides a detailed depth profile around the shallow décollement, which is difficult to observe using core sample measurements. Such information is important for understanding the structural geology to consider the controls on the location of the shallow décollement.

CONCLUSIONS

Determining sediment lithologies, especially the clay mineral fraction, in accretionary prisms is important, because it is essential information that can be used for a variety of geological studies. The clay mineral fraction is usually measured using cores acquired during drilling. The sand–clay ratio is also qualitatively evaluated using NGR logs, because radioactive thorium and uranium tend to concentrate in the clay minerals. However, NGR logs have not been quantitatively related to the clay mineral fraction, because the concentrations of radioactive elements are strongly dependent on lithology. This study focused on the Nankai Trough and established an empirical conversion function from the NGR logs to the clay mineral fraction of the sediment.

In this new method, NGR logs compensated for the influence of porosity were empirically converted into the clay mineral fractions. Porosities were estimated from the electrical resistivity logs using the method of Yabe et al. (2019). As a byproduct of the porosity estimations, estimations of thermal conductivity and temperature depth profiles were obtained at all NanTroSEIZE drilling sites, where logging data are available. To establish the empirical conversion function, NGR logs compensated for the influence of porosity were compared with the clay mineral fractions measured from core samples. We found a linear relationship between these two values, with some exceptions. The exceptions were due to high concentrations of volcanic ash in the slope apron facies younger than ~ 1.5 Ma and in Unit 1A of the USB sediments. The empirical conversion function established in this study is for the Nankai trough, though our method is applicable to other subduction zones if enough log data and clay fraction data are available.

This new method and its results are useful in various ways to understand seismogenesis and evolutionary process of the shallow accretionary prism. For example, when deviations are observed between core sample measurements and estimations using the new method, irregular components are present in the sediments, which is important for determining the sedimentology. Continuously measured logging data provide a more detailed depth profile of the clay mineral fraction than discrete core sample measurements. Such a detailed profile is important for interpreting the subsurface structural geology. As electrical resistivity logs and NGR logs are essential measurements conducted during downhole logging, this new method can easily be applied to future scientific drilling expeditions in the Nankai Trough.

Acknowledgements

This study was made possible by participating in the IODP workshop "Core-Log-Seismic Integration at Sea" and IODP Expedition 358 on D/V Chikyu. The authors appreciate the shipboard assistance of the Marine Works Japan laboratory technicians, as well as Mantle Quest Japan Ltd. for operating D/V Chikyu. We thank Prof. Kyuichi Kanagawa, Drs. Takehiro Hirose, Manami Kitamura, and Yoshinori Sanada for their helpful comments on this manuscript. Some of the figures were produced using the GMT software package (Wessel et al. 2013).

Contributions

All authors contributed toward the conception of this study. SY analyzed data and drafted the manuscript. All authors contributed to the revision of the manuscript. All authors read and approved the final manuscript.

ABBREVIATIONS

BSR: Bottom simulating reflector
IODP: International Ocean Discovery Project
LWD: Logging while drilling
MAD: Moisture and density
NGR: Natural gamma radiation
PSP: Philippine Sea Plate
USB/MSB/LSB: Upper/Middle/Lower Shikoku Basin
XRD: X-ray diffraction

REFERENCES

1. Archie GE (1942) The electrical resistivity log as an aid in determining some reservoir characteristics, JPT. J Pet Technol 5:1–8

2. Clift PD, Carter A, Nicholson U, Masago H (2013) Zircon and apatite thermochronology of the Nankai Trough accretionary prism and trench, Japan: Sediment transport in an active and collisional margin setting. Tectonics 32:377–395. https://doi.org/10.1002/tect.20033

3. DeMets C, Gordon RG, Argus DF (2010) Geologically current plate motion. Geophys J Int 181(1):1–80. https://doi.org/10.1111/j.1365-246X.2009.04491.x

4. Expedition 314 Scientists (2009a) Expedition 314 methods. In: Kinoshita M, Tobin H, Ashi J, Kimura G, Lallement S, Screaton EJ, Curewitz D, Masago H, Moe KT, and the Expedition 314/315/316 Scientists, Proc. IODP, 314/315/316: Washington, DC (Integrated Ocean Drilling Program Management International, Inc.). https://doi.org/10.2204/iodp.proc.314315316.112.2009

5. Expedition 314 Scientists (2009b) Expedition 314 Site C0006. In: Kinoshita M, Tobin H, Ashi J, Kimura G, Lallemant S, Screaton EJ, Curewitz D, Masago H, Moe KT, and the Expedition 314/315/316 Scientists, Proc. IODP, 314/315/316: Washington, DC (Integrated Ocean Drilling Program Management International, Inc.). https://doi.org/10.2204/iodp.proc.314315316.118.2009

6. Expedition 314 Scientists (2009c) Expedition 314 Site C0001. In: Kinoshita M, Tobin H, Ashi J, Kimura G, Lallement S, Screaton EJ, Curewitz D, Masago H, Moe KT, and the Expedition 314/315/316 Scientists, Proc. IODP, 314/315/316: Washington, DC (Integrated Ocean Drilling Program Management International, Inc.). https://doi.org/10.2204/iodp.proc.314315316.113.2009

7. Expedition 314 Scientists (2009d) Expedition 314 Site C0004. In: Kinoshita M, Tobin H, Ashi J, Kimura G, Lallement S, Screaton EJ, Curewitz D, Masago H, Moe KT, and the Expedition 314/315/316 Scientists, Proc. IODP, 314/315/316: Washington, DC (Integrated Ocean Drilling Program Management International, Inc.). https://doi.org/10.2204/iodp.proc.314315316.116.2009

8. Expedition 314 Scientists (2009e). Expedition 314 Site C0002. In: Kinoshita M, Tobin H, Ashi J, Kimura G, Lallement S, Screaton EJ, Curewitz D, Masago H, Moe KT, and the Expedition 314/315/316 Scientists, Proc. IODP, 314/315/316: Washington, DC (Integrated Ocean Drilling Program Management International, Inc.). https://doi.org/10.2204/iodp.proc.314315316.114.2009

9. Expedition 315 Scientists (2009a) Expedition 315 methods. In: Kinoshita M, Tobin H, Ashi J, Kimura G, Lallemant S, Screaton EJ, Curewitz D, Masago H, Moe KT, and the Expedition 314/315/316 Scientists, Proc. IODP, 314/315/316: Washington, DC (Integrated Ocean Drilling Program Management International, Inc.). https://doi.org/10.2204/iodp.proc.314315316.122.2009

10. Expedition 315 Scientists (2009b) Expedition 315 Site C0001. In: Kinoshita M, Tobin H, Ashi J, Kimura G, Lallemant S, Screaton EJ, Curewitz D, Masago H, Moe KT, and the Expedition 314/315/316 Scientists, Proc. IODP, 314/315/316: Washington, DC (Integrated Ocean Drilling Program Management International, Inc.). https://doi.org/10.2204/iodp.proc.314315316.123.2009

11. Expedition 315 Scientists (2009c). Expedition 315 Site C0002. In: Kinoshita M, Tobin H, Ashi J, Kimura G, Lallemant S, Screaton EJ, Curewitz D, Masago H, Moe KT, and the Expedition 314/315/316 Scientists, Proc. IODP, 314/315/316: Washington, DC (Integrated Ocean Drilling Program Management International, Inc.). https://doi.org/10.2204/iodp.proc.314315316.124.2009

12. Expedition 316 Scientists (2009a) Expedition 316 Site C0006. In: Kinoshita M, Tobin H, Ashi J, Kimura G, Lallemant S, Screaton EJ, Curewitz D, Masago H, Moe KT, and the Expedition 314/315/316 Scientists, Proc. IODP, 314/315/316: Washington, DC (Integrated Ocean Drilling Program Management International, Inc.). https://doi.org/10.2204/iodp.proc.314315316.134.2009
13. Expedition 316 Scientists (2009b) Expedition 316 Site C0004. In: Kinoshita M, Tobin H, Ashi J, Kimura G, Lallemant S, Screaton EJ, Curewitz D, Masago H, Moe KT, and the Expedition 314/315/316 Scientists, Proc. IODP, 314/315/316: Washington, DC (Integrated Ocean Drilling Program Management International, Inc.). https://doi.org/10.2204/iodp.proc.314315316.133.2009
14. Expedition 316 Scientists (2009c) Expedition 316 Site C0007. In: Kinoshita M, Tobin H, Ashi J, Kimura G, Lallemant S, Screaton EJ, Curewitz D, Masago H, Moe KT, and the Expedition 314/315/316 Scientists, Proc. IODP, 314/315/316: Washington, DC (Integrated Ocean Drilling Program Management International, Inc.). https://doi.org/10.2204/iodp.proc.314315316.135.2009
15. Expedition 319 Scientists (2010a) Site C0009. In: Saffer D, McNeill L, Byrne T, Araki E, Toczko S, Eguchi N, Takahashi K, and the Expedition 319 Scientists, Proc. IODP, 319: Tokyo (Integrated Ocean Drilling Program Management International, Inc.). https://doi.org/10.2204/iodp.proc.319.103.2010
16. Expedition 319 Scientists (2010b) Methods. In: Saffer D, McNeill L, Byrne T, Araki E, Toczko S, Eguchi N, Takahashi K, and the Expedition 319 Scientists, Proc. IODP, 319: Tokyo (Integrated Ocean Drilling Program Management International, Inc.). https://doi.org/10.2204/iodp.proc.319.102.2010
17. Expedition 319 Scientists (2010c) Site C0010. In: Saffer D, McNeill L, Byrne T, Araki E, Toczko S, Eguchi N, Takahashi K, and the Expedition 319 Scientists, Proc. IODP, 319: Tokyo (Integrated Ocean Drilling Program Management International, Inc.). https://doi.org/10.2204/iodp.proc.319.104.2010
18. Expedition 322 Scientists (2010a) Site C0011. In: Saito S, Underwood MB, Kubo Y, and the Expedition 322 Scientists, Proc. IODP, 322: Tokyo (Integrated Ocean Drilling Program Management International, Inc.). https://doi.org/10.2204/iodp.proc.322.103.2010
19. Expedition 322 Scientists (2010b) Site C0012. In: Saito S, Underwood MB, Kubo Y, and the Expedition 322 Scientists, Proc. IODP, 322: Tokyo (Integrated Ocean Drilling Program Management International, Inc.). https://doi.org/10.2204/iodp.proc.322.104.2010
20. Expedition 333 Scientists (2012a) Site C0011. In: Henry, P., Kanamatsu, T., Moe, K., and the Expedition 333 Scientists, Proc. IODP, 333: Tokyo (Integrated Ocean Drilling Program Management International, Inc.). https://doi.org/10.2204/iodp.proc.333.104.2012
21. Expedition 333 Scientists (2012b) Site C0012. In: Henry P, Kanamatsu T, Moe K, and the Expedition 333 Scientists, Proc. IODP, 333: Tokyo (Integrated Ocean Drilling Program Management International, Inc.). https://doi.org/10.2204/iodp.proc.333.105.2012
22. Expedition 333 Scientists (2012c) Site C0018. In: Henry P, Kanamatsu T, Moe K, and the Expedition 333 Scientists, Proc. IODP, 333: Tokyo (Integrated Ocean Drilling Program Management International, Inc.). https://doi.org/10.2204/iodp.proc.333.103.2012
23. Faure G (1986). Principles of Isotope Geology, second ed. Wiley, (ISBN: 0471864129)
24. Guo J, Underwood MB (2012) Data report: clay mineral assemblages from the Nankai Trough accretionary prism and the Kumano Basin, IODP Expeditions

315 and 316, NanTroSEIZE Stage 1. In: Kinoshita M, Tobin H, Ashi J, Kimura G, Lallemant S, Screaton EJ, Curewitz D, Masago H, Moe KT, and the Expedition 314/315/316 Scientists, Proc. IODP, 314/315/316: Washington, DC (Integrated Ocean Drilling Program Management International, Inc.). https://doi.org/10.2204/iodp.proc.314315316.202.2012

25. Hampson GJ, Davies W, Davies SJ, Howell JA, Adamson KR (2005) Use of spectral gamma-ray data to refine subsurface fluvial stratigraphy: late Cretaceous strata in the Book Cliffs, Utah, USA. J Geol Soc 162(4):603–621. https://doi.org/10.1144/0016-764904-059
26. Harris RN, Schmidt-Schierhorn F, Spinelli G (2011) Heat flow along the NanTroSEIZE transect: Results from IODP Expeditions 315 and 316 offshore of the Kii Peninsula, Japan, Geochem Geophys Geosyst 12, Q0AD16. https://doi.org/10.1029/2011gc003593
27. Hüpers A, Ikari MJ, Underwood MB, Kopf AJ (2015) Origin of a zone of anomalously high porosity in the subduction inputs to Nankai Trough. Mar Geol 361:147–162. https://doi.org/10.1016/j.margeo.2015.01.004
28. Hüpers A, Torres ME, Owari S, McNeill LC, Dugan B, Henstock TJ, Milliken KL, Petronotis KE, Backman J, Bourlange S, Chemale F, Chen W, Colson TA, Frederik MCG, Guèrin G, Hamahashi M, House BM, Jeppson TN, Kachovich S, Kenigsberg AR, Kuranaga M, Kutterolf S, Mitchison FL, Mukoyoshi H, Nair N, Pickering KT, Pouderoux HFA, Shan Y, Song I, Vannucchi P, Vrolijk PJ, Yang T, Zhao X (2017) Release of mineral-bound water prior to subduction tied to shallow seismogenic slip off Sumatra. Science 356(6340):841–844. https://doi.org/10.1126/science.aal3429
29. Ishii T, Sato H, Machida S, Haraguchi S, Usui A, Ishizuka O, Taniguchi H, Yagi K (2000) Geological and petrological studies of the Kinan and Izu-Ogasawara back arc enchelon seamount chains. Bull Geol Surv Jpn 51:615–630
30. Ishizuka O, Taylor RN, Yuasa M, Ohara Y (2011) Making and breaking an island arc: a new perspective from the Oligocene Kyushu-Palau arc, Philippine Sea. Geochem Geophys Geosyst 12:Q05005. https://doi.org/10.1029/2010GC003440
31. Kimura G, Kitamura Y, Hashimoto Y, Yamaguchi A, Shibata T, Ujiie K, Okamoto S (2007) Transition of accretionary wedge structures around the up-dip limit of the seismogenic subduction zone. Earth Planet Sci Lett 255:471–484. https://doi.org/10.1016/j.epsl.2007.01.005
32. Kimura G, Moore GF, Strasser M, Screaton E, Curewitz D, Streiff C, Tobin H (2011) Spatial and temporal evolution of the megasplay fault in the Nankai Trough. Geochem Geophys Geosyst. https://doi.org/10.1029/2010gc003335
33. Kimura G, Koge H, Tsuji T (2018) Punctuated growth of an accretionary prism and the onset of a seismogenic megathrust in the Nankai Trough. Progr Earth Planet Sci 5:78. https://doi.org/10.1186/s40645-018-0234-1
34. Kitajima H, Hirose T, Ikari M, Kanagawa K, Kimura G, Kinoshita M, Saffer D, Tobin H, Yamaguchi A, Eguchi N, Maeda L, Toczko S, Bedford J, Chiyonobu S, Colson TA, Conin M, Cornard PH, Dielforder A, Doan M-L, Dutilleul J, Faulkner DR, Fukuchi R, Guérin G, Hamada Y, Hamahashi M, Hong W-L, Ijiri A, Jaeger D, Jeppson T, Jin Z, John BE, Kitamura M, Kopf A, Masuda H, Matsuoka A, Moore GF, Otsubo M, Regalla C, Sakaguchi A, Sample J, Schleicher A, Sone H, Stanislowski K, Strasser M, Toki T, Tsuji T, Ujiie K, Underwood MB, Yabe S, Yamamoto Y, Zhang J, Sanada Y, Kido Y, Le Ber E, Saito S (2020) Site C0002. With contributions from T. Kanamatsu. In: Tobin H, Hirose T, Ikari M, Kanagawa K, Kimura G, Kinoshita M, Kitajima H, Saffer D, Yamaguchi A, Eguchi N, Maeda L, Toczko S, and Expedition 358 Scientists, NanTroSEIZE Plate Boundary Deep Riser 4: Nankai Seismogenic/Slow Slip Megathrust. In: Proceedings

of the International Ocean Discovery Program, 358: College Station, TX (International Ocean Discovery Program). https://doi.org/10.14379/iodp.proc.358.103.2020

35. Miyazaki S, Heki K (2001) Crustal velocity field of southwest Japan: subduction and arc–arc collision. J Geophys Res 106:4305–4326. https://doi.org/10.1029/2000JB900312

36. Moore GF, Shipley TH (1988) Mechanisms of sediment accretion in the Middle America Trench off Mexico. J Geophys Res 93:8911–8927

37. Moore JC, Roeske SM, Lundberg N, Schoonmaker J, Cowan DS, Gonzales E, Lucas SE (1986) Scaly fabrics from the deep sea drilling project cores from forearcs. In: Moore JC (Ed.), Structural Fabric in Deep Sea Drilling Project Cores from Forearcs: Mem. Geol. Soc. Am., 166, pp 55–73

38. Moore GF, Shipley TH, Stoffa PL, Karig DE, Taira A, Kuramoto S, Tokuyama H, Suyehiro K (1990) Structure of the Nankai Trough accretionary zone from multichannel seismic reflection data. J Geophys Res 95:8753–8765

39. Moore GF, Boston BB, Strasser M, Underwood MB, Ratliff RA (2015a) Evolution of tectono-sedimentary systems in the Kumano Basin, Nankai Trough forearc. Mar Petrol Geol 67:604–616. https://doi.org/10.1016/j.marpetgeo.2015.05.032

40. Moore JC, Plank TA, Chester FM, Polissar PJ, Savage HM (2015b) Sediment provenance and controls on slip propagation: lessons learned from the 2011 Tohoku and other great earthquakes of the subducting northwest Pacific Plate. Geosphere 11(3):533–541

41. Ohde A, Otsuka H, Kioka A, Ashi J (2018) Distribution and depth of bottom-simulating reflectors in the Nankai subduction margin. Earth Planets Space 70:60. https://doi.org/10.1186/s40623-018-0833-5

42. Okino K, Shimakawa Y, Nagaoka S (1994) Evolution of the Shikoku Basin. J Geomagn Geoelectr 46:463–479. https://doi.org/10.5636/jgg.46.463

43. Okino K, Ohara Y, Kasuga S, Kato Y (1999) The Philippine Sea: new survey results reveal the structure and history of the marginal basins. Geophys Res Lett 26:2287–2290. https://doi.org/10.1029/1999GL900537

44. Oohashi K, Hirose T, Takahashi M, Tanikawa W (2015) Dynamic weakening of smectite-bearing faults at intermediate velocities: implications for subduction zone earthquakes. J. Geophys. Res. Solid Earth 120:1572–1586. https://doi.org/10.1002/2015JB011881

45. Pickering KT, Underwood MB, Saito S, Naruse H, Kutterolf S, Scudder R, Park J-O, Moore GF, Slagle A (2013) Depositional architecture, provenance, and tectonic/eustatic modulation of Miocene submarine fans in the Shikoku Basin: results from the Nankai Trough Seismogenic Zone Experiment. Geochem Geophys Geosyst 14:1722–1739. https://doi.org/10.1002/ggge.20107

46. Saffer D, Kopf A, Toczko S, Araki E, Carr S, Kimura T, Kinoshita C, Kobayashi R, Machida Y, Rösner A, Wallace LM (2017) Site C0010. With contributions from Chiyonobu S, Kanagawa K, Kanamatsu T, Kimura G, Underwood MB. In: Saffer D, Kopf A, Toczko S, and Expedition 365 Scientists, NanTroSEIZE Stage 3: Shallow Megasplay Long-Term Borehole Monitoring System. Proceedings of the International Ocean Discovery Program, 365: College Station, TX (International Ocean Discovery Program). https://doi.org/10.14379/iodp.proc.365.103.2017

47. Sawai M, Hirose T, Kameda J (2014) Frictional properties of incoming pelagic sediments at the Japan Trench: implications for large slip at a shallow plate boundary during the 2011 Tohoku earthquake. Earth Planets Space 66:65. https://doi.org/10.1186/1880-5981-66-65

48. Sdrolias M, Roest WR, Müller RD (2004) An expression of Philippine Sea plate rotation: the Parece Vela and Shikoku Basins. Tectonophysics 394:69–86. https://doi.

org/10.1016/j.tecto.2004.07.061

49. Strasser M, Moore GF, Kimura G, Kitamura Y, Kopf AJ, Lallemant S, Park J-O, Screaton EJ, Su X, Underwood MB (2009) Origin and evolution of a splay fault in the Nankai accretionary wedge. Nat Geosci 2:648–652. https://doi.org/10.1038/ngeo609
50. Strasser M, Dugan B, Kanagawa K, Moore GF, Toczko S, Maeda L, Kido Y, Moe KT, Sanada Y, Esteban L, Fabbri O, Geersen J, Hammerschmidt S, Hayashi H, Heirman K, Hüpers A, Jurado Rodriguez MJ, Kameo K, Kanamatsu T, Kitajima H, Masuda H, Milliken K, Mishra R, Motoyama I, Olcott K, Oohashi K, Pickering KT, Ramirez SG, Rashid H, Sawyer D, Schleicher A, Shan Y, Skarbek R, Song I, Takeshita T, Toki T, Tudge J, Webb S, Wilson DJ, Wu H-Y, Yamaguchi A (2014a) Site C0022. In Strasser M, Dugan B, Kanagawa K, Moore GF, Toczko S, Maeda L, and Expedition 338 Scientists, Proc. IODP, 338: Yokohama (Integrated Ocean Drilling Program). https://doi.org/10.2204/iodp.proc.338.107.2014
51. Strasser M, Dugan B, Kanagawa K, Moore GF, Toczko S, Maeda L, Kido Y, Moe KT, Sanada Y, Esteban L, Fabbri O, Geersen J, Hammerschmidt S, Hayashi H, Heirman K, Hüpers A, Jurado Rodriguez MJ, Kameo K, Kanamatsu T, Kitajima H, Masuda H, Milliken K, Mishra R, Motoyama I, Olcott K, Oohashi K, Pickering KT, Ramirez SG, Rashid H, Sawyer D, Schleicher A, Shan Y, Skarbek R, Song I, Takeshita T, Toki T, Tudge J, Webb S, Wilson DJ, Wu H-Y, Yamaguchi A (2014b). Site C0012. In ', Proc. IODP, 338: Yokohama (Integrated Ocean Drilling Program). https://doi.org/10.2204/iodp.proc.338.104.2014
52. Strasser M, Dugan B, Kanagawa K, Moore GF, Toczko S, Maeda L, Kido Y, Moe KT, Sanada Y, Esteban L, Fabbri O, Geersen J, Hammerschmidt S, Hayashi H, Heirman K, Hüpers A, Jurado Rodriguez MJ, Kameo K, Kanamatsu T, Kitajima H, Masuda H, Milliken K, Mishra R, Motoyama I, Olcott K, Oohashi K, Pickering KT, Ramirez SG, Rashid H, Sawyer D, Schleicher A, Shan Y, Skarbek R, Song I, Takeshita T, Toki T, Tudge J, Webb S, Wilson DJ, Wu H-Y, Yamaguchi A (2014c) Site C0018. In: Strasser M, Dugan B, Kanagawa K, Moore GF, Toczko S, Maeda L, and Expedition 338 Scientists, Proc. IODP, 338: Yokohama (Integrated Ocean Drilling Program). https://doi.org/10.2204/iodp.proc.338.105.2014
53. Strasser M, Dugan B, Kanagawa K, Moore GF, Toczko S, Maeda L, Kido Y, Moe KT, Sanada Y, Esteban L, Fabbri O, Geersen J, Hammerschmidt S, Hayashi H, Heirman K, Hüpers A, Jurado Rodriguez MJ, Kameo K, Kanamatsu T, Kitajima H, Masuda H, Milliken K, Mishra R, Motoyama I, Olcott K, Oohashi K, Pickering KT, Ramirez SG, Rashid H, Sawyer D, Schleicher A, Shan Y, Skarbek R, Song I, Takeshita T, Toki T, Tudge J, Webb S, Wilson DJ, Wu H-Y, Yamaguchi A (2014d) Site C0021. In: Strasser M, Dugan B, Kanagawa K, Moore GF, Toczko S, Maeda L, and Expedition 338 Scientists, Proc. IODP, 338: Yokohama (Integrated Ocean Drilling Program). https://doi.org/10.2204/iodp.proc.338.106.2014
54. Strasser M, Dugan B, Kanagawa K, Moore GF, Toczko S, Maeda L, Kido Y, Moe KT, Sanada Y, Esteban L, Fabbri O, Geersen J, Hammerschmidt S, Hayashi H, Heirman K, Hüpers A, Jurado Rodriguez MJ, Kameo K, Kanamatsu T, Kitajima H, Masuda H, Milliken K, Mishra R, Motoyama I, Olcott K, Oohashi K, Pickering KT, Ramirez SG, Rashid H, Sawyer D, Schleicher A, Shan Y, Skarbek R, Song I, Takeshita T, Toki T, Tudge J, Webb S, Wilson DJ, Wu H-Y, Yamaguchi A (2014e) Site C0002. In: Strasser M, Dugan B, Kanagawa K, Moore GF, Toczko S, Maeda L, and Expedition 338 Scientists, Proc. IODP, 338: Yokohama (Integrated Ocean Drilling Program). https://doi.org/10.2204/iodp.proc.338.103.2014
55. Sugihara T, Kinoshita M, Araki E, Kimura T, Kyo M, Namba Y, Kido Y, Sanada Y, Thu M (2014) Reevaluation of temperature at the updip limit of locked portion of Nankai

megasplay inferred from IODP Site C0002 temperature observatory. Earth, Planets and Space 66:107. https://doi.org/10.1186/1880-5981-66-107

56. Tobin HJ, Kinoshita M (2006) NanTroSEIZE: the IODP Nankai Trough seismogenic zone experiment. Sci Drill 2:23–27. https://doi.org/10.2204/iodp.sd.2.06.2006

57. Tobin H, Kinoshita M, Moe KT, and the Expedition 314 Scientists (2009). Expedition 314 summary. In Kinoshita M, Tobin H, Ashi J, Kimura G, Lallemant S, Screaton EJ, Curewitz D, Masago H, Moe KT, and the Expedition 314/315/316 Scientists, Proc. IODP, 314/315/316: Washington, DC (Integrated Ocean Drilling Program Management International, Inc.). https://doi.org/10.2204/iodp.proc.314315316.111.2009

58. Tobin H, Hirose T, Saffer D, Toczko S, Maeda L, Kubo Y, Boston B, Broderick A, Brown K, Crespo-Blanc A, Even E, Fuchida S, Fukuchi R, Hammerschmidt S, Henry P, Josh M, Jurado MJ, Kitajima H, Kitamura M, Maia A, Otsubo M, Sample J, Schleicher A, Sone H, Song C, Valdez R, Yamamoto Y, Yang K, Sanada Y, Kido Y, Hamada Y (2015a) Site C0002. In Tobin H, Hirose T, Saffer D, Toczko S, Maeda L, Kubo Y, and Expedition 348 Scientists, Proc. IODP, 348: College Station, TX (Integrated Ocean Drilling Program). https://doi.org/10.2204/iodp.proc.348.103.2015

59. Tobin H, Hirose T, Saffer D, Toczko S, Maeda L, Kubo Y, Boston B, Broderick A, Brown K, Crespo-Blanc A, Even E, Fuchida S, Fukuchi R, Hammerschmidt S, Henry P, Josh M, Jurado MJ, Kitajima H, Kitamura M, Maia A, Otsubo M, Sample J, Schleicher A, Sone H, Song C, Valdez R, Yamamoto Y, Yang K, Sanada Y, Kido Y, Hamada Y (2015b) Methods. In: Tobin H, Hirose T, Saffer D, Toczko S, Maeda L, Kubo Y, and Expedition 348 Scientists, Proc. IODP, 348: College Station, TX (Integrated Ocean Drilling Program). https://doi.org/10.2204/iodp.proc.348.102.2015

60. Tobin H, Hirose T, Ikari M, Kanagawa K, Kimura G, Kinoshita M, Kitajima H, Saffer, D Yamaguchi A, Eguchi N, Maeda L, Toczko S, Bedford J, Chiyonobu S, Colson TA, Conin M, Cornard PH, Dielforder A, Doan M-L, Dutilleul J, Faulkner DR, Fukuchi R, Guérin G, Hamada Y, Hamahashi M, Hong W-L, Ijiri A, Jaeger D, Jeppson T, Jin Z, John BE, Kitamura M, Kopf A, Masuda H, Matsuoka A, Moore GF, Otsubo M, Regalla C, Sakaguchi A, Sample J, Schleicher A, Sone H, Stanislowski K, Strasser M, Toki T, Tsuji T, Ujiie K, Underwood MB, Yabe S, Yamamoto Y, Zhang J, Sanada Y, Kido Y, Le Ber E, Saito S (2020) Expedition 358 summary. With contributions by Kanamatsu T. In: Tobin H, Hirose T, Ikari M, Kanagawa K, Kimura G, Kinoshita M, Kitajima H, Saffer D, Yamaguchi A, Eguchi N, Maeda L, Toczko S, and the Expedition 358 Scientists, NanTro- SEIZE Plate Boundary Deep Riser 4: Nankai Seismogenic/ Slow Slip Megathrust. Proceedings of the International Ocean Discovery Program, 358: College Station, TX (International Ocean Discovery Program). https://doi.org/10.14379/iodp.proc.358.101.2020

61. Ujiie K, Hisamitsu T, Taira A (2003) Deformation and fluid pressure variation during initiation and evolution of the plate boundary décollement zone in the Nankai accretionary prism. J Geophys Res Solid Earth 108(B8):2398. https://doi.org/10.1029/2002JB002314

62. Underwood MB (2017a) Data report: clay mineral assemblages in slope basin sediments and mass-transport deposits at Sites C0018 and C0021, IODP Expeditions 333 and 338. In Strasser M, Dugan B, Kanagawa K, Moore GF, Toczko S, Maeda L, and the Expedition 338 Scientists, Proceedings of the Integrated Ocean Drilling Program, 338: Yokohama (Integrated Ocean Drilling Program). https://doi.org/10.2204/iodp.proc.338.207.2017

63. Underwood MB (2017b) Data report: clay mineral assemblages in cuttings from Hole C0002N, IODP Expedition 348, Nankai Trough accretionary prism. In: Tobin

H, Hirose T, Saffer D, Toczko S, Maeda L, Kubo Y, and the Expedition 348 Scientists, Proceedings of the Integrated Ocean Drilling Program, 348: College Station, TX (Integrated Ocean Drilling Program). https://doi.org/10.2204/iodp.proc.348.203.2017

64. Underwood MB (2017c) Data report: clay mineral assemblages and illite/smectite diagenesis in cuttings from Hole C0002P, IODP Expedition 348, Nankai Trough accretionary prism. In: Tobin H, Hirose T, Saffer D, Toczko S, Maeda L, Kubo Y, and the Expedition 348 Scientists, Proceedings of the Integrated Ocean Drilling Program, 348: College Station, TX (Integrated Ocean Drilling Program). https://doi.org/10.2204/iodp.proc.348.204.2017

65. Underwood MB (2018) The origin of strata within the inner accretionary prism of Nankai Trough: Evidence from clay mineral assemblages along the NanTroSEIZE transect. The Island Arc, e12252. https://doi.org/10.1111/iar.12252

66. Underwood MB, Guo J (2013) Data report: clay mineral assemblages in the Shikoku Basin, NanTroSEIZE subduction inputs, IODP Sites C0011 and C0012. In: Saito S, Underwood MB, Kubo Y, and Expedition 322 Scientists, Proc. IODP 322: Tokyo (Integrated Ocean Drilling Program Management International, Inc.). https://doi.org/10.2204/iodp.proc.322.202.2013

67. Underwood MB, Guo J (2017) Data report: clay mineral assemblages in the upper Shikoku Basin, results from IODP Expedition 333, Sites C0011 and C0012. In: Henry P, Kanamatsu T, Moe K, and Expedition 333 Scientists, Proceedings of the Integrated Ocean Drilling Program, 333: Tokyo (Integrated Ocean Drilling Program Management International, Inc.). https://doi.org/10.2204/iodp.proc.333.205.2017

68. Underwood MB, Guo J (2018) Clay-mineral assemblages across the Nankai-Shikoku subduction system, offshore Japan: A synthesis of results from the NanTroSEIZE project: Geosphere, 14, 5, 2009–2043. https://doi.org/10.1130/ges01626.1

69. Underwood MB, Song C (2016) Data report: clay mineral assemblages in cuttings from Hole C0002F, IODP Expedition 338, upper Nankai Trough accretionary prism. In: Strasser M, Dugan B, Kanagawa K, Moore GF, Toczko S, Maeda L, and the Expedition 338 Scientists, Proceedings of the Integrated Ocean Drilling Program, 338: Yokohama (Integrated Ocean Drilling Program). https://doi.org/10.2204/iodp.proc.338.206.2016

70. Usman MO, Masago H, Winkler W, Strasser M (2014) Mid-Quaternary decoupling of sediment routing in the Nankai Forearc revealed by provenance analysis of turbiditic sands. Int J Earth Sci 103:1141–1161. https://doi.org/10.1007/s00531-014-1011-z

71. Vrolijk P (1990) On the mechanical role of smectite in subduction zones. Geology 18(8):703–707

72. Wessel P, Smith WHF, Scharroo R, Luis J, Wobbe F (2013) Generic mapping tools: improved version released. Eos Trans AGU 94(45):409. https://doi.org/10.1002/2013EO450001

73. Yabe S, Fukuchi R, Hamada Y, Kimura G (2019) Simultaneous estimation of in situ porosity and thermal structure from core sample measurements and resistivity log data at Nankai accretionary prism. Earth Planets Space 71:116. https://doi.org/10.1186/s40623-019-1097-4

74. Yamaguchi A, Hirose T, Ikari M, Kanagawa K, Kimura G, Kinoshita M, Kitajima H, Saffer D, Tobin H, Eguchi N, Maeda L, Toczko S, Bedford J, Chiyonobu S, Colson TA, Conin M, Cornard PH, Dielforder A, Doan M-L, Dutilleul J, Faulkner DR, Fukuchi R, Guérin G, Hamada Y, Hamahashi M, Hong W-L, Ijiri A, Jaeger D, Jeppson T, Jin Z, John BE, Kitamura M, Kopf A, Masuda H, Matsuoka A, Moore GF, Otsubo M, Regalla C, Sakaguchi A, Sample J, Schleicher A, Sone H, Stanislowski K, Strasser M, Toki T, Tsuji T, Ujiie K, Underwood MB, Yabe S, Yamamoto Y, Zhang J, Sanada Y, Kido Y, Le Ber E,

Saito S (2020) Site C0024. With contributions from Kanamatsu T. In: Tobin H, Hirose T, Ikari M, Kanagawa K, Kimura G, Kinoshita M, Kitajima H, Saffer D, Yamaguchi A, Eguchi N, Maeda L, Toczko S, and Expedition 358 Scientists, NanTroSEIZE Plate Boundary Deep Riser 4: Nankai Seismogenic/Slow Slip Megathrust. In: Proceedings of the International Ocean Discovery Program, 358: College Station, TX (International Ocean Discovery Program). https://doi.org/10.14379/iodp.proc.358.104.2020

3

IDENTIFYING AND MAPPING ALTERATION MINERALS USING HYSPEX AIRBORNE HYPERSPECTRAL DATA AND RANDOM FOREST ALGORITHM

Shanshan Wang[1,2,3,4], Kefa Zhou[1,2,3], Jinlin Wang[1,2,3,4] and Jie Zhao[2]

[1]State Key Laboratory of Desert and Oasis Ecology, Xinjiang Institute of Ecology and Geography, Chinese Academy of Sciences, Urumqi, China
[2]Xinjiang Key Laboratory of Mineral Resources and Digital Geology, Xinjiang Institute of Ecology and Geography, Chinese Academy of Sciences, Urumqi, China
[3]Xinjiang Research Centre for Mineral Resources, Chinese Academy of Sciences, Urumqi, China
[4]University of Chinese Academy of Sciences, Beijing, China

ABSTRACT

Airborne hyperspectral remote sensing data provide rapid, non-destructive, and near laboratory quality reflectance spectra for mineral mapping and lithological discrimination, thereby ushering an innovative era of remote sensing. In this study, NEO HySpex cameras, which comprise 504 spectral channels in the spectral ranges of 0.4–1.0 μm and 1.0–2.5 μm, were mounted on a delta wing XT-912 aircraft. The designed flexibility and modular nature of the HySpex aircraft hyperspectral imaging system made it relatively easy to test, transport, install, and remove the system multiple times before the acquisition flights. According to the design fight plan, including the route distance, length, height, and flight speed, we acquired high spectral and spatial resolutions airborne hyperspectral images of Yudai porphyry Cu (Au, Mo) mineralization in Kalatag District, Eastern Tianshan terrane, Northwest China. By comparing the features of the HySpex hyperspectral data and standard spectra data from the United States Geological Survey database, endmember pixels of spectral signatures for most alteration mineral assemblages (goethite, hematite, jarosite, kaolinite, calcite, epidote, and chlorite) were extracted. After a HySpex data processing workflow, the distribution of alteration mineral assemblages (iron oxide/

hydroxide, clay, and propylitic alterations) was mapped using the random forest (RF) algorithm. The experiments demonstrated that the workflow for processing data and RF algorithm is feasible and active, and show a good performance in classification accuracy. The overall classification accuracy and Kappa classification of alteration mineral identification were 73.08 and 65.73%, respectively. The main alteration mineral assemblages were primarily distributed around pits and grooves, consistent with field-measured data. Our results confirm that HySpex airborne hyperspectral data have potential application in basic geology survey and mineral exploration, which provide a viable alternative for mineral mapping and identifying lithological units at a high spatial resolution for large areas and inaccessible terrains.

Keywords: *Airborne hyperspectral imaging, HySpex sensor, Alteration mineral, Random forest algorithm, Image classification*

INTRODUCTION

Hyperspectral imaging is a highly relevant topic in remote sensing research owing to the characteristics of high spectral resolution, strong wavelength continuity and large amount of spectral information (Grove et al., 1992; Clark et al., 2003; van der Meer et al., 2012; Guha, 2020).

Each pixel in a hyperspectral image has its continuous spectrum, the technique has wide geological applications, including the identification and classification of mineralogical alteration assemblages and correlation of geological units (Gomez et al., 2018; Xie B. et al., 2022; Xie B. S. et al., 2022). For decades, hyperspectral imaging has been widely used in geological prospecting, such as porphyry-type copper deposits (Batchelor et al., 2016; Dalm et al., 2017), hydrothermal gold deposits (Osterloo et al., 2012), sulphide deposits (Iyakwari et al., 2016), ore deposits (Murphy and Monteiro, 2013; Metelka et al., 2015), uranium Ore deposits (Reath and Ramsey, 2013), and rare earth deposits (Riaza et al., 2001). The most popular hyperspectral sensors for mineral exploration from spaceborne and airborne platforms are summarized in Table 1.

Airborne hyperspectral imagers have a major role in geological exploration and mineralogy owing to the advantages of spatial resolution and performance capabilities in a cloudy atmosphere, compared with the images of spaceborne airborne platforms.

Airborne hyperspectral sensors, such as AVIRIS, HyMAP, CASI/SASI and HySpex open new possibilities for the detection and mapping of mineral deposits (Rowan et al., 2000; Rowan et al., 2004; Jing et al., 2014). Compared with the airborne sensor parameters in Table 1, HySpex provides high-performance parameters such as spectral range (0.4–2.5 μm), sampling interval of 2.8 nm/5.45 nm, 504 spectral channels, a flexible spatial resolution, and light weight (10.2 kg) to be mounted on the aircraft platform and ground platform.

In 1995, the hyperspectral sensor of HySpex is a "pushbroom" scanner produced by NEO (Norsk Elektro Optikk, Oslo, Norway). The camera scans an object and collects a hyperspectral image or "cube" with two spatial dimensions and one spectral dimension. In recent years, numerous studies have identified the detection capabilities of HySpex sensors for various applications (Baissa et al., 2011; Steffens and Buddenbaum, 2013; Mathieu et al., 2017).

Table 1. Widely used hyperspectral sensors

	Sensors	Spectral range (µm)	No. of bands	Spectral resolution (nm)	Ground sample distance (GSD; m per pixel)	Year of launch	References
Satellite	Hyperion	0.40–2.50	220	10	30	2000	Salimi et al., (2018)
	EnMAP	0.42–2.40	228	5.25–12.5	30	2006	Rogge et al., (2014)
	MODIS	0.40–1.45	36	10	250/500/1000	199-2002	Sun et al., (2011)
	DESIS	0.40–2.50	180	3.3	30	2018	Eckardt et al., (2015)
	GF-5	0.40–2.50	330	5	30/60	2018	Sun et al. (2011)
	PRISMA	0.40–2.50	239	12	30	2019	Vangi et al., (2021)
Airborne	AVIRIS	0.36–2.45	224	10	20	1986	Tripathi et al., (2020)
	HyMAP	0.45–2.50	126	15	5	1997	Tan et al., (2021)
	CASI/SASI	0.36–1.05	144	2.4	2.5	1989	Jing et al., (2014)
	HYDICE	0.40–2.50	210	10.2	7	1994	Rickard et al., (1993)
	HySpex	0.40–2.50	504	2.8/5.45	/	1995	Cristóbal et al., (2021)

For example, Cristóbal et al. (2021) demonstrated the viability of HySpex airborne hyperspectral wetland mapping in the high latitudes of Alaska. Baissa et al. (2011) identified and mapped hand specimens of Jurassic carbonate facies via a HySpex SWIR-320 m hyperspectral sensor in the Agadir Basin. Buckley et al. (2013) analyzed and mapped geological outcrops by merging LiDAR and HySpex imaging data.

Using a HySpex sensor in a laboratory with a spectral range from 0.4 to 0.99 μm, Steffens and Buddenbaum (2013) developed a fast and effective imaging method that enables the spatially accurate classification of diagnostic horizons as well as the mapping of different contents of mineral elements. Ali et al. (2017) developed a rapid and efficient method to determine hyperspectral specific leaf area by leaf and canopy spectral measurements using HySpex airborne imagery.

HySpex imagery is useful when detecting target objects in hundreds of narrow bands that are only a few nanometers wide from the visible and near-infrared (VNIR) to the short-wave infrared region (SWIR). The HySpex airborne hyperspectral sensors are more effective in the advanced stage of exploration where there is a need to focus on small ground objects with spectral data at high spectral and spatial resolutions (Paoletti et al., 2019; Guha, 2020). Furthermore, the designed flexibility and modular nature of the HySpex aircraft system made it relatively easy to test, transport, install, and remove the system multiple times before the acquisition flights. Besides, the flexible setup also reduced cost by not committing aircraft use solely to data acquisition. In contrast, compared with conventional mineral mapping based on geological fieldwork spectral surveys, HySpex hyperspectral sensors can provide and collect easily accessible mineral information over a large spatial area and are effective in mapping spectrally conspicuous host rocks of mineral deposits, residual enrichment deposits, and surface mineralogical proxies occupying a larger spatial domain (Guha, 2020). HySpex imaging can provide airborne hyperspectral data that satisfy the high spatial and spectral resolution requirements of many geoscience remote sensing applications, offering a near laboratory quality spectrum and representing a fundamental development in the field of hyperspectral remote sensing. However, airborne hyperspectral imaging using HySpex has not been used as commonly as ground- and laboratory-based imaging and has rarely been applied for mineral mapping and discrimination of lithologic units.

Using hyperspectral images on our own HySpex airborne flight, we extracted and identified alteration mineral assemblages of the Yudai porphyry Cu (Au, Mo) mineralization (Kalatag District, northwest China). The main objectives of this study were to 1) acquire HySpex airborne hyperspectral images of the Yudai Porphyry Cu (Au, Mo) mineralization, 2) determine a workflow for processing HySpex images, and 3) identify alteration minerals using a random forest (RF) algorithm and a comprehensive field survey.

STUDY AREA

The Eastern Tianshan in Xinjiang, northwest China, is a vital component of the Central Asian Orogenic Belt as well as the Altaids. The Kalatag district is located on the southern margin of the Turpan–Hami Basin and hosts the Yudai, Hongshi,

Hongshan, South Meiling, and Meiling deposits. The Yudai Cu (Mo, Au) deposit is a newly discovered porphyry Cu deposit in the northwest part of Kalatag (Figure 1); this is a key area of early Palaeozoic rocks in the Eastern Tianshan (Chen et al., 2017). The deposit is a typical porphyry Cu deposit and occupies an area of nearly 30 km². Alteration and mineralization have formed distinct zones around the quartz diorite porphyry, with disseminated, vein, or veinlet Cu (Mo, Au) mineralization types mainly related to potassic alteration such as biotite, magnetite, and quartz (Sun et al., 2017).

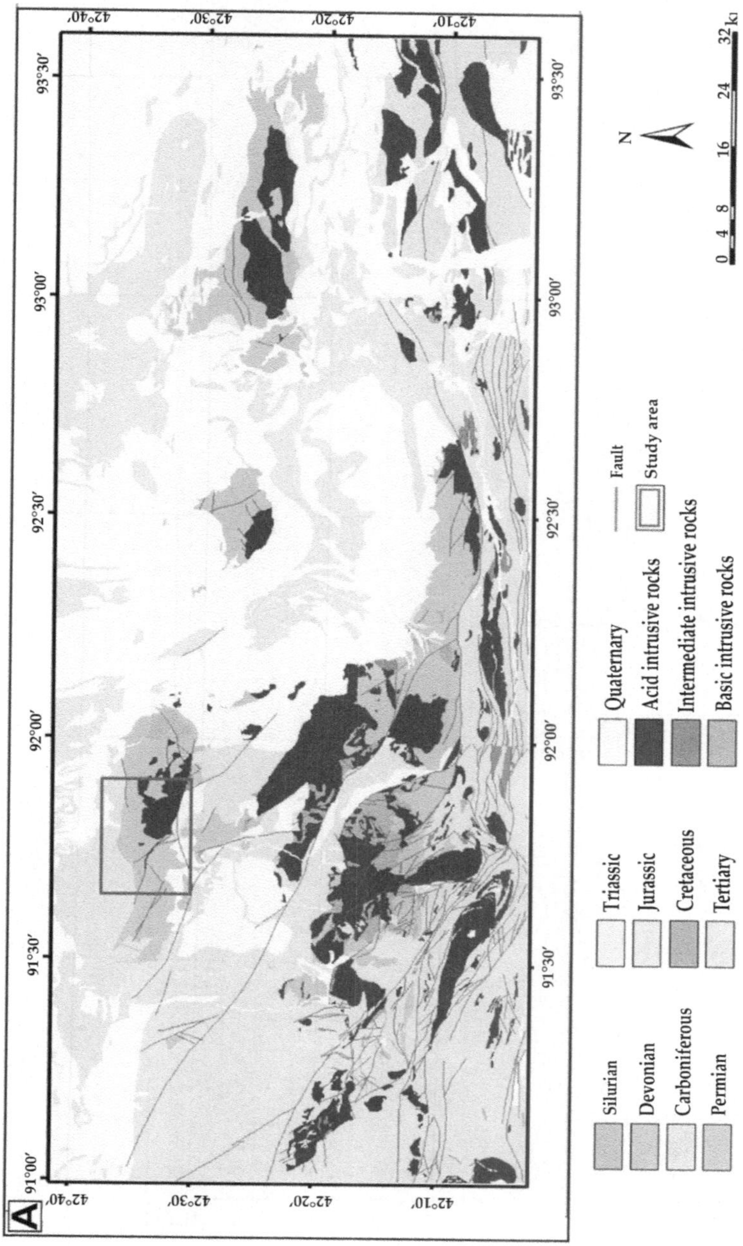

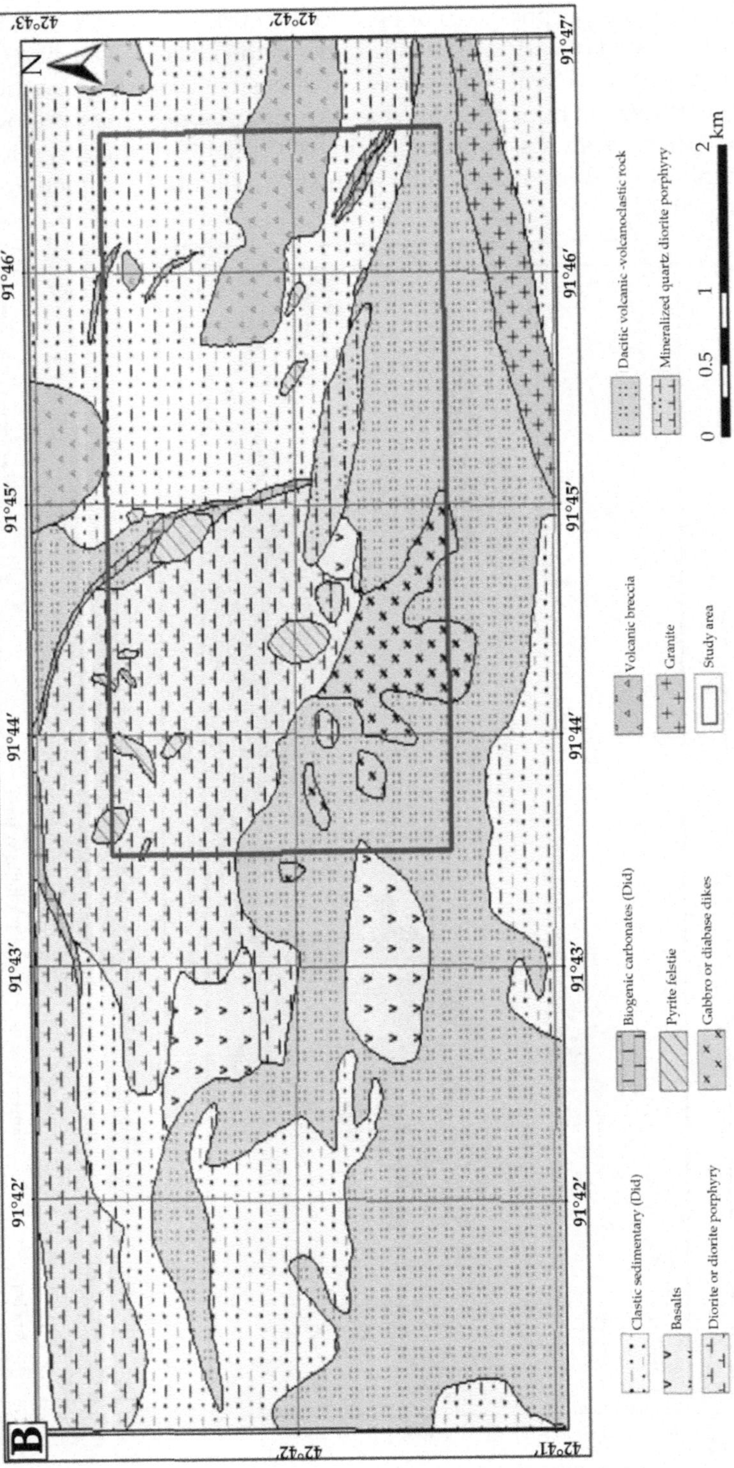
B
N
42°43′
42°42′
42°41′
91°42′
91°43′
91°44′
91°45′
91°46′
91°47′
Clastic sedimentary (Did)
Basalts
Diorite or diorite porphyry
Biogenic carbonates (Did)
Pyrite felstie
Gabbro or diabase dikes
Volcanic breccia
Granite
Study area
Dacitic volcanic -volcanoclastic rock
Mineralized quartz diorite porphyry
0
0.5
1
2
km

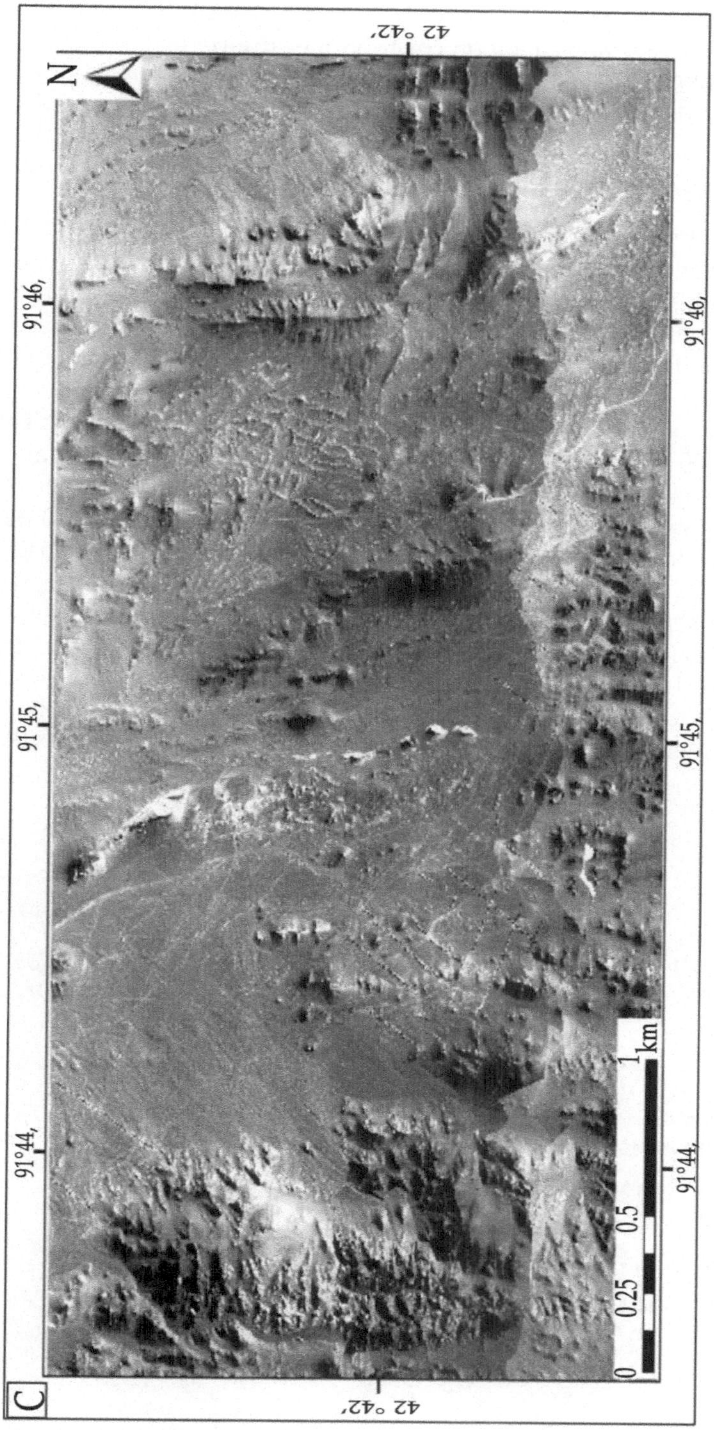

Figure 1. Schematic diagram of study area (A) Overview map of Kalatag district; (B) generalized geological map of Yudai; (C) hyperspectral images from the HySpex airborne flight.

The Xinjiang Karatag Jade Zone mining area is an arid area with large amounts of evaporation and no vegetation cover on the surface, making it suitable for the use of remote sensing data for large-area mineral resources exploration work. Therefore, airborne hyperspectral remote sensing provides a viable and reliable alternative for a wide range of mineral mapping in this area.

HYSPEX AIRBORNE HYPERSPECTRAL IMAGING SYSTEM

HySpex hyperspectral sensors were mounted on a delta wing XT-912 aircraft platform, a light vehicle with good gliding performance, low cost, simple structure, ability to fold, and ease of transport and storage. The assembly instrument, an airborne flight tool for HySpex data, included HySpex sensors, a computer assembly, an inertial navigation set, and a battery pack and was mounted on the delta wing XT-912 aircraft (Figure 2).

The take-off and landing ground sliding range of the delta wing XT-912 aircraft is between 50 and 150 m, the flight altitude range is 50–4,000 m, and the flight speed range is 45–110 km/h. According to the study area terrain, detection target, and spatial resolution, the flight height of the HySpex airborne is 500–3,000 m. The airborne hyperspectral sensor of HySpex consists of Sensor 1 and Sensor 2 imaging spectrometers, which have spectral ranges of 0.4–1.0 and 0.93–2.5 μm, respectively. It also consists of 504 spectral channels with spectral sampling resolutions of 2.8 and 5.45 nm for Sensor 1 and Sensor 2, respectively. The sensors of VNIR and SWIR are pushbroom scanners that use an angular recording image width of 15 m. The field of view (FOV) can be doubled to 30° by a field expander lens. Table 2 lists the spectrometer parameters of the HySpex sensors of VNIR and SWIR.

Figure 2. Photographs of the HySpex system for airborne acquisition. (A) Delta wing XT-912 aircraft, (B) HySpex sensors in aircraft, (C) HySpex sensors assembly, (D) main computer assembly, (E) 4 m × 4 m white reference panel, and (F) a geographical reference correction point. Details of HySpex sensors and computer assembly: ① HySpex VNIR sensor (Sensor 1), ② HySpex SWIR sensor (Sensor 2), ③ inertial navigation set, ④ battery pack, ⑤ main computer case, ⑥ air-to-ground communication equipment and cables.

Table 2. HySpex spectrometer parameters

	VNIR (Sensor 1)	SWIR (Sensor 2)
Spectral sampling (nm)	2.8	5.45
Spectral range (μm)	0.4–1.0	0.93–2.5
Number of spectral bands	216	288
Spatial pixels	1024	384
Max frame rate (fps)	350	450
FOV across track (°)	17	14
Pixel FOV across/along track (mrad)	0.28/0.56	0.73/0.73
Digitization (bits)	12	16
Sensor head power consumption (W)	6	30
Detector material	Si CCD	Hg Cd Te
Sensor head weight (kg)	4.5	5.7

VNIR, visible and near-infrared; SWIR, short-wave infrared region.

The HySpex instrument is equipped with an inertial navigation system that applies a correction of geometric errors and georeferencing. The inertial navigation system can record the posture data of the aircraft (i.e., the sensor) in real-time to provide data support for the geometric coarse correction of the image data in the later stage. Data are transmitted and read through antennas between the hyperspectral sensor and the inertial navigation system. The antenna of the inertial navigation system should be aligned to the sky and the control antenna to the ground. Ground-based field tasks were synchronously carried out during flight operations. A 4 m × 4 m white reference panel of identified reflectivity

(~80%) was placed on the ground in the test area. Several geographical reference correction points were also placed on the ground and measured by real-time kinematic navigation.

HYSPEX DATA PROCESSING WORKFLOW

Figure 3 shows a workflow of HySpex data processing, including data acquisition, data preprocessing, RF methods, and accuracy evaluation. The afore mentioned stages are presented in more detail below.

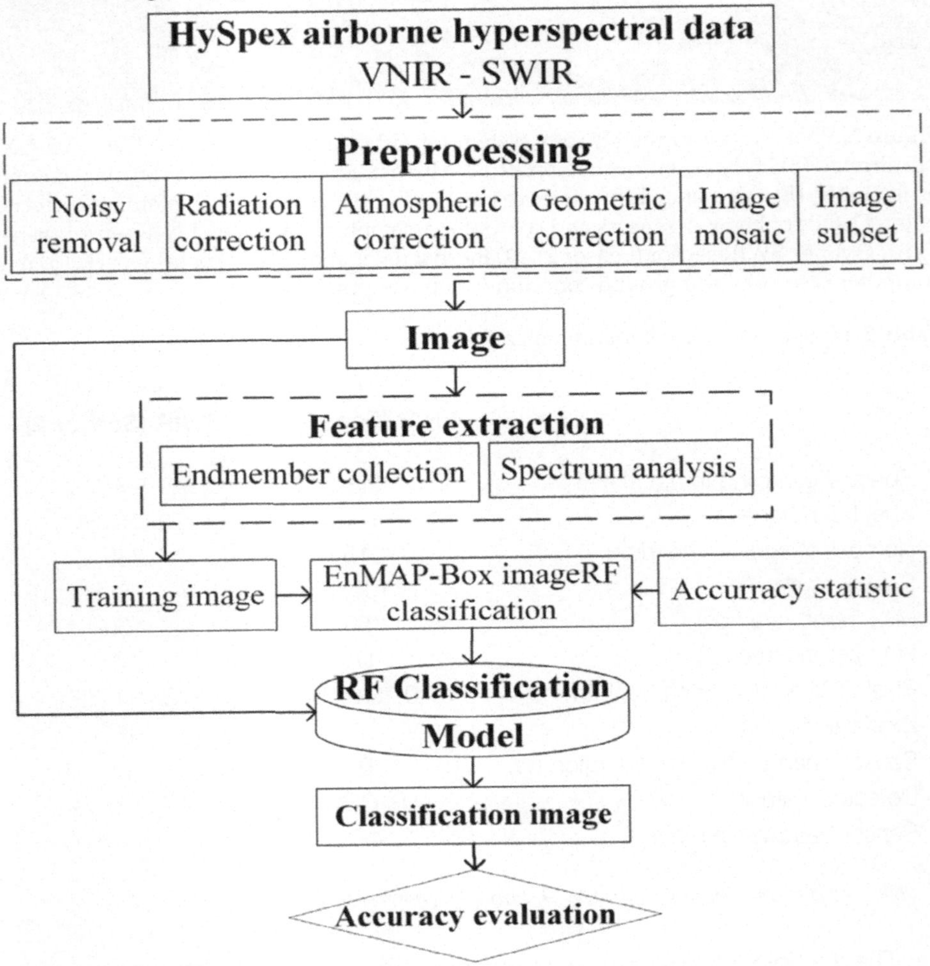

Figure 3. Workflow of HySpex data processing.

Flight Planning and Data Acquisition

Before the flight experiments, the route distance, length, height, and flight speed were determined to ensure that repetition, a low signal-to-noise ratio, and a spatial resolution of high spectral imaging data were achieved. The relationship between route spacing, flight height, and image repetition rate is as follows:

$$\tan\frac{A}{2} = \frac{W}{2H} \tag{1}$$

$$D = 2\tan\frac{A}{2} \times H \times (1 - x) \tag{2}$$

where D represents the route spacing, A represents the instantaneous FOV angle of the spectrometer sensor, W represents the scan width of a route, H represents the flight height, and x represents the image repetition rate.

Flight experiments were conducted from October 9 to 11, 2017 between 09:00 and 12:15 local time in the area of Yudai. The weather conditions were as follows: sunny, cloudless, 15°C—25°C, and wind speed < 4 levels. To ensure that data splicing did not produce voids and affect the quality of the data, the repetition rate of the image data was set to approximately 40%. The instantaneous FOV angle of the HySpex two sensors was 32°. When the delta wing XT-912 aircraft was flown over the test site at an average altitude of 1 km above sea level, the average ground spatial resolutions were 1 and 1.2 m for the VNIR and SWIR images, respectively. Each route length was designed to be 5 km, and each route had a scan width of 286.7 m. Depending on the flight height and repetition rate, the route distance was approximately 400 m. Figure 4 shows the route designed for flight planning, which included the designed route of the flight location and flight endurance (Figure 4A), the flight tracks obtained using HySpex AIR software (Figure 4B), and a photo taken in-flight (Figure 4C).

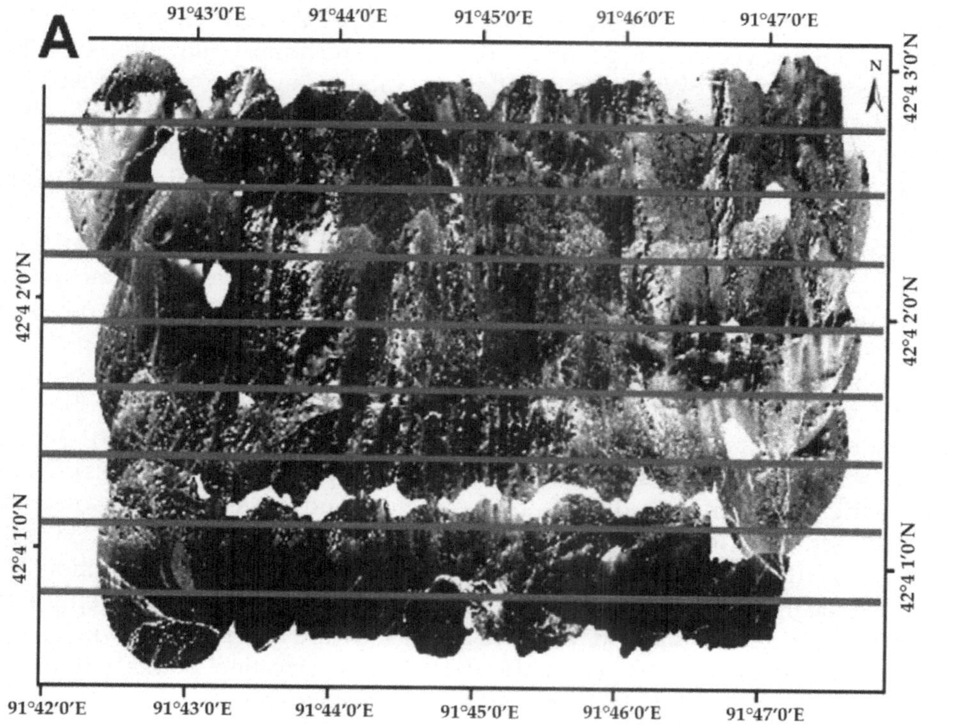

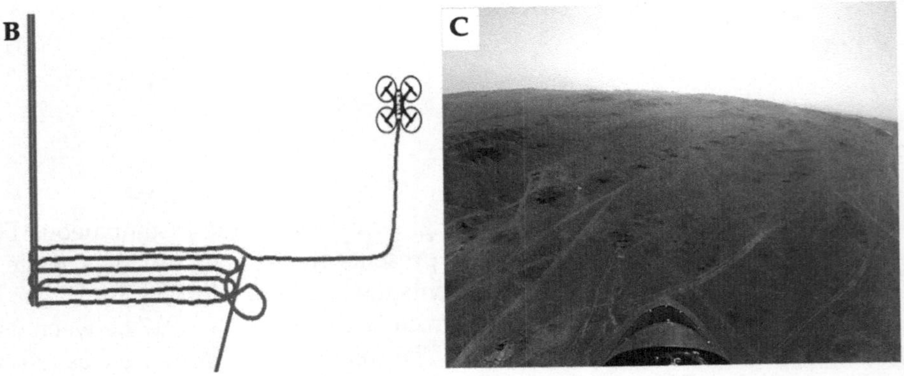

Figure 4. Route design for flight planning. (A) Designed route of the flight location and flight endurance, (B) flight tracks of HySpex AIR software, and (C) a photo taken in-flight.

For this study, we selected a small area as the method training site. In the training site, we collected 26 hand specimens of alteration minerals, and each sample was ground and mounted on a glass holder for thin section preparation. We analyzed the thin sections using photomicrographs to determine the alteration and mineral composition of each specimen. The HySpex image processing used the professional remote sensing software ENVI 5.3 (Environment for Visualizing Images 5.3).

Hyperspectral Images Preprocessing

The obtained HySpex hyperspectral raw data were the original digital number (DN) value data without any physical significance; the data were without coordinate systems but had very strong geometric distortions and atmospheric effects. For the most recent HySpex SWIR instrument, the so called "keystone" effect (related to spatial mistration for different bands within a pixel) and the spectral smile effect (the spectral curvature effect is a spectral distortion) are considered during the data correction process to remove spectral and spatial misregistrations (Koloniatis et al., 2020).

Data preprocessing forms the basis of alteration mineral mapping analysis, and mainly consists of radiation calibration, geometric correction, atmospheric correction, and bad band removal. First, the original DN value was converted to a radiation brightness value to obtain the attitude data of the flight platform synchronized with the images using HySpex RAD and HySpex NAV software, which was brought by the instrument manufacturer of NEO A/S. Second, atmospheric correction of the HySpex image data was performed by the atmospheric correction ATCOR4 software (ReSe Applications LLC, Swiss). Third, geometric correction, including coarse corrections and precision corrections, was carried out. PARGE (PARametric Geocoding, ReSe Applications LLC, Swiss)[1], HySpex's airborne post-processing software, was used for orthorectification and georeferencing with a standard digital elevation model. PARGE used an empirical view angle correction method available in the software to remove the view angle effects. Approximately 10 geographical reference correction points within the study area were used for precision geometric correction.

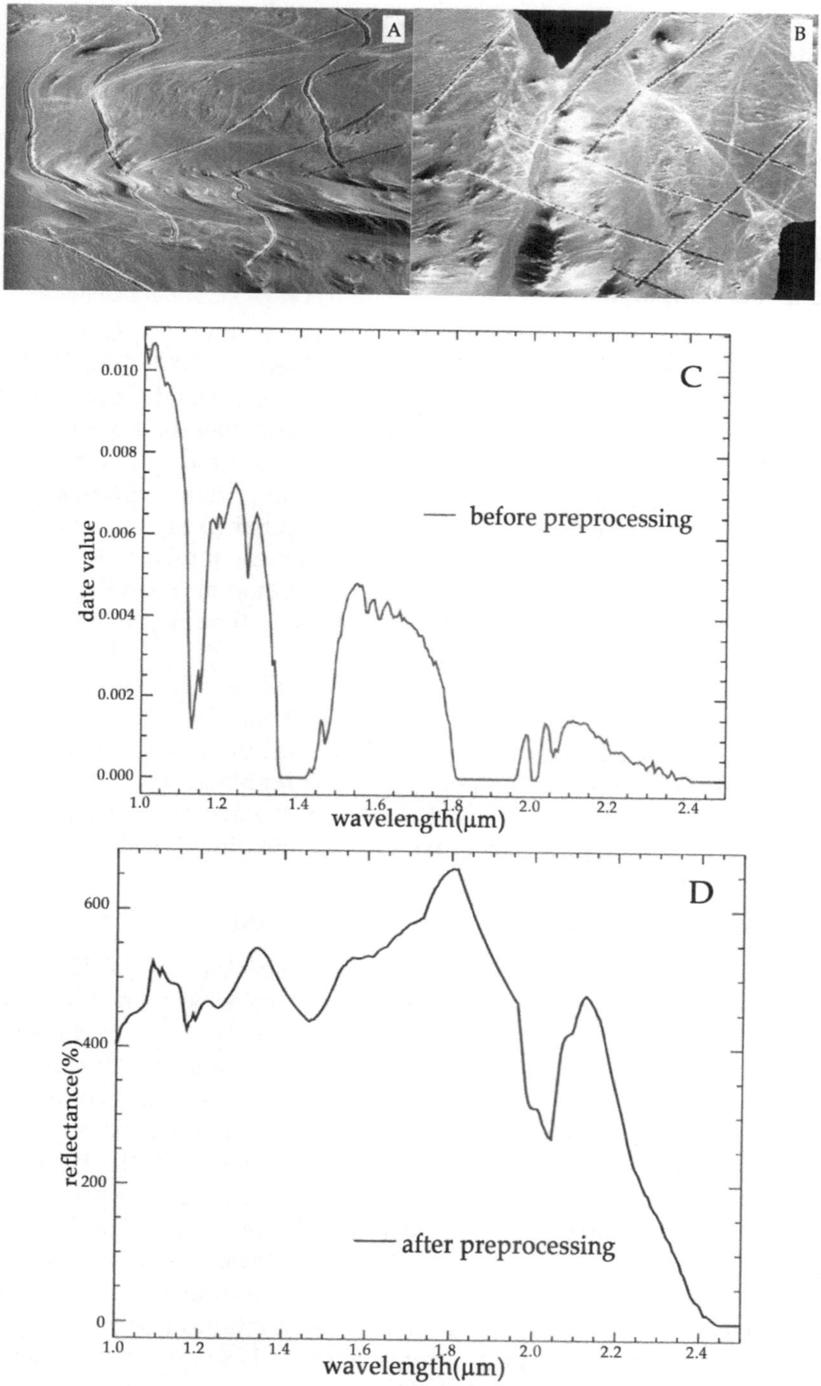

Figure 5. Geometric correction, atmospheric correction, and bad band removal (A) before geometric correction preprocessing, (B) after geometric correction preprocessing, (C) before atmospheric correction and bad band removal preprocessing, and (D) after atmospheric correction and bad band removal preprocessing.

The resulting geometric accuracy was > 0.50 m. Furthermore, a certified reflectance standard white reference panel of known reflectivity was recorded with each image and used as reference material to convert the obtained spectral data into reflectance data. To further analyze the HySpex spectral data, image registration, noise removal, mosaic, and mask were used by ENVI.

The efficacy of the geometric and atmospheric corrections was assessed by comparing the spectra extracted from the original and preprocessed image from the same location (Figure 5).

Random Forest Algorithm

RF is an ensemble learning method for data classification, regression, and other tasks based on classification and regression trees (Breiman, 2001). The RF algorithm performs well with fewer training sample sets, less computation time, and more accuracy through its out-of-bag error estimation than other methods (Demarchi et al., 2014). Moreover, it offers a cross-validation-like accuracy measure through its out-of-bag error estimation and indicates variable importance through the loss in accuracy when feature values are randomly permuted (Breiman, 2001; Waske et al., 2012). Hence, when remotely sensed spectra represent mixtures of a variety of materials, RF is a highly robust discrimination method and can improve overall prediction accuracy. In this study, RF was run using the ImageRF toolbox of the EnMAP-Box v2.2.1 software. ImageRF is an Interactive Data Language-based tool for RF classification and regression analysis of remote sensing imagery. A four-step image analysis workflow consisting of model parameterization and application is shown in Figure 4. First, the RF classification (RFC) model was made by a reference dataset for training and internal validation. Second, the RFC model was used to perform image classification. Third, the accuracy of the model was estimated by comparing it with an independent validation. Finally, the model was applied to the target image data.

Endmember Spectra Collection and Spectrum Analysis

The collection and analysis of endmember spectra are the first stages of classification because they are used as references for subsequent processing and directly affect the mapping results. Each endmember spectrum is the average of 10 typical endmember spectra obtained by manual selection from images collected with the VNIR and SWIR sensors. We analyzed the spectral absorption positions and overall spectral shapes based on expert knowledge to select possible alteration mineral endmembers. In addition, the spectra extracted from the HySpex image were compared with existing reference library standard spectra from the United States Geological Survey (USGS) spectral library (Kokaly et al., 2017). The extracted spectral information was evaluated by interactive visual inspection and ENVI 5.3. To complement this analysis, continuum removal was used to find differences between varying spectra (Clark and Roush, 1984). Continuum removal is generally normalized in the range of spectral reflectance (0–1) to enhance the absorption features and intensity of characteristic peaks.

In this study, images acquired by the VNIR and SWIR sensors were used to map different alteration minerals. Alteration minerals of metal ores (e.g., Fe, Cu, and Mn) can be mapped using the VNIR image of Sensor 1 (0.4–1.0 μm), and the

OH-group and carbonate minerals can be identified using the SWIR image of Sensor 2 (0.93–2.5 μm) (Sabins, 1999; Clark et al., 2003; Kruse, 2012). Wavelengths outside the atmospheric windows of ~1.45 and ~1.91 μm of SWIR were eliminated from further consideration because atmospheric water vapor causes strong disturbances in these regions (Vane and Goetz, 1988; Magendran, 2014).

According to geological map and field references, endmember extraction yielded average spectral features for most alteration minerals in the study area, such as hematite, goethite, jarosite, calcite, epidote, kaolinite, and chlorite (Figure 7). The standard spectra, which were used for comparison with the endmember spectra, were obtained from the USGS spectral library. The common major absorption areas are highlighted by color blocks corresponding to the chemical components' letters and arrows.

RESULTS

Alteration Mineral Mapping

Iron Oxide Minerals

In the visible light bands, the endmember 1 spectrum displays a significant absorption feature at 0.51 μm, weak absorption valleys near 0.74 and 0.78 μm, and a distinct band centered at 0.91 μm (Figure 7A).

A previous study on visible light spectra of minerals and rocks found that crystal field effects are the most common electronic processes and are associated with transition elements such as ferrous (Fe^{2+}) and ferric (Fe^{3+}) iron (Hunt, 1971). Minerals containing Fe^{2+} have a wide absorption range centered near 1.0–1.1 μm with a longwave slope that can extend to ~1.8 μm. For example, Fe^{3+} minerals in jarosite have crystal field absorptions near 0.51 μm (Hunt and Ashley, 1979). The position of the ~0.91 μm absorption in hematite shifts to longer wavelengths in goethite samples, and this can be detected by hyperspectral imaging sensors.

The ferric iron minerals goethite (α-FeOOH) and hematite (α-Fe_2O_3) have distinct spectral curves in the region of 0.35–0.90 μm owing to absorption induced by crystal field transitions near 0.46, 0.65, and 0.85–0.95 μm (Murphy and Monteiro, 2013). The general chemical formula of jarosite is $AB_3(SO_4)_2(OH)_6$, in which A represents K^+ and B represents Fe^{3+} for typical jarosite. The groups of jarosite minerals are important indicators of aqueous, acidic, and oxidizing formation conditions (Singh et al., 2016).

The spectra assigned to goethite and hematite show strong absorption at 0.52 μm with matching bands at 0.42 μm (Figure 7C). In addition, the spectra of goethite and jarosite display a broad absorption feature at 0.91 μm with matching band-189 of HySpex VNIR. The absorption of spectra between 0.91 and 0.93 μm (bands 187–96) can, therefore, be assigned to hematite (Figure 7C). The wavelength position of the primary Fe^{3+} absorption band (~ 0.92 μm) and the shapes of the absorption edge and shoulder near 0.48 and 0.55 μm, respectively, are most consistent with hematite, goethite, jarosite, or their combinations (Figure 7C). However, features of the endmember 1 spectrum show slight deviations and are relatively weaker than the standard spectrum. There are two possible reasons for this behavior: first, a mixing of the spectra of hematite, goethite, and jarosite may have occurred, which also mixes the spectra of primary rock-forming minerals and secondary weathering minerals (Metelka et al., 2015); second, there may be potential masking by pyrophyllite and chlorite owing to Fe–OH and OH stretching and bending (Hosseinjani Zadeh et al., 2014). The iron oxide minerals (red-colored pixels in Figure 7) such as hematite, goethite, and jarosite occur in a mostly longitudinal distribution and coincide with mine pits and grooves in the Yudai area, reflecting the secondary alteration of iron-rich minerals. Under oxygen-rich conditions, oxidation is the main iron mineral alteration, and secondary weathering minerals dominate the reflectance spectra (Singh et al., 2016). Iron absorptions associated with both primary rock-forming minerals and secondary weathering minerals (hematite and goethite) were detected in most samples (Metelka et al., 2015).

Clay Alteration Minerals

The spectral profile of endmember 2 has absorption bands centered at 1.43, 1.91, 2.18, and 2.21 μm (Figure 7D), corresponding to H_2O, OH, Al-OH, and Mg-OH vibrations, respectively. In the 2.0–2.5 μm region (Figure 7E), the endmember 2 spectrum shows characteristic absorption at 2.21 μm and minor changes in the intensity of reflectance near 2.16 μm, which is an easily identifiable bimodal feature. This strong absorption with discrete absorption features at 2.21 μm is characteristic of Al-OH minerals (Hosseinjani Zadeh et al., 2014). Kaolinite predictably exhibited the strongest water overtones at 1.43 μm, in addition to water combination bands at 1.90 μM, with strong OH combination bands from 2.1 to 2.4 μm (Figures 7E,F). Most importantly, Al-OH clays (kaolin-group minerals) have diagnostic absorption features between 2.18 and 2.21 μm as a result of a combination of Al-OH bending and OH stretching vibrations (Clark et al., 2003). In our data, spectra with diagnostic absorption features at 2.16 μm (band of 440) and 2.21 μm (band of 449) attributable to Al-OH vibration are indicators of kaolinite (Figure 7E).

Kaolinite, smectite, and illite are mainly the result of clay alteration; this hydrothermal alteration of wall rock that contains the alteration minerals of kaolinite alteration (light-green-colored pixels in Figure 7) is concentrated near pits and grooves.

Propylitic Alteration Minerals

The reflectance change from the 2.0–2.5 μm region is lower, and the endmember three spectrum tends to be similar to that of calcite, epidote, and chlorite (Figure 7I). The SWIR spectrum displays a weak absorption valley close to 2.26 μm and a distinct band centered near 2.35 μm. In addition, Fe-OH molecules lead to absorption bands at 2.44 μm (Murphy and Monteiro, 2013). Figures 6H,I show the reflectance spectra of chlorite, calcite, and epidote between 2.0 and 2.5 μm in the near-infrared that. The diagnostic infrared absorption feature of calcite ($CaCO_3$) is focused at 2.34 μm, as is those of epidote $[Ca_2(Fe^{3+}, Al)_3(SiO_4)_3(OH)]$ and chlorite $[(Mg, Fe^{2+})_5 Al (Si_3Al) O_{10}(OH)_8]$ (Dalton et al., 2004). Even though we found a small shoulder near 2.34 μm in the main absorption valley, there is no corresponding absorption feature in the spectrum of calcite. Calcite shows a weak and small absorption valley near 2.15 μm, that is, not detected for other minerals (Dalton et al., 2004). The secondary absorption feature (~2.26 μm) of the chlorite spectrum would have changed and obscured the tendency and strength of the primary absorption feature (~2.34 μm; Figures 7H, 5I).

The spectral signature of epidote has nearly the same features as that of chlorite. The spectral relationships help with the identification of these minerals using remote sensing data. The spectrum of endmember 3, which was assigned to calcite, epidote, and chlorite, shows broad and strong absorption at 2.35 μm, matching band 475 (Figure 7I). In addition, the spectra with absorptions at 2.26 μm (band 459) can be assigned to epidote and chlorite.

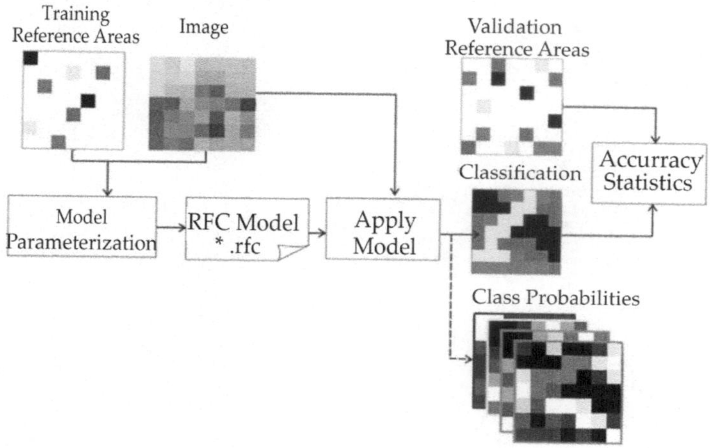

Figure 6. Workflow of parameterization, application, and assessment of the random forest classification (RFC) model in the ImageRF toolbox of EnMAP-Box v2.2.1.

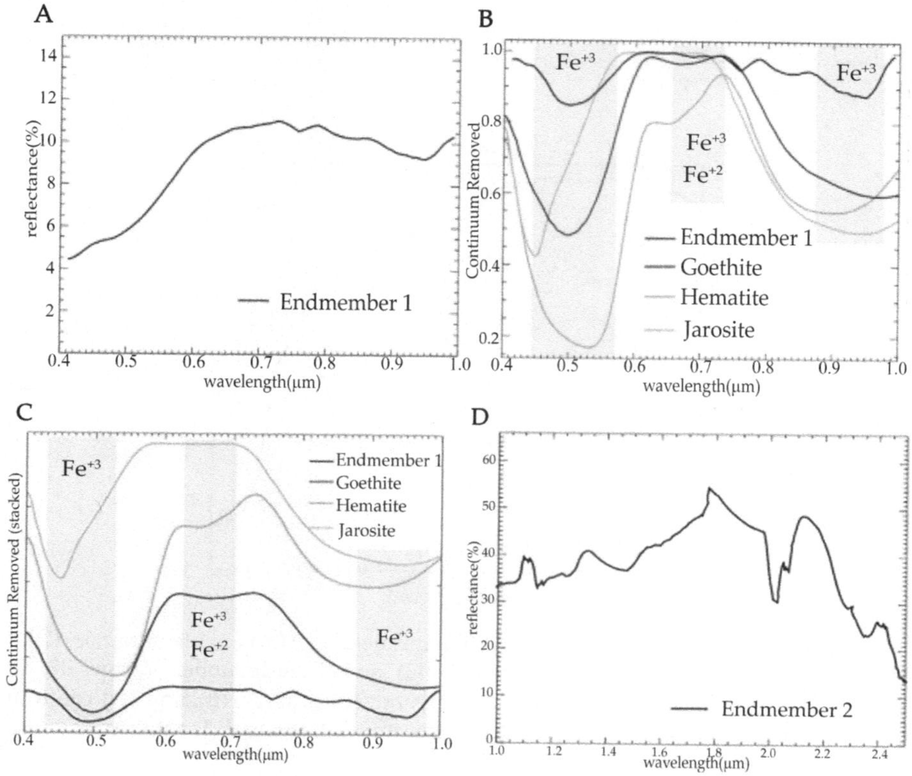

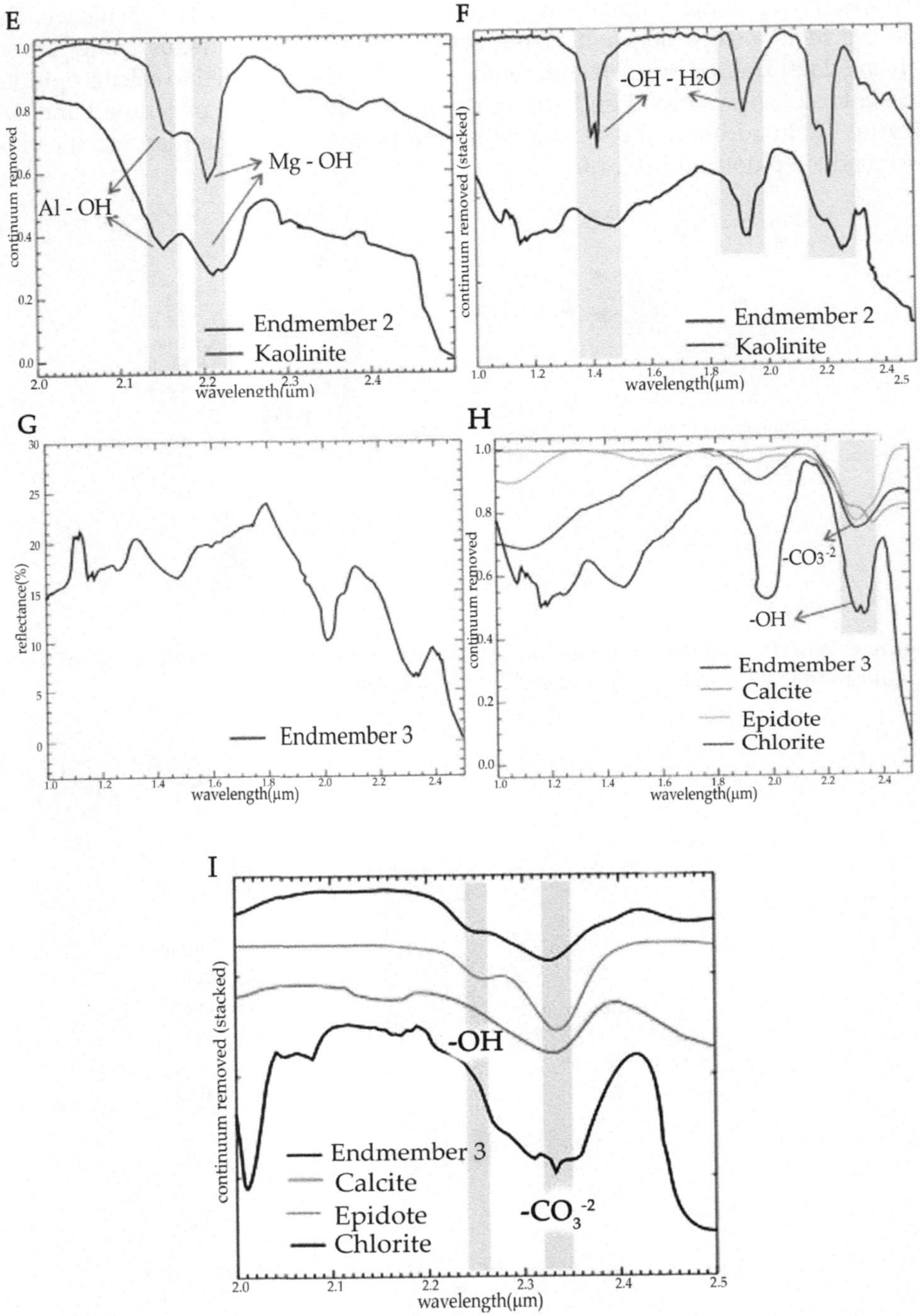

Figure 7. Endmember spectra. (A), (B), and (C) Endmember 1/iron oxide minerals; (D), (E), and (F) endmember 2/clay alteration; and (G), (H), and (I) endmember 3/propylitic alteration; (A), (D), and (G) reflectance of the endmember spectra; (B), (E), and (H) continuum removal spectra of endmembers and standard spectra; (C), (F), and (I) continuum removal spectra (stacked) of endmembers and standard spectra.

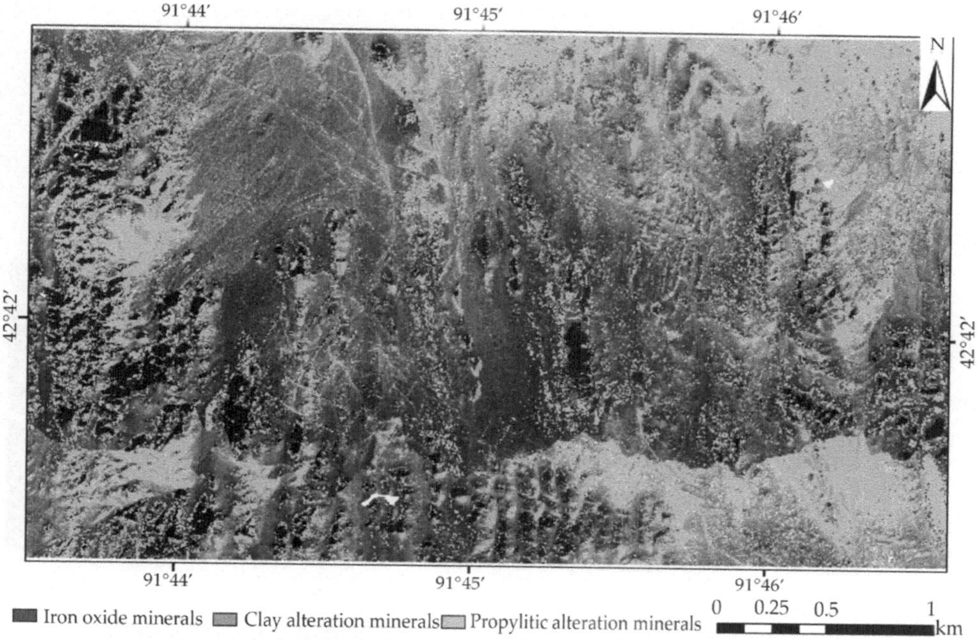

Figure 8. Alteration mineral map of the mineralization of Yudai.

Propylitic alteration is the chemical alteration of rock, that is, rich in biotite or amphibole owing to iron- and magnesium-bearing hydrothermal fluids. Pixels indicating propylitic alteration (light-blue-colored pixels in Figure 8) are the most common class of pixels on the map of Yudai, demonstrating the dominance of propylitic alteration over porphyry copper bodies. Chlorite, calcite, and epidote can be identified through the method; however, the three endmember minerals create discrepancies in areas where they are combined and have different particle sizes.

Accuracy Evaluation

The accuracy of our mineral identification was evaluated via the mineralogical analysis of samples collected during fieldwork (Figure 9). We used the common classification comparison method, overall accuracy, and Kappa classification accuracy. The classification accuracy of iron oxide minerals, clay alteration minerals and propylitic alteration minerals was 82.60, 62.16, and 70.59%, respectively. The overall classification accuracy and Kappa classification were 73.08 and 65.73%, respectively. The experiments demonstrate that the workflow for processing data and using RF algorithm is feasible and active, and show a good performance in classification accuracy.

The mineralogy of each HySpex image pixel identified using the proposed method generally matched our field observations (Foody, 2002). There was excellent consistency in the alteration minerals observed through visual inspection (Figure 9). The main alteration mineral assemblages were primarily distributed around pits and grooves. However, minor differences were noticed, most of which were owing to co-occurrences of alteration minerals in the study area. Furthermore, the poor effect of classification includes some other environmental reasons, such as shadows of mountains, valleys, and rock walls; the footprint of wheels in the study area; and white holes caused by sun's violent reflection.

Figure 9. Accuracy evaluation map, hand specimen photographs, and thin section photomicrographs at 10-fold magnification. (A) Iron oxide minerals, (B) clay alteration minerals, (C) clay alteration minerals, (D) propylitic alteration minerals, (E) propylitic alteration minerals, and (F) iron oxide minerals. Qtz, quartz; Pl, plagioclase; Ser, sericite; Py, pyrite; Ep, epidote.

DISCUSSION

HySpex Airborne Hyperspectral Image Data Acquisition

In this study, a HySpex aircraft system including HySpex sensors, computer assembly, inertial navigation set, and battery pack was successfully mounted on an aircraft platform. Using this system, we designed a route and performed flight test acquisition, geometric correction, and atmospheric correction based on the collected hyper spectroscopic image to provide basic data for corrosion information extraction. The final remote sensing data were obtained through radiation correction, bad band removal, spectral reconstruction, cutting, and splicing.

With recent fundamental developments in remote sensing, sensor design, processing rate, computer power, and speed, this technology has wide-ranging applications including mineral identification with high-quality spatial resolution. Recent research has focused on HySpex data from laboratory and field-based research. Most studies have shown how spectra can be interpreted and processed to produce mineralogical information using an array of HySpex data, such as mine faces, diamond cores, and drill chips (Kurz et al., 2008, 2009; Buckley et al., 2013; Qiu et al., 2017). However, compared with ground-based data, airborne acquisition of HySpex hyperspectral data has several advantages in terms of convenience and efficiency of data collection over large areas. In addition, the images include data on the geological context and textures of a rock unit and can detect faults and fractures as well as evaluate the degree of veining or dissemination of minerals (Qiu et al., 2017).

Specifically, using HySpex airborne hyperspectral data in large mining areas offers three distinct advantages. First, HySpex airborne hyperspectral data connect both spectral and spatial imaging methods, creating the illustration of objects in 504 contiguous and narrow spectral bands, with spectral resolutions of 2.8 and 5.45 nm, respectively. Second, SWIR sensors (0.93–2.5 μm) are used in combination with sensors operating in the VNIR (0.4–1.0 μm) range, allowing the user to separate and identify different alteration mineral assemblages (Dalponte et al., 2013). Third, a very high spatial resolution can be obtained using aerial platforms, which supports mapping at different scales. The spatial resolution is based on the performance parameter of the sensor, the flying altitude, and weather conditions during acquisition. With a delta wing XT-912 aircraft, spatial resolutions of 1 and 1.2 m were achieved for the VNIR and SWIR images, respectively, enabling the accurate identification and mapping of alteration minerals.

RF Algorithm

The RF algorithm refers to decision tree modeling of each bootstrap sample using bootstrap resampling methods by taking multiple samples from the original sample. When image classification is performed using the RF method, the final category of each image element is determined by the joint voting after each decision tree. Therefore, the RF classifier has a high-dimensional processing capability and can solve problems that occur as a result of high spectra (e.g., few training samples and class imbalance). To differentiate the ensemble predictors, not all available features are used to determine the optimum split point in a node. Instead, only a predetermined number of randomly selected features are used to avoid overfitting the model.

In this study, the identification of corrosion mineral information differed from the conventional classification; it is difficult to identify and extract the target information from the algorithm in the absence of prior knowledge. Moreover, the alternation mineral information has its own characteristics; therefore, extracting information from complex ground object spectra, making the property judgment, and then extracting the training samples were all required. For the applicability of the RF algorithm in imaging for the identification of high spectral erosion mineral information, combining the RF algorithm with the spectral feature matching technique was the proposed method. During the experiment, the training samples overcame the difficulty of extracting corrosion mineral information, and the training model of RF was constructed to apply the RF algorithm to hyperspectral corrosion mineral information identification.

Identification and Mapping Alteration Minerals

The main absorption features of rocks in VNIR are affected by the presence of various iron minerals, in addition to mixtures of ferric oxides, ferric hydroxides, and other minerals (Ramanaidou et al., 2007; Salles et al., 2017). Ferric oxides are recognized by their various spectral properties; for example, an absorption edge at 0.52 μm, an absorption band or shoulder at 0.74 μm, and a strong absorption band between 0.86 and 0.98 μm (endmember 1). The SWIR region (0.1–2.5 μm) is key to identifying carbonates and hydrous minerals (Clark, 1999; Guo et al., 2017) that are often products of hydrothermal alteration. Spectral features at 1.0–2.5, 2.2–2.5, and 1.9 μm are attributable to O–H vibrations, hydroxyl groups, and the water content of the rocks, respectively (Velasco et al., 2005; Milton et al., 2009). All spectra exhibit several typical absorption bands, including bound water resulting in absorption bands generally at 1.4 and 1.9 μm, and structural water absorption bands at 1.4, 2.2, 2.3, and 2.4 μm. The geometries of sites occupied by water affect the shapes, intensities, and widths of the 1.4 and 1.9 μm bands (Clark et al., 2003; Dalton et al., 2004). We did not consider features located around 1.45 and 1.91 μm as features in these regions are attributable to atmospheric water vapor (Vane and Goetz, 1988).

This study provides an example of the airborne acquisition and utilization of HySpex hyperspectral data for mineral mapping and the identification of lithologic units. By overcoming problems regarding safety, flight technology, field wind impact, and image processing, the preliminary results indicate that hyperspectral images of HySpex provide excellent and effective data for separating alteration mineral assemblages in the Yudai porphyry mineralization. The spectral curve characteristics of Cu mineralization show that the strong iron oxide/hydroxide characteristic absorption at 0.51 μm and the weak absorption valleys near 0.74 and 0.78 μm are caused by goethite and hematite; the absorption bands centered at 2.18 and 2.21 μm are caused by Al-OH and Mg-OH vibrations; and the weak absorption valley close to 2.26 μm and a distinct band centered near 2.35 μm may be the reflectance spectra of propylitic alteration minerals. According to the type and distribution characteristics of alteration mineral assemblages by hyperspectral remote sensing and geological characteristics of the study area, the hyperspectral remote sensing prospecting model of the alteration mineral assemblages of the Yudai copper deposit was determined as iron oxide/hydroxide, clay, and propylitic alteration.

Notably, high classification accuracy with a low time-cost for identification of alteration mineral assemblages in the mine pits and grooves was obtained. However, the reflectance spectra did not provide the complete identification and mapping of the surface mineralogy. In the field investigation, physical characteristics of minerals and rocks revealed that there is a stain of iron oxide/hydroxide, clay, and propylitic alteration minerals on the rocks and the surface of the mine pits, whereas the grooves are colored yellow but change to red with increasing depth. These changes occur mainly because of hydrothermally altered minerals but can also be generated through redox potentials at the time of weathering, and even during secondary weathering (Tripathi et al., 2020). This limitation is most likely related to the spectral resolution of diagnostic absorption features of the field reflectance spectra, the spectral resolution of the image, the spatial extent of the ground target, the spatial resolution of the image, noise in the image, and the accuracy of image preprocessing (Kodikara et al., 2012). Moreover, geologic surfaces are always partly shielded with non-geologic materials or composed of mixtures of minerals with different grain sizes and varying grades of compaction and weathering. The remote spectral measurements and the limit on the number of pixels that can be classified and mapped were greatly influenced by these factors. By comparing the spectra of rock and ore samples from hyperspectral images with standard spectra from the USGS database and the open integrated rock spectral library, the principle spectral absorption characteristics of the altered rock units were comprehensively understood, laying a theoretical foundation for mapping alteration products (Xie B. et al., 2022). The total shape of some reference spectra differed from the extracted spectra, most likely because the reference spectra are based on pure materials, which are unusual in natural environments. In addition, residual atmospheric absorption features, grain size variation, rock/desert varnish of rock surfaces, and calibration errors of laboratory spectrometers and/or hyperspectral instruments may have contributed to these differences. Therefore, image spectra, instead of library reference spectra, should be used as reference spectra in the RF algorithm.

The applicability of the algorithm was verified experimentally and provides a technical reference for subsequent work. However, this study has some limitations. For classification, the problem of mixed image decomposition may affect high spectral images. Information identification in this study was based on image endmembers, which directly affected the classification accuracy and caused the results to show discrete points and discontinuities in graph spots. Meanwhile, some environmental factors, such as shadows, ground trajectory, and violent reflection also affect the classification accuracy. Consequently, subsequent studies should focus on extending our analyses over a wider range of spatial scales, including field and laboratory scales, using hyperspectral data with high spatial and spectral resolutions. HySpex sensors can also be used in the field and laboratory to acquire ground spectra of altered minerals, considering weathered surfaces, grain size variation, and other natural variations. To increase our understanding of the differences in spectra between mixtures of minerals and pure minerals, spectral tests of such minerals and mineral assemblages are required, both at field and laboratory scales. In the meantime, we will expand our analyses to better understand the connection between spectral features and mineral alteration in addition to the mineral structure.

CONCLUSION

This study investigated the practicability of using HySpex airborne hyperspectral imaging for the detection and mapping of alteration zones in the Yudai porphyry mineralization area of Kalatag district, northwest China. The results show that HySpex airborne hyperspectral data can be used to detect alteration zones. The main conclusions are summarized follows:

1) The study presents an effective sequence for analyzing HySpex airborne hyperspectral data for lithological discrimination and mineral mapping. The analytical sequence consists of geometric correction, atmospheric correction, bad band removal, spectral reconstruction, the selection of a classification method (RF algorithm), spectrum analysis, and alteration mineral mapping.

2) Using alteration mineral spectra, expert knowledge, and an RF algorithm, the distribution of alteration minerals was mapped. Endmember pixel extraction yields average spectral signatures for most alteration mineral assemblages (goethite, hematite, jarosite, kaolinite, calcite, epidote, and chlorite). As remotely sensed spectra generally represent mixtures of materials, the RF method performs well.
3) Alteration mineral assemblages of iron oxide/hydroxide, clay, and propylitic alteration were found, and their distributions generally coincide with field observations. Although endmember minerals can be identified using this proposed scheme, discrepancies arise in areas where minerals occur as complex mixtures. The shape of the spectral curve of endmembers can change based on several factors, such as the replacement of iron by aluminum, particle size variations causing trans-opaque spectral effects in different wavelengths of the spectrum, and particle shape.

HySpex sensors mounted on a delta wing XT-912 aircraft have the advantage of producing integrated images and spectra. The high spatial resolution produces images at a large scale, and the high spectral resolution can facilitate the accurate identification of alteration minerals. Hence, this study demonstrates that airborne hyperspectral imaging can be widely applied to basic geological research and mineral exploration.

AUTHOR CONTRIBUTIONS

WSS was responsible for most of the experimental analysis and paper writing; ZKF and ZJ was responsible for sampling and experimental processing; WSS and WJL was responsible for the experimental design and the paper revision.

ACKNOWLEDGMENTS

The authors would like to thank the Xinjiang Laboratory of Mineral Resources and Digital Geology, CAS, Urumqi 830011, China, for their contribution in the collection of information for this study.

REFERENCES

1. Ali, A. M., Darvishzadeh, R., Skidmore, A. K., and van Duren, I. (2017). Specific Leaf Area Estimation from Leaf and Canopy Reflectance through Optimization and Validation of Vegetation Indices. *Agric. For. Meteorology* 236, 162–174. doi:10.1016/j.agrformet.2017.01.015
2. Baissa, R., Labbassi, K., Launeau, P., Gaudin, A., and Ouajhain, B. (2011). Using Hyspex Swir-320m Hyperspectral Data for the Identification and Mapping of Minerals in Hand Specimens of Carbonate Rocks from the Ankloute Formation (Agadir Basin, Western Morocco). *J. Afr. Earth Sci.* 61, 1–9. doi:10.1016/j.jafrearsci.2011.04.003
3. Batchelor, A. R., Ferrari-John, R. S., Katrib, J., Udoudo, O., Jones, D. A., Dodds, C., et al. (2016). Pilot Scale Microwave Sorting of Porphyry Copper Ores: Part 1 - Laboratory Investigations. *Miner. Eng.* 98, 303–327. doi:10.1016/j.mineng.2016.07.007
4. Breiman, L. (2001). Random Forests. *Mach. Learn.* 45, 5–32. doi:10.1023/a:1010933404324
5. Buckley, S. J., Kurz, T. H., Howell, J. A., and Schneider, D. (2013). Terrestrial Lidar and Hyperspectral Data Fusion Products for Geological Outcrop Analysis. *Comput. Geosciences* 54, 249–258. doi:10.1016/j.cageo.2013.01.018
6. Chen, L., Wang, J.-B., Bagas, L., Wu, X.-B., Zou, H.-Y., Zhang, H.-Q., et al. (2017).

Significance of Adakites in Petrogenesis of Early Silurian Magmatism at the Yudai Copper Deposit in the Kalatag District, NW China. *Ore Geol. Rev.* 91, 780–794. doi:10.1016/j.oregeorev.2017.08.022

7. Clark, R. N., and Roush, T. L. (1984). Reflectance Spectroscopy: Quantitative Analysis Techniques for Remote Sensing Applications. *J. Geophys. Res.* 89, 6329–6340. doi:10.1029/jb089ib07p06329
8. Clark, R. N. (1999). "Spectroscopy of Rocks and Minerals, and Principles of Spectroscopy," in *Remote Sensing for the Earth Sciences: Manual of Remote Sensing* (New York: John Wiley & Sons), 3–58.
9. Clark, R. N., Swayze, G. A., Livo, K. E., Kokaly, R. F., Sutley, S. J., Dalton, J. B., et al. (2003). Imaging Spectroscopy: Earth and Planetary Remote Sensing with the Usgs Tetracorder and Expert Systems. *J.-Geophys.-Res.* 108, 5131. doi:10.1029/2002je001847
10. Cristóbal, J., Graham, P., Prakash, A., Buchhorn, M., Gens, R., Guldager, N., et al. (2021). Airborne Hyperspectral Data Acquisition and Processing in the Arctic: A Pilot Study Using the Hyspex Imaging Spectrometer for Wetland Mapping. *Remote Sens.* 13, 1178.
11. Dalm, M., Buxton, M. W. N., and van Ruitenbeek, F. J. A. (2017). Discriminating Ore and Waste in a Porphyry Copper Deposit Using Short-Wavelength Infrared (SWIR) Hyperspectral Imagery. *Miner. Eng.* 105, 10–18. doi:10.1016/j.mineng.2016.12.013
12. Dalponte, M., Orka, H. O., Gobakken, T., Gianelle, D., and Naesset, E. (2013). Tree Species Classification in Boreal Forests with Hyperspectral Data. *IEEE Trans. Geosci. Remote Sens.* 51, 2632–2645. doi:10.1109/tgrs.2012.2216272
13. Dalton, J. B., Bove, D. J., Mladinich, C. S., and Rockwell, B. W. (2004). Identification of Spectrally Similar Materials Using the USGS Tetracorder Algorithm: the Calcite-Epidote-Chlorite Problem. *Remote Sens. Environ.* 89, 455–466. doi:10.1016/j.rse.2003.11.011
14. Demarchi, L., Canters, F., Cariou, C., Licciardi, G., and Chan, J. C.-W. (2014). Assessing the Performance of Two Unsupervised Dimensionality Reduction Techniques on Hyperspectral APEX Data for High Resolution Urban Land-Cover Mapping. *ISPRS J. Photogrammetry Remote Sens.* 87, 166–179. doi:10.1016/j.isprsjprs.2013.10.012
15. Eckardt, A., Horack, J., Lehmann, F., Krutz, D., Drescher, J., Whorton, M., et al. (2015). "Desis (DLR Earth Sensing Imaging Spectrometer for the ISS-MUSES Platform)," in *2015 IEEE International Geoscience and Remote Sensing Symposium*, 1457–1459. doi:10.1109/igarss.2015.7326053
16. Foody, G. M. (2002). Status of Land Cover Classification Accuracy Assessment. *Remote Sens. Environ.* 80, 185–201. doi:10.1016/s0034-4257(01)00295-4
17. Gomez, C., Adeline, K., Bacha, S., Driessen, B., Gorretta, N., Lagacherie, P., et al. (2018). Sensitivity of Clay Content Prediction to Spectral Configuration of VNIR/SWIR Imaging Data, from Multispectral to Hyperspectral Scenarios. *Remote Sens. Environ.* 204, 18–30. doi:10.1016/j.rse.2017.10.047
18. Grove, C. I., Hook, S. J., and Paylor Ii, E. D. (1992). *Compilation of Laboratory Reflectance Spectra of 160 minerals, 0.4 to 25 Micrometers.* Washington, DC: Jet Propulsion Laboratory, National Astronautics and Space Administration.
19. Guha, A. (2020). "Mineral Exploration Using Hyperspectral Data," in *Hyperspectral Remote Sensing*. Editors P. C. Pandey, P. K. Srivastava, H. Balzter, B. Bhattacharya, and G. P. Petropoulos (Elsevier), 15, 293–318. doi:10.1016/b978-0-08-102894-0.00012-7
20. Guo, N., Thomas, C., Tang, J., Tong, Q., Guo, N., Cudahy, T., et al. (2019). Mapping

White Mica Alteration Associated with the Jiama Porphyry-Skarn Cu Deposit, Central Tibet Using Field Swir Spectrometry. *Ore Geol. Rev.* 108, 147–157. doi:10.1016/j.oregeorev.2017.07.027

21. Hosseinjani Zadeh, M., Tangestani, M. H., Roldan, F. V., and Yusta, I. (2014). Sub-Pixel Mineral Mapping of a Porphyry Copper Belt Using EO-1 Hyperion Data. *Adv. Space Res.* 53, 440–451. doi:10.1016/j.asr.2013.11.029
22. Hunt, G. R., and Ashley, R. P. (1979). Spectra of Altered Rocks in the Visible and Near Infrared. *Econ. Geol.* 74, 1613–1629. doi:10.2113/gsecongeo.74.7.1613
23. Hunt, G. R. (1971). Visible and Near-Infrared Spectra of Minerals and Rocks: II Carbonates. *Mod. Geol.* 2, 23–30.
24. Iyakwari, S., Glass, H. J., Rollinson, G. K., and Kowalczuk, P. B. (2016). Application of Near Infrared Sensors to Preconcentration of Hydrothermally-Formed Copper Ore. *Miner. Eng.* 85, 148–167. doi:10.1016/j.mineng.2015.10.020
25. Jing, C., Bokun, Y., Runsheng, W., Feng, T., Yingjun, Z., Dechang, L., et al. (2014). Regional-scale Mineral Mapping Using ASTER VNIR/SWIR Data and Validation of Reflectance and Mineral Map Products Using Airborne Hyperspectral CASI/SASI Data. *Int. J. Appl. Earth Observation Geoinformation* 33, 127–141. doi:10.1016/j.jag.2014.04.014
26. Kodikara, G. R. L., Woldai, T., van Ruitenbeek, F. J. A., Kuria, Z., van der Meer, F., Shepherd, K. D., et al. (2012). Hyperspectral Remote Sensing of Evaporate Minerals and Associated Sediments in Lake Magadi Area, kenya. *Int. J. Appl. Earth Observation Geoinformation* 14, 22–32. doi:10.1016/j.jag.2011.08.009
27. Kokaly, R. F., Clark, R. N., Swayze, G. A., Livo, K. E., and Hoefen, T. M. (2017). *USGS Spectral Library Version 7*. U.S. Geological Survey, Reston, Virginia
28. Koloniatis, K., Andronis, V., and Karathanassi, V. (2020). "Spectral Smile Correction for Airborne Imaging Spectrometers," in *Hyperspectral Remote Sensing*. Editors P. C. Pandey, P. K. Srivastava, H. Balzter, B. Bhattacharya, and G. P. Petropoulos (Elsevier), 2, 23–44. doi:10.1016/b978-0-08-102894-0.00003-6
29. Kruse, F. A. (2012). Mapping Surface Mineralogy Using Imaging Spectrometry. *Geomorphology* 137, 41–56. doi:10.1016/j.geomorph.2010.09.032
30. Kurz, T. H., Buckle, S. J., Howell, J. A., and Schneider, D. (2009). Close Range Hyperspectral and Lidar Data Integration for Geological Outcrop Analysis. The Workshop on Hyperspectral Image & Signal Processing: Evolution in. *Remote Sens.* 1–4.
31. Kurz, T. H., Buckley, S. J., Howell, J. A., and Schneider, D. (2008). *Geological Outcrop Modelling and Interpretation Using Ground Based Hyperspectral and Laser Scanning Data Fusion*, 37.
32. Magendran, T. (2014). *Statistical and Neural Network Analysis of Hyperspectral Radiometric Data to Characterise Hematite of Singbhum Iron Ore Belt*. (SPIE India).
33. Mathieu, M., Roy, R., Launeau, P., Cathelineau, M., and Quirt, D. (2017). Alteration Mapping on Drill Cores Using a Hyspex Swir-320m Hyperspectral Camera: Application to the Exploration of an Unconformity-Related Uranium Deposit (Saskatchewan, Canada). *J. Geochem. Explor.* 172, 71–88. doi:10.1016/j.gexplo.2016.09.008
34. Metelka, V., Baratoux, L., Jessell, M. W., and Naba, S. (2015). Visible and Infrared Properties of Unaltered to Weathered Rocks from Precambrian Granite-Greenstone Terrains of the West African Craton. *J. Afr. Earth Sci.* 112, 570–585. doi:10.1016/j.jafrearsci.2015.10.003
35. Milton, E. J., Schaepman, M. E., Anderson, K., Kneubühler, M., and Fox, N. (2009). Progress in Field Spectroscopy. *Remote Sens. Environ.* 113, S92–S109. doi:10.1016/j.rse.2007.08.001

36. Murphy, R. J., and Monteiro, S. T. (2013). Mapping the Distribution of Ferric Iron Minerals on a Vertical Mine Face Using Derivative Analysis of Hyperspectral Imagery (430-970nm). *ISPRS J. Photogrammetry Remote Sens.* 75, 29–39. doi:10.1016/j.isprsjprs.2012.09.014

37. Osterloo, M. M., Hamilton, V. E., and Anderson, F. S. (2012). A Laboratory Study of the Effects of Roughness on the Thermal Infrared Spectra of Rock Surfaces. *Icarus* 220, 404–426. doi:10.1016/j.icarus.2012.04.020

38. Paoletti, M. E., Haut, J. M., Plaza, J., and Plaza, A. (2019). Deep Learning Classifiers for Hyperspectral Imaging: A Review. *ISPRS J. Photogrammetry Remote Sens.* 158, 279–317. doi:10.1016/j.isprsjprs.2019.09.006

39. Qiu, J.-T., Zhang, C., Yu, Z.-F., Xu, Q.-J., Wu, D., Li, W.-W., et al. (2017). Subsetting Hyperspectral Core Imaging Data Using a Graphic-Identification-Based IDL Program. *Comput. Geosciences* 106, 68–76. doi:10.1016/j.cageo.2017.05.009

40. Ramanaidou, E., Wells, M. A., Belton, D., Verral, D., and Ryan, C. (2007). Mineralogical and Microchemical Methods for the Characterization of High-Grade BIF Derived Iron Ore. *Econ. Geol.* 15, 129–156.

41. Reath, K. A., and Ramsey, M. S. (2013). Exploration of Geothermal Systems Using Hyperspectral Thermal Infrared Remote Sensing. *J. Volcanol. Geotherm. Res.* 265, 27–38. doi:10.1016/j.jvolgeores.2013.08.007

42. Riaza, A., Strobl, P., Beisl, U., Hausold, A., and Müller, A. (2001). Spectral Mapping of Rock Weathering Degrees on Granite Using Hyperspectral DAIS 7915 Spectrometer Data. *Int. J. Appl. Earth Observation Geoinformation* 3, 345–354. doi:10.1016/s0303-2434(01)85042-x

43. Rickard, L. J., Basedow, R. W., Zalewski, E. F., Silverglate, P. R., and Lander, L. M. (1993). *HYDICE: An Airborne System for Hyperspectral Imaging*. International Society for Optical Engineering, Imaging Spectrometry of the Terrestrial Environment, Orlando, FL, United States.

44. Rogge, D., Rivard, B., Segl, K., Grant, B., and Feng, J. (2014). Mapping of NiCu-PGE Ore Hosting Ultramafic Rocks Using Airborne and Simulated EnMAP Hyperspectral Imagery, Nunavik, Canada. *Remote Sens. Environ.* 152, 302–317. doi:10.1016/j.rse.2014.06.024

45. Rowan, L. C., Crowley, J. K., Schmidt, R. G., Ager, C. M., and Mars, J. C. (2000). Mapping Hydrothermally Altered Rocks by Analyzing Hyperspectral Image (AVIRIS) Data of Forested Areas in the Southeastern United States. *J. Geochem. Explor.* 68, 145–166. doi:10.1016/s0375-6742(99)00081-3

46. Rowan, L. C., Simpson, C. J., and Mars, J. C. (2004). Hyperspectral Analysis of the Ultramafic Complex and Adjacent Lithologies at Mordor, NT, Australia. *Remote Sens. Environ.* 91, 419–431. doi:10.1016/j.rse.2004.04.007

47. Sabins, F. F. (1999). Remote Sensing for Mineral Exploration. *Ore Geol. Rev.* 14, 157–183. doi:10.1016/s0169-1368(99)00007-4

48. Salimi, A., Ziaii, M., Amiri, A., Hosseinjani Zadeh, M., Karimpouli, S., and Moradkhani, M. (2018). Using a Feature Subset Selection Method and Support Vector Machine to Address Curse of Dimensionality and Redundancy in Hyperion Hyperspectral Data Classification. *Egypt. J. Remote Sens. Space Sci.* 21, 27–36. doi:10.1016/j.ejrs.2017.02.003

49. Salles, R. d. R., de Souza Filho, C. R., Cudahy, T., Vicente, L. E., and Monteiro, L. V. S. (2017). Hyperspectral Remote Sensing Applied to Uranium Exploration: A Case Study at the Mary Kathleen Metamorphic-Hydrothermal U-REE Deposit, NW, Queensland, Australia. *J. Geochem. Explor.* 179, 36–50. doi:10.1016/j.gexplo.2016.07.002

50. Singh, M., Rajesh, V. J., Sajinkumar, K. S., Sajeev, K., and Kumar, S. N. (2016). Spectral and Chemical Characterization of Jarosite in a Palaeolacustrine Depositional Environment in Warkalli Formation in Kerala, South India and its Implications. *Spectrochimica Acta Part A Mol. Biomol. Spectrosc.* 168, 86–97. doi:10.1016/j.saa.2016.05.035

51. Steffens, M., and Buddenbaum, H. (2013). Laboratory Imaging Spectroscopy of a Stagnic Luvisol Profile - High Resolution Soil Characterisation, Classification and Mapping of Elemental Concentrations. *Geoderma* 195-196, 122–132. doi:10.1016/j.geoderma.2012.11.011

52. Sun, Y., Wang, J., Li, Y., Wang, Y., Yu, M., Long, L., et al. (2017). Recognition of Late Ordovician Yudai Porphyry Cu (Au, Mo) Mineralization in the Kalatag District, Eastern Tianshan Terrane, NW China: Constraints from Geology, Geochronology, and Petrology. *Ore Geol. Rev.* 100, 220–236. doi:10.1016/j.oregeorev.2017.07.011

53. Sun, Z., Chang, N.-B., Opp, C., and Hennig, T. (2011). Evaluation of Ecological Restoration through Vegetation Patterns in the Lower Tarim River, china with MODIS NDVI Data. *Ecol. Inf.* 6, 156–163. doi:10.1016/j.ecoinf.2010.10.002

54. Tan, K., Ma, W., Chen, L., Wang, H., Du, Q., Du, P., et al. (2021). Estimating the Distribution Trend of Soil Heavy Metals in Mining Area from HyMap Airborne Hyperspectral Imagery Based on Ensemble Learning. *J. Hazard. Mater.* 401, 123288. doi:10.1016/j.jhazmat.2020.123288

55. Tripathi, M. K., Govil, H., and Chattoraj, S. L. (2020). Identification of Hydrothermal Altered/weathered and Clay Minerals through Airborne AVIRIS-NG Hyperspectral Data in Jahajpur, India. *Heliyon* 6, e03487. doi:10.1016/j.heliyon.2020.e03487

56. van der Meer, F. D., van der Werff, H. M. A., van Ruitenbeek, F. J. A., Hecker, C. A., Bakker, W. H., Noomen, M. F., et al. (2012). Multi- and Hyperspectral Geologic Remote Sensing: A Review. *Int. J. Appl. Earth Observation Geoinformation* 14, 112–128. doi:10.1016/j.jag.2011.08.002

57. Vane, G., and Goetz, A. F. H. (1988). Terrestrial Imaging Spectroscopy. *Remote Sens. Environ.* 24, 1–29. doi:10.1016/0034-4257(88)90003-x

58. Vangi, E., D'Amico, G., Francini, S., Giannetti, F., Lasserre, B., Marchetti, M., et al. (2021). The New Hyperspectral Satellite PRISMA: Imagery for Forest Types Discrimination. *Sensors* 21, 1182. doi:10.3390/s21041182

59. Velasco, F., Alvaro, A., Suarez, S., Herrero, J.-M., and Yusta, I. (2005). Mapping Fe-Bearing Hydrated Sulphate Minerals with Short Wave Infrared (SWIR) Spectral Analysis at San Miguel Mine Environment, Iberian Pyrite Belt (SW Spain). *J. Geochem. Explor.* 87, 45–72. doi:10.1016/j.gexplo.2005.07.002

60. Waske, B., van der Linden, S., Oldenburg, C., Jakimow, B., Rabe, A., and Hostert, P. (2012). imageRF - A User-Oriented Implementation for Remote Sensing Image Analysis with Random Forests. *Environ. Model. Softw.* 35, 192–193. doi:10.1016/j.envsoft.2012.01.014

61. Xie, B. S., Zhou, S. Y., Wu, L. X., Mao, W. F., and Wang, W. (2022b). RockSL: An Integrated Rock Spectral Library for Better Global Shared Services. *Big Earth Data,* 1–21. doi:10.1080/20964471.2021.2017111

62. Xie, B., Wu, L., Mao, W., Zhou, S., and Liu, S. (2022a). An Open Integrated Rock Spectral Library (RockSL) for a Global Sharing and Matching Service. *Minerals* 12, 118. doi:10.3390/min12020118

4

INSIGHTS INTO THE CRUSTAL EVOLUTION AND TUNGSTEN MINERALIZATION OF THE WEST CATHAYSIA BLOCK: CONSTRAINTS FROM THE INHERITED ZIRCONS FROM THE MESOZOIC DENGFUXIAN AND PALEOZOIC TANGHU PLUTONS, SOUTH CHINA

Jingya Cao[1,2,3], Youyue Lu[1], Lei Liu[4], Jianming Fu[1], Guofeng Xu[4], Qianhong Wu[4], Shengxiong Yang[2], Xiaofei Qiu[1] and Zunzun Zhang[1]

[1]Research Center for Petrogenesis and Mineralization of Granitoid Rocks, China Geological Survey, Wuhan 430205, China
[2]Southern Marine Science and Engineering Guangdong Laboratory (Guangzhou), Guangzhou 511458, China
[3]Department of Ocean Science, The Hong Kong University of Science & Technology, Hong Kong, China
[4]Key Laboratory of Metallogenic Prediction of Nonferrous Metals and Geological Environment Monitoring, Ministry of Education, Central South University, Changsha 410083, China

ABSTRACT

The formation and evolution of the ancient continental crust are crucial issues in solid-earth geology which are commonly associated with global tectonic events and the formation of economically valuable magmatic-hydrothermal ore deposits. The Cathaysia Block, one of the ancient continents in Southeast Asia, can be subdivided into two parts: the West Cathaysia Block and the East Cathaysia Block. Unlike the East Cathaysia Block, no Precambrian rocks are exposed in the West Cathaysia Block, constraining further understanding of the formation and evolution of this block. In this study, a total of four hundred and thirty-three zircon U-Pb dating analyses and two hundred and eighteen Lu-Hf isotopic analyses on zircon grains from the Jurassic Dengfuxian granites and Ordovician Tanghu granites, Nanling Range, were carried out. LA-ICP-MS zircon U-Pb dating yields mean average $^{206}Pb/^{238}U$ *ages of 152.6 ± 2.2 Ma (MSWD = 1.6) and 442.4 ± 1.7 Ma (MSWD = 3.8), which are regarded as the rock-forming age for the Jurassic Dengfuxian granites and Ordovician Tanghu granites, respectively. The* $^{207}Pb/^{206}Pb$ *ages of the inherited zircons from the Jurassic Dengfuxian granites and Ordovician Tanghu granites range from 522 Ma to 3395 Ma, hosting two major peaks at the 0.9–1.0 Ga and 2.4–2.5 Ga. In contrast to the East Cathaysia Block, the West Cathaysia Block lacks the age peak of 1.8–1.9 Ga, indicating that the West Cathaysia Block was not influenced by the assembly of the Columbia supercontinent in the Paleo-Proterozoic. In combination with the Lu-Hf isotopes, we proposed that the crust evolution of the West Cathaysia Block in Archean is dominated by juvenile crustal growth events, and dominated by the crustal reworking since the Proterozoic. The long duration of crustal reworking in the West Cathaysia Block resulted in the enrichment of lithophile elements (e.g., W, Sn, Nb, and Ta) in the crust of that region. Therefore, the Jurassic granites in the Nanling Range, which are mainly derived from the partial melting of Proterozoic basement rocks, became associated with large-scale tungsten polymetallic mineralization.*

Keywords: *U-Pb ages, Lu-Hf isotopes, Inherited zircons, Crustal evolution, South China*

INTRODUCTION

Understanding crustal generation and reworking, essential for the formation and evolution of the global continental crust, commonly depends on the basement rocks of each continent. The nature of these rocks can provide key information, not only on the formation and evolution of the crust but also on the origin of polygenetic granites and related mineral deposits [1,2,3,4,5,6,7,8]. However, these rocks in some early Precambrian cratons might be overprinted and/or obscured, caused by the intensive thermo-tectonic events [9,10]. The Cathaysia Block, one of the important parts of the South China Block, is well-known for its intensive polygenetic tectonic-magmatism and related rare metal mineralization [11,12,13,14,15,16,17,18,19,20]. To date, the Paleoproterozoic Badu Group exposed in the south-western Zhejiang province and north-western Fujian province are well accepted as the oldest rock unit in the Cathaysia Block [21,22,23,24,25,26]. Hence, the Precambrian crust evolution of the Cathaysia Block is poorly understood due to the absence of Archean rocks. Although no Archean rocks were found in the Cathaysia Block, abundant Archean and even Hadean ages of the inherited zircon grains in igneous and detrital zircon grains in sedimentary rocks and/or river sediments were reported [4,22,23,26,27,28,29]. Therefore, several geologists argued that the Archean basement rocks of "Cathaysia old continent" might exist in the Cathaysia Block, however, they might be broken up into several microplates and/or reworked by the subsequent tectonic-magmatic events [9,30]. Further studies indicated that the Cathaysia Block can be subdivided into the West Cathaysia Block and East Cathaysia Block, which consist of the Yunkai-Nanling and Wuyi terranes, respectively [4,11]. The Precambrian rocks are well exposed in the East Cathaysia Block, including the Paleo-Proterozoic metasedimentary rocks and granites [21,22,23,31]. Conversely, rare Precambrian rocks are identified in the West Cathaysia Block, constraining the understanding of the formation and evolution of the Precambrian crust of the West Cathaysia Block.

It is currently known that granitic rocks not only play an extremely important role in the evolution of the Earth's crust [32,33,34], they also are commonly accompanied by migration and/or enrichment of elements and the formation of intense tungsten polymetallic mineralization [34,35]. The inherited zircons from the granites, which are commonly derived by partial melting of the deep crustal rocks, will elucidate the ages and source of these deep crust basement rocks and might provide a new perspective for geologists to unravel the missing geological records of the formation and evolution of the continental crust [36,37,38,39]. Nanling range is one of the important tungsten–tin polymetallic ore belts in South China and a world-famous tungsten mineralization province, hosting a series of large to super large tungsten deposits, e.g., Shizhuyuan, Xihuashan, Taoxikeng, Xiangdong, and Yaogangxian tungsten–tin deposits. In addition, these deposits are genetically linked to the granitic rocks both in time and space. Therefore, LA-ICP-MS U-Pb dating and LA-MC-ICPMS Lu-Hf isotopic analyses were carried out on these inherited zircons from the Mesozoic Dengfuxian and Paleozoic Tanghu plutons, Nanling Range, aiming to provide some insights into the generation and reworking history, and the relationship between the crust evolution and tungsten polymetallic mineralization of the West Cathaysia Block.

REGIONAL GEOLOGY

The South China Block (SCB), composed of the Yangtze Block and the Cathaysia Block, is one of the largest Precambrian blocks in East Asia and might be formed by the amalgamation of these two blocks at ca. 820 Ma [12]. It was surrounded by the Qinling-Dabei orogenic belt to the north, the Songpan-Garze orogenic to the northwest, and Sibumasu-Indochina orogenic belt to the southwest (Figure 1a). Two blocks were proposed to be bounded by the Qinzhou-Hangzhou Belt, which starts from Hangzhou Bay to the northeast and Qinzhou Bay to the southwest (Figure 1b). The Yangtze Block, located at the northwest SCB, hosts a series of Mesoproterozoic-Archean igneous rocks, e.g., Kongling Group, Dongling Group, Douling Group, and Cuoke Group [40,41,42,43,44,45]. The Kongling Group, commonly regarded as the oldest exposed rock unit in the Yangtze Block, contains the Archean tonalite-trondhjemite-granodiorite (TTG) gneiss with ages of 3.2–3.3 Ga [41,42,43]. In addition, two-stage depleted mantle Hf model ages of zircon grains in the TTG gneiss show a significant peak at 3.3–3.5 Ga. Therefore, it is well accepted that an Archean basement might exist in the Yangtze Block [41,42,43]. Meanwhile, the Neoproterozoic igneous rocks are dominated by the arc-related basic and granitic rocks (820–850 Ma) which are mainly distributed in the margin area of the Yangtze Block, especially in the Jiangnan Orogen [46,47,48,49]. It is proposed that the Neoproterozoic oceanic subduction and/or arc-continent collision event might occur around the margin area of the Yangtze Block, which might cause the collision between the Yangtze and Cathysia blocks [12,24,50,51]. The Mesozoic granodiorite and granites in the Yangtze Block, mainly emplaced in Late-Jurassic and Early-Cretaceous and related to the regional Cu-Au-W polymetallic mineralization, are involved in the Pacific subduction event toward the South China Block, for example, the Cu-Au deposits of the Tongling ore cluster and the Chating ore cluster [52,53,54,55,56,57].

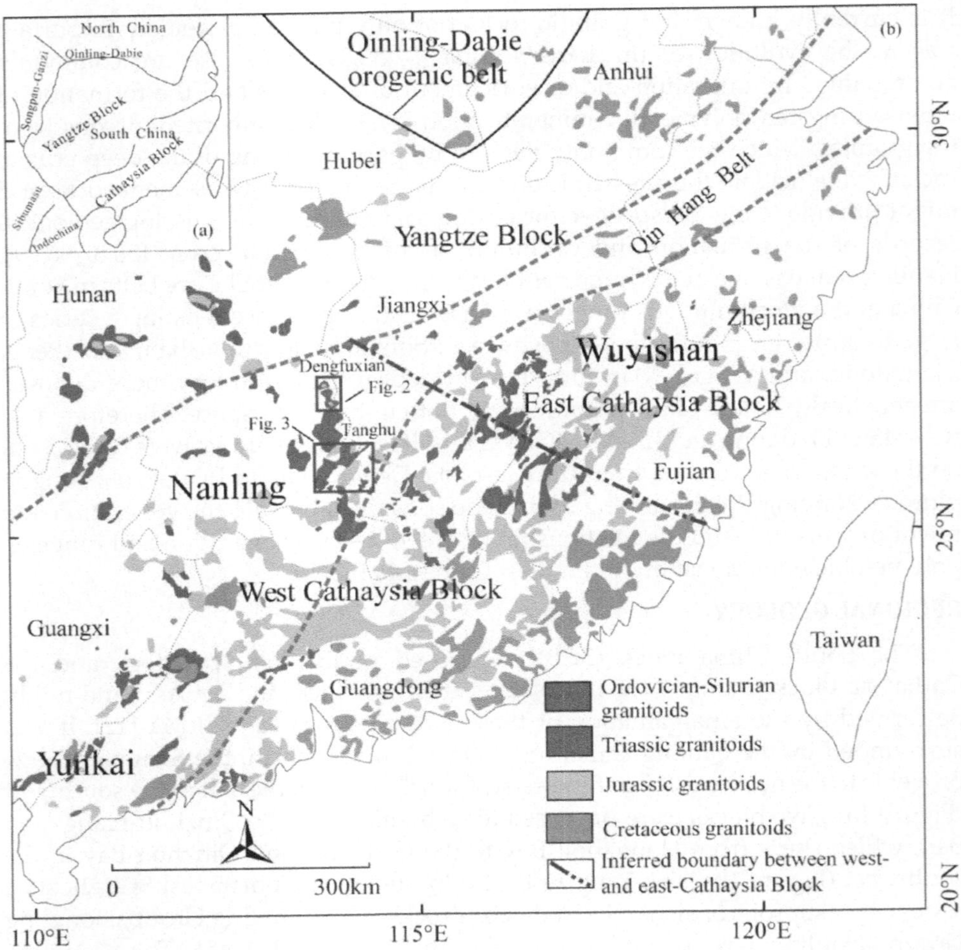

Figure 1. The sketch tectonic location of the South China Block (a) and the Geological sketch map of the South China Block (b), modified from [58]. The inferred boundary between the West and East Cathaysia Block is modified from [4]. The red box refers to the study regions, whereas the area between the blue dashed lines refers to the Qian-Hang Belt of SE China. The grey dushed lines refer to the provincial boundaries.

Unlike the Yangtze Block, the Archean basement rocks are absent in the Cathysia Block, however, a series of Archean-Hadean detrital zircon ages were reported, indicating that the Archean basement and/or Hadean fragments might exist in the Cathysia Block [4,21,26,27,28,29]. The Precambrian rocks exposed in this block are mostly distributed between the NE-trending Qin-Hang Belt and Zhenghe-Dapu fault (Figure 1b), e.g., the Paleoproterozoic Badu Group and Mesoproterozoic Longquan Group in the Wuyishan area, the East Cathysia Block, and Neoproterozoic Sibao Group in the Nanling Range, and the West Cathysia Block. The Badu Group mainly comprises the mica schists, gneisses, and migmatites, which were mainly formed at ca. 2.5 Ga and subsequently reworked in ca. 1.8 Ga and 250–230 Ma, respectively [21,59]. The Longquan Group is mainly composed of mica-quartz schist, epidote amphibolite, actinolite schist, and fine-grained biotite gneiss, showing that the protoliths of mafic volcanic rocks and pelitic–arenaceous–calcareous sedimentary rocks experienced greenschist to low amphibolite facies metamorphism [21]. Sm-Nd isotopic dating suggests that rocks of the Longquan Group might be formed at 1.4–1.0 Ga, however, detailed age data are lacking [59,60]. The Sibao Group, mainly composed of epimetamorphic shale and volcaniclastic rocks, metamorphosed basalts, spilite, and keratophyre, might have been formed at ca. 830 Ma [61]. The Paleozoic strata are widespread in the Cathysia Block, including a series of marine to continental sedimentary rocks. However, they have experienced various degrees of deformation due to the following tectonic events.

The oldest igneous rocks, mainly S- and/or A-type granites, were distributed in the Wuyishan area, Eastern Cathaysia Block, with ages of 1.8–1.9 Ga [22,23]. However, rare Paleoproterozoic igneous rocks were identified in the Nanling-Yunkai region. The igneous rocks in the Cathaysia Block are dominated by the Mesozoic granitoid with few Middle Paleozoic granitoids (Figure 1b). The Middle Paleozoic granitoids, related to uranium mineralization, are mainly located in the Nanling Range with peak ages of 444–419 Ma [62]. The Mesozoic granitoids, especially Jurassic granitoids, are widespread in the Cathaysia Block and are associated with the regional W-Sn polymetallic mineralization. These Jurassic granites, mainly emplaced at 165–150 Ma, are related to the intense magmatic magmatism which was derived by the partial melting of the Proterozoic basement rocks of the Cathaysia Block and triggered by the Pacific subduction event [17].

SAMPLING AND ANALYTICAL METHODS

Sampling and Description

Two fresh samples were collected from the Nanling Range, West Cathaysia Block, including a two-mica monzonitic granite from the Dengfuxian composite pluton (Figure 2a) and a biotite monzonitic granite from the Tanghu pluton (Figure 3a). The Dengfuxian composite pluton, located in Eastern Hunan, is composed of the Triassic biotite monzonitic granite and Jurassic two-mica monzonitic granite, with ages of 225.7 ± 1.6 Ma and 151.1 ± 1.5 Ma, respectively [63,64]. A two-mica monzonitic granite sample was collected from the Dengfuxian composite pluton, and the sampling location was represented in Figure 2a. In addition, the Jurassic granites, genetically related to the large-scale tungsten mineralization, belong to the S-type granite and originated from the partial melting of the Precambrian basement rocks of the Cathaysia Block [65,66]. The Jurassic two-mica monzonitic granites are mostly grey with a fine-grained texture, consisting of K-feldspar (~25%), plagioclase (~20%), quartz (~35%), muscovite (~12%), and biotite (~8%) (Figure 2b,c). These granites also contain accessory minerals of apatite, zircon, and titanite.

The Tanghu pluton, part of the Zhuguangshan composite pluton which is composed of Ordovician, Triassic, and Jurassic granites, is located at the junction region between Jiangxi and Guangdong [68]. The Tanghu pluton is composed of biotite monzonitic granite and granodiorite with a precise U-Pb age of ca. 440 Ma, indicating that the formation of these Ordovician granites has a close link with the Caledonian orogeny in South China [68]. A biotite monzonitic granite sample was collected from the Tanghu pluton, and the sampling location was shown in Figure 3a. The Tanghu biotite monzonitic granite is mostly grey with a porphyritic-like texture (Figure 3b). Phenocrysts are mainly composed of subhedral tabular plagioclase, which commonly hosts a zonary structure (Figure 3c). Mineral assemblages are composed of K-feldspar (15%~25%), plagioclase (35%~45%), quartz (20%~25%), and biotite (5%), with accessory minerals of magnetite, apatite, zircon, and titanite.

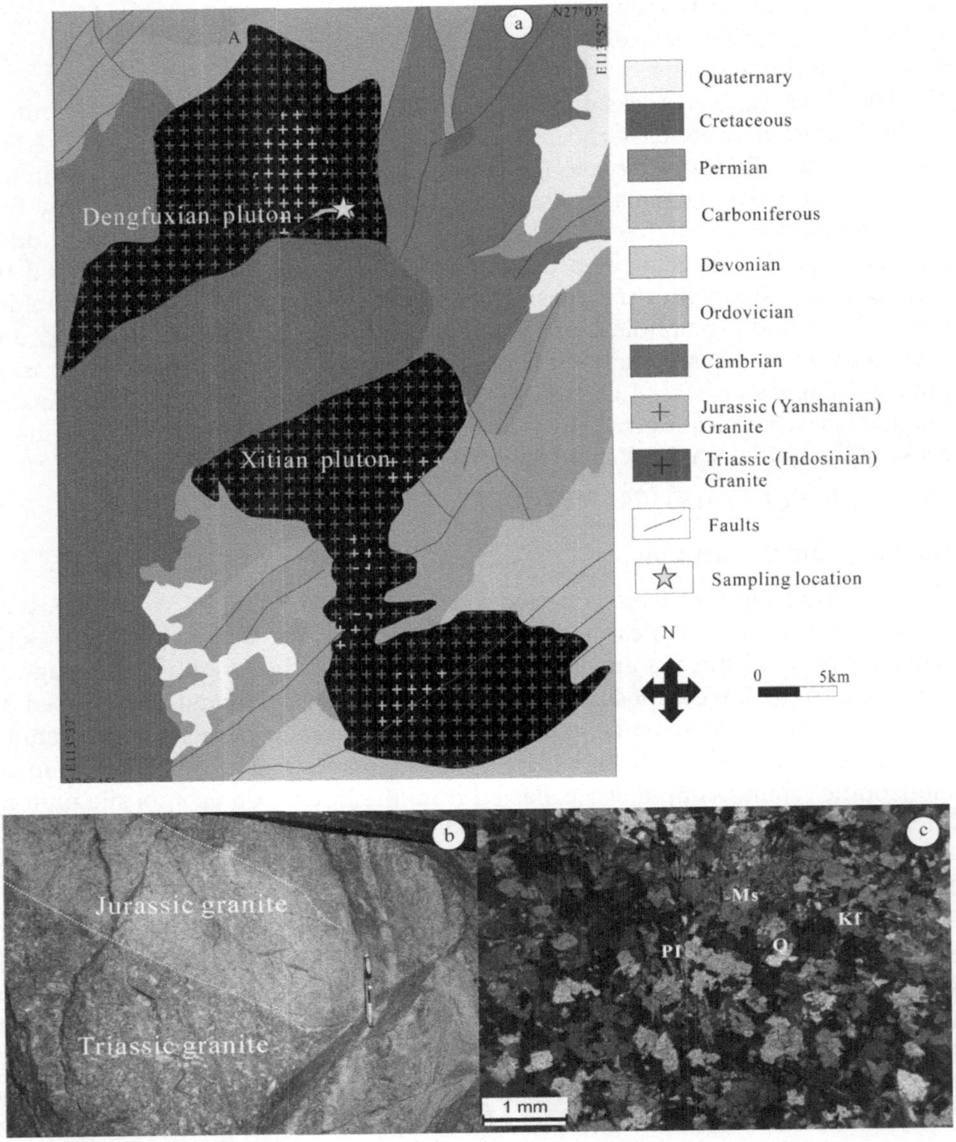

Figure 2. Geological sketch map of the Xitian and Dengfuxian plutons (a), modified from [67]; Outcrop (b) and orthogonal polarizing micrograph (c) of the Dengfuxian Jurassic granite. The location of Figure 2a has been shown in Figure 1b.

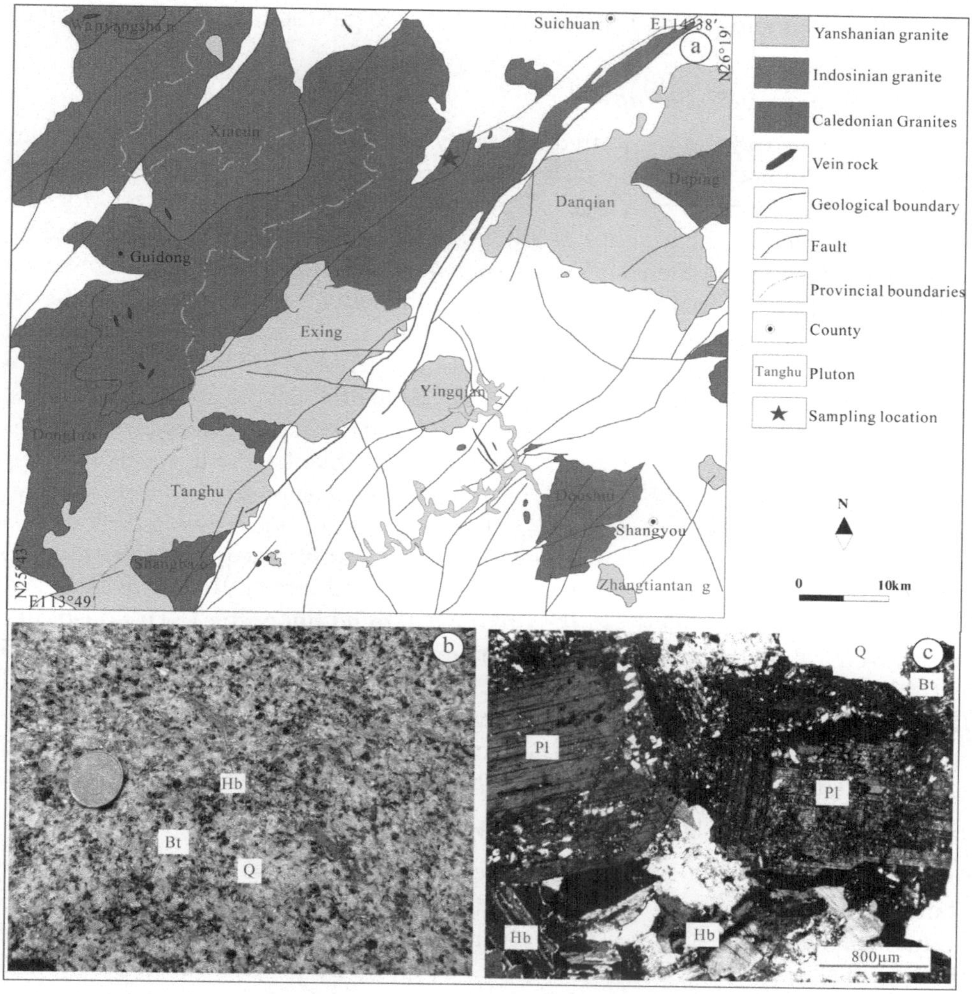

Figure 3. Geological sketch map of the north Zhuguangshan area (a); Outcrop (b) and orthogonal polarizing micrograph (c) of the Tanghu granite. The location of Figure 3a has been shown in Figure 1b.

Analytical Methods

Selection and CL Imaging of Zircon

Based on the conventional magnetic and heavy liquid techniques, zircon grains were first separated from the Dengfuxian Jurassic granites (sample no. 0326–2s2) and Tanghu Ordovician granites (sample no. 13D40). Second, they were handpicked under a binocular microscope. Third, they were mounted into epoxy resin blocks and polished to obtain flat surfaces. Before the LA-ICP-MS U-Pb dating analyses, these zircon grains were photographed and observed under an electron microscope to avoid inclusions and fractures in these grains. Finally, these zircon grains were photographed through a scanning electron microscope (TESCAN MIRA 3 LMH FE-SEM, TESCAN, Brno, Czech Republic) at the Sample Solution

Analytical Technology Co., Ltd, Wuhan, China, to get the Cathodoluminescence (CL) images.

LA-ICP-MS Zircon U-Pb Dating

Ninety-two and three hundred and forty-two U-Pb dating and trace element analyses of zircons were collectively conducted by LA-ICP-MS at the Laboratory of Isotope Geochemistry, Wuhan Center of China Geological Survey for the Dengfuxian and Tanghu granites, respectively. The LA-ICP-MS were assembled by an icapQ ICP-MS instrument and a resolution laser ablation system. The former was used to obtain ion-signal intensities, whereas the latter was used to denudate the samples. The helium and argon were used as the carrier gas and the make-up gas, respectively., which were optimized by ablating NIST SRM 610 to obtain maximum signal intensity for ^{238}Uand to keep low ThO/Th (0.01%–0.03%) and Ba^{2+}/Ba^{+} ratios (0.02%–0.03%) to reduce the matrix-induced interferences. Average intensities of ^{202}Hg, ^{204}(Pb + Hg), ^{206}Pb, ^{207}Pb, and ^{208}Pb of the gas blank were lower than 50 cps at a sensitivity of 0.4 × 106 cps ^{238}U. In this analysis, a spot size of 29 μm and an energy density of 4 J/cm^2 were applied. The detailed procedure was as follows: (1) Each analysis contained approximately 15 s of background acquisition (gas blank) followed by 45 s of data acquisition; (2) The standard zircon 91500 was analyzed twice in every eight analyses of the unknowns. In addition, the U-Th-Pb isotopic ratios were calibrated relative to this standard zircon 91500 whose age is ca. 1064.5 ± 0.6 Ma [69]. Common Pb correcting was followed by using the model proposed by [70]. Further data processing for zircon U-Pb dating and trace elements were performed by the software ICPMSDataCal 10.7 [71], while concordia diagrams and age calculations were obtained by using the software Isoplot v.3.0 [72].

LA-MC-ICP-MS Zircon Lu-Hf Isotope Analysis

Ninety-two and one hundred and twenty-seven in situ Lu-Hf isotope ratio analyses of zircons were conducted using a Neptune Plus MC-ICP-MS combined with a RESOlution laser ablation system at the Laboratory of Isotope Geochemistry, Wuhan Center of China Geological Survey for the Dengfuxian and Tanghu granites, respectively. Helium and argon were used as the carrier gases during the process. Masses of ^{171}Yb, ^{173}Yb, ^{175}Lu, ^{176}(Hf + Yb + Lu), ^{177}Hf, ^{178}Hf, ^{179}Hf, and ^{180}Hf were measured by a 60s of ablation signal acquisition and followed by 30s of wash time. The size of the laser spot diameter and the energy density were set by 43 μm and 6 J/cm^2, respectively. The detailed correcting process of the interference of ^{176}Yb/^{176}Hf and ^{176}Yb/^{177}Hf were following the methods proposed by Thirlwall et al. (2004) [73]. The ^{176}Lu/^{177}Hf ratio was calculated by using the recommended ^{176}Lu/^{175}Lu = 0.02656 [74]. Offline data processing was performed using the software ICPMSDataCal 10.7 [71]. Reference standard zircons 91500 and Plĕsovice were analyzed as "unknowns" to control the reproducibility and instrument stability. In this analysis, the ^{176}Hf/^{177}Hf of standard zircons 91500 and Plĕsovice were 0.282297 ± 0.000002 (MSWD = 1.5, n = 126) and 0.282479 ± 0.000001 (MSWD = 1.4, n = 134), respectively, and both agree with the reference values of 0.282307 ± 0.000031 [75] and 0.282482 ± 0.000013 [76], respectively. The decay constant value ^{176}Lu was 1.867 × $10^{-11\ a-1}$ for [77], while the $\varepsilon_{Hf}(t)$ values were calculated by using the present-day chondritic ratios of ^{176}Hf/^{177}Hf = 0.282785 and $^{176}Lu/^{177}Hf$ = 0.0336 [78]. The crustal model age (T_{DMC}) calculation was assuming that the parental magma was produced from the average continental crust ($^{176}Lu/^{177}Hf$ = 0.015) that was primarily derived from the depleted mantle [79]. A detailed analytical procedure was described by [75].

RESULTS

Zircon Characteristics

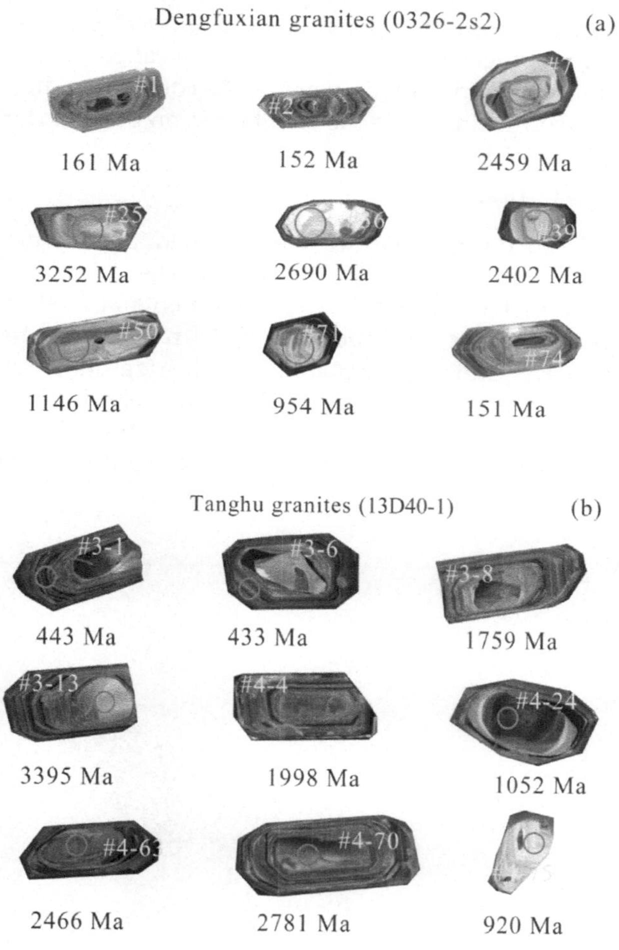

Figure 4. Cathodoluminescence (CL) images of zircons from the Dengfuxian (a) and Tanghu granites (b). The red circle is the location for the LA-ICP-MS dating. The red circle diameter, marked on zircons, is ca. 29 μm. The numbers under the zircon grains represent the U-Pb ages of the corresponding zircon. The selected ages in this study followed the procedure that the young ages (lower than 1000 Ma) and old ages (higher than 1000 Ma) used $^{206}Pb/^{238}U$ ages and $^{207}Pb/^{206}Pb$ ages, respectively.

Zircon grains from the Dengfuxian granites (0326–2s2) are mostly characterized by regular to irregular prisms with light yellow to colorless, a length of 100–250 μm, a width of 80–120 μm and aspect ratios of 3:1 to 1:1. Most of these zircon grains exhibit a clear zoning structure with no cores, however, some zircon grains show a rim-core structure (Figure 4a). These zircon grains with a rim-core structure contain a clear boundary between the rim and core with a zoning structure for the rims and a weak zonation for the cores, respectively. In addition, most of the cores are characterized by bright fluorescence under the CL images.

Zircon grains from the Tanghu granites (13D40) are mostly stubby prismatic with light yellow to colorless, a length of 100–400 μm, a width of 100–200 μm, and aspect ratios of 4:1 to 1:1. Most of these zircon grains exhibit a rim-core structure with a clear zoning structure for the rims and a weak zoning structure for the cores in the CL images (Figure 4b). In general, the cores show different features to the Dengfuxian zircon grains, since the Tanghu zircon grains mainly show a dark fluorescence under the CL images.

Zircon U-Pb Isotopic Ages

Ninety-two U-Pb dating analyses were carried out on ninety-two zircon grains from the Dengfuxian granites (0326–2s2). These zircon grains from the Dengfuxian granites have variable U and Th contents of 18–1294 ppm and 27–3320 ppm, respectively. The corresponding Th/U ratios range from 0.03 to 1.60 (mean = 0.59), however, most of them are over 0.1, suggesting a magmatic origin for most of these zircon grains (Figure 5) [80]. After rejecting a small number of analyses (concordance < 90%), the isotopic ages vary from 137 to 3252 Ma, hosting five major peaks at 150–160 Ma, 900–1000 Ma, 1100–1200 Ma, 1500–1600 Ma and 2300–2500 Ma.

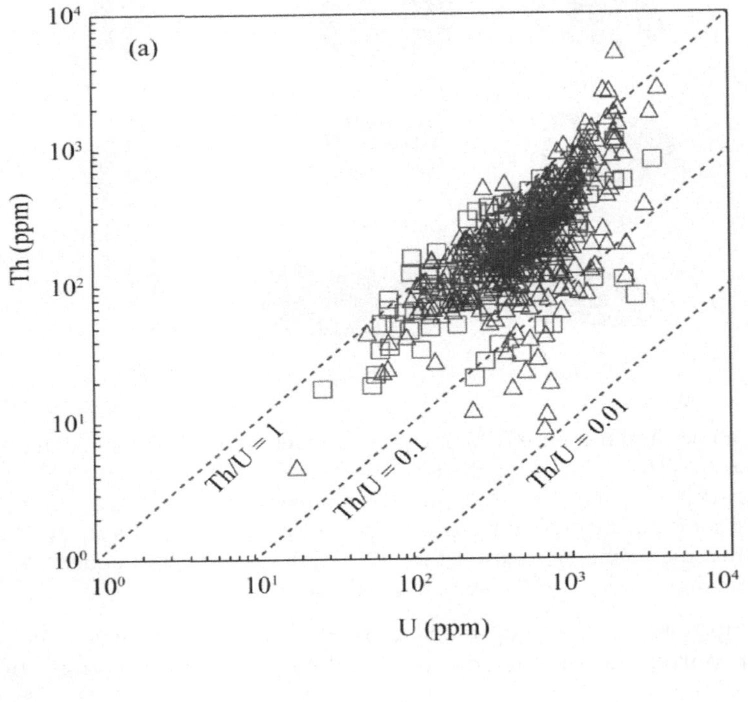

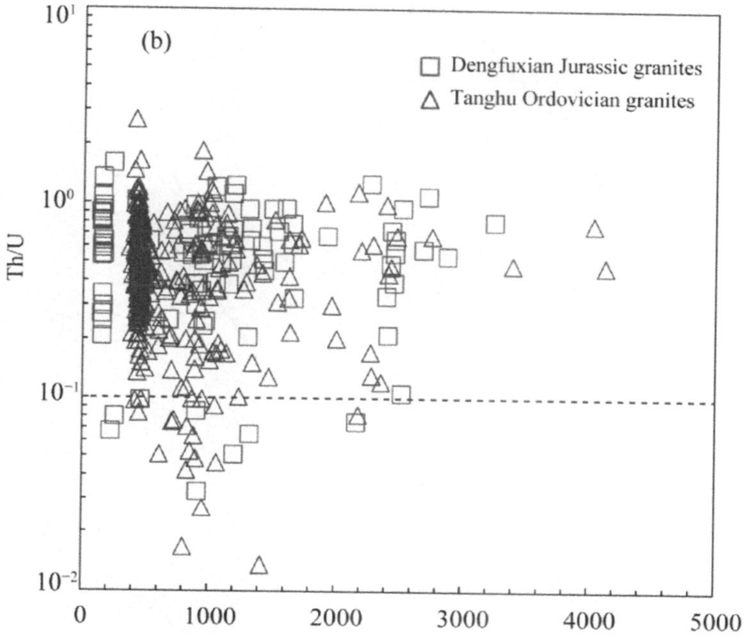

Figure 5. U versus Th (a) and age versus Th/U diagrams (b) of zircons from the Dengfuxian and Tanghu granites.

In addition, these zircon grains are mostly plotted on or near the concordant curve and can be subdivided into five populations which have ages of 152.6 ± 2.2 Ma (MSWD = 1.6), 966 ± 12 Ma (MSWD = 1.2), 1115 ± 23 Ma (MSWD = 1.7), 1558 ± 76 Ma (MSWD = 3.2) and 2382 ± 72 Ma (MSWD = 1.4), respectively (Figure 6). These prismatic zircon grains with an age of 152.6 ± 2.2 Ma, which is the same as the previously reported ages, are regarded to be formed during the emplacement of the Jurassic Dengfuxian granites.

A total of three hundred and thirty-two zircon U-Pb dating analyses were carried out on three hundred and thirty-two zircon grains from the Tanghu granites (13D40-1).

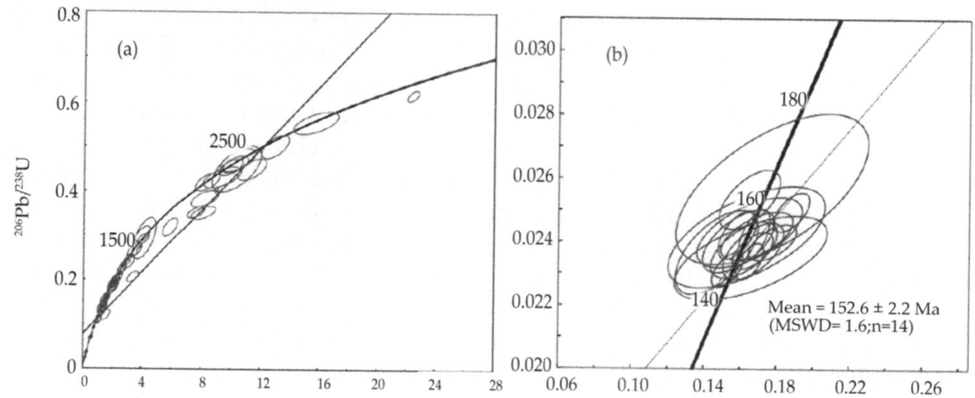

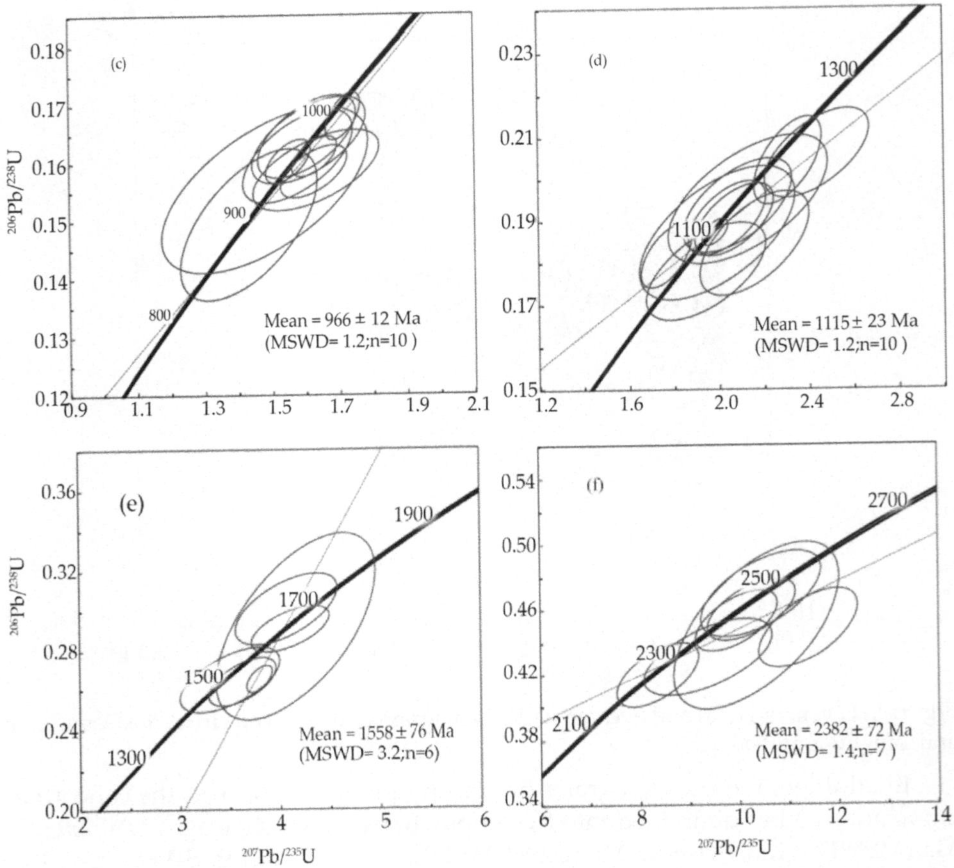

Figure 6. Concordia plots of zircon dating results for the Dengfuxian granites. (a) shows concordia plot of the whole data; (b,c) show concordia plot and mean average ages of Jurassic and Neoproterozoic, respectively. (d,e) show concordia plot and mean average ages of Mesoproterozoic zircons of ca. 1115 Ma and 1558 Ma, respectively; (f) show concordia plot and mean average age of Paleoproterozoic zircons. The red circle, bule solid lines and bule dashed lines represent the radius error for a single spot, concordant curve and unconcordant curve, respectively.

The U and Th contents for the zircon grains from the Tanghu granites were 19–3600 ppm and 5–5214 ppm, respectively, with corresponding Th/U ratios of 0.01–2.67 (mean = 0.48), indicating a magmatic origin for most of these zircon grains (Figure 5) [80]. After rejecting a small number of analyses (concordance < 90%), the isotopic ages vary from 367 Ma to 3395 Ma. Three major peaks at 400–500 Ma, 800–900 Ma, and 1000–1100 Ma were identified. In addition, these zircon grains are mostly plotted on or near the concordant curve (Figure 7a), yielding weighted mean ages of 442.4 ± 1.7 Ma (MSWD = 3.8), 896.2 ± 8.8 Ma (MSWD = 0.31) and 1043 ± 18 Ma (MSWD = 2.5), respectively (Figure 7b,d). These prismatic zircon grains with an age of 442.4 ± 1.7 Ma, which is the same as the previously reported ages, are regarded to be formed during the emplacement of the Tanghu Ordovician granites.

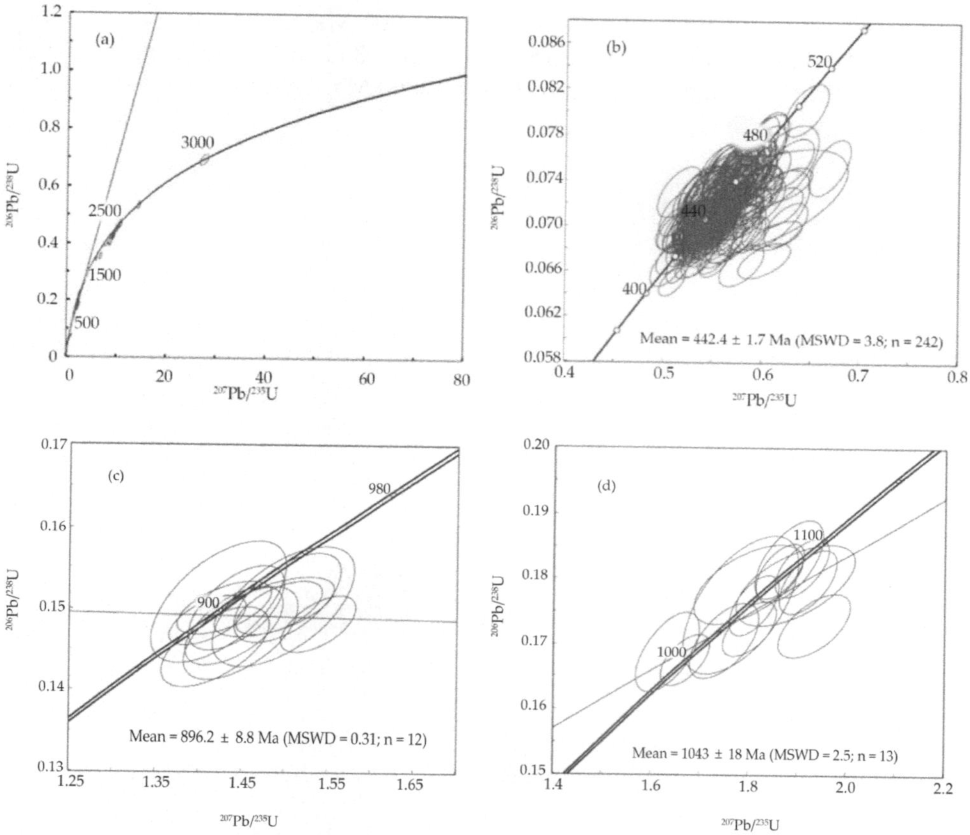

Figure 7. Concordia plots of zircon dating results for the Tanghu granites. (a) shows concordia plot of the whole data; (b–d) show concordia plot and mean average ages of Ordovician, Neoproterozoic and Mesoproterozoic zircons, respectively. The red circle, bule solid lines and bule dashed lines represent the radius error for a single spot, concordant curve and unconcordant curve, respectively.

Zircon Lu-Hf Isotopes

Ninety-two Lu-Hf isotopic analyses were carried out on these zircon grains from the Dengfuxian granites (0326-2s2) which have been analyzed by U-Pb dating, and the results. In general, these zircon grains have variable $^{176}Hf/^{177}Hf$ ratios of 0.28087–0.28256 with corresponding $\varepsilon_{Hf}(t)$ values of −12.9 to 10.8 and crustal model ages (T_{DMC}) of 1081–3583 Ma, respectively (Figure 8 and Figure 9). Based on age populations of 150–160 Ma, 900–1000 Ma, 1100–1200 Ma, 1500–1600 Ma, and 2300–2500 Ma, they have $\varepsilon_{Hf}(t)$ values of −11.3 to −5.8, −10.9 to 10.8, −12.9 to 10.5, −11.6 to 9.1, and −10.0 to 8.9 Ma, respectively. Their corresponding T_{DMC} ages are 1572–1918 Ma, 1081–2336 Ma, 1301–2779 Ma, 1615–3055 Ma, and 2602–3583 Ma, respectively. One hundred and twenty-eight zircon Lu-Hf isotopic analyses were carried out on these zircon grains from the Tanghu granites (13D40-1) which have been analyzed by U-Pb dating, and the results. These zircon grains also have variable $^{176}Hf/^{177}Hf$ ratios of 0.280223–0.282509 with corresponding $\varepsilon_{Hf}(t)$ values of −35.2 to 13.8 and crustal model ages (T_{DMC}) of 1306–4234 Ma, respectively (Figure 8 and Figure 9).

Based on age populations of 400–500 Ma, 800–900 Ma, and 1000–1100 Ma, they have $\varepsilon_{Hf}(t)$ values of −10.1 to 0.06, −6.8 to 4.02, and −35.2 to 10.5, respectively. Their corresponding T_{DMC} ages range from 1431 to 2060 Ma, 1596 to 2189 Ma, and 1494 to 4061 Ma, respectively.

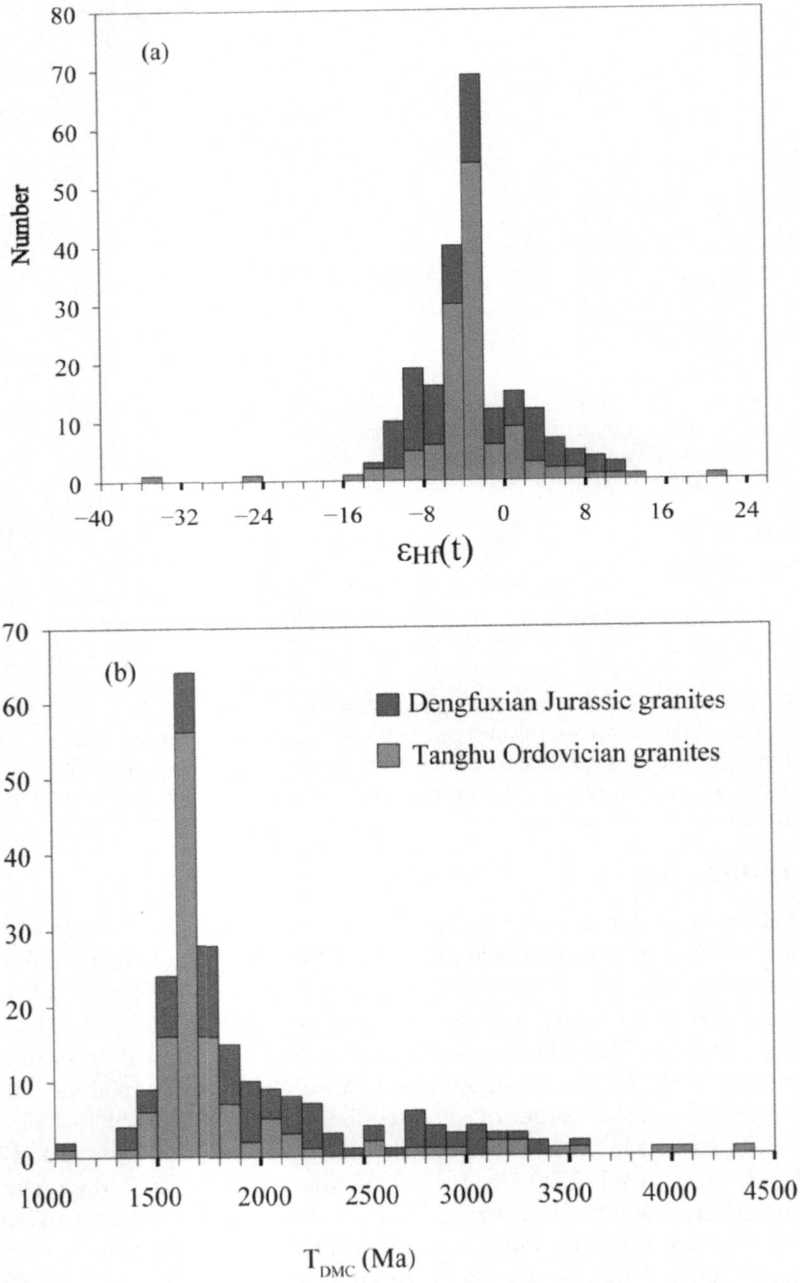

Figure 8. Histograms of zircon $\varepsilon_{Hf}(t)$ values (a) and Hf model ages (T_{DMC}) (b) for the Dengfuxian and Tanghu granites.

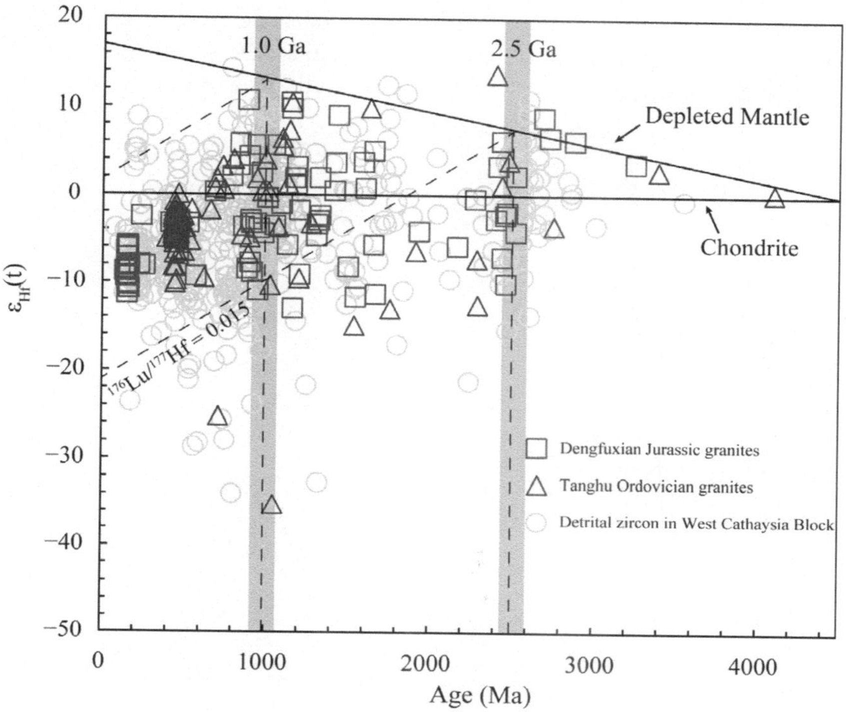

Figure 9. Zircon $\varepsilon_{Hf}(t)$ values versus U-Pb ages diagram. The data of the detrital zircons in the west Cathaysia Block are from [4,29,81,82], respectively.

DISCUSSION

Precambrian Crust Evolutionary History of the West Cathaysia Block

Archean (4000~2500 Ma)

Although no Archean rocks were exposed in the Cathaysia Block, geologists tend to believe that an Archean basement might be unexposed, although limited information on the Archean rocks was identified in the Cathaysia Block [9,23,30]. Recently, more and more geologists regarded that the Neo- and Meso-Archean (3200~2500 Ma) might be one of the key crustal generation periods of the Cathaysia Block since plentiful detrital zircon grains have an apparent age peak at ca. 2500 Ma in both the East and West Cathaysia Blocks (Figure 10) [10,15,28]. In addition, most of these zircon grains have positive $\varepsilon_{Hf}(t)$ values, indicating a mantle origin for the magma from which these zircons crystallized (Figure 9). Meanwhile, the difference between the zircon Hf model ages and the U–Pb ages, defined as the crustal incubation time [83,84], also have a cluster at ca. 2500 Ma, unraveling that the juvenile materials were involved in these zircons (Figure 11). In this study, both the Dengfuxian and Tanghu granites contain Neo- and Meso-Archean zircons with ages over 2500 Ma, positive $\varepsilon_{Hf}(t)$ values, and the consistent Hf model ages, further supporting the perspective of a juvenile crust growth event for the West

Cathaysia Block during the Neo- and Meso-Archean. However, it's worth noting that a considerable amount of Neo- and Meso-Archean detrital zircon grains have negative $\varepsilon_{Hf}(t)$ values and Hf model ages of 2600~3600 Ma, signifying that crustal reworking is also a crucial aspect of the West Cathaysia Block crust evolution (Figure 9).

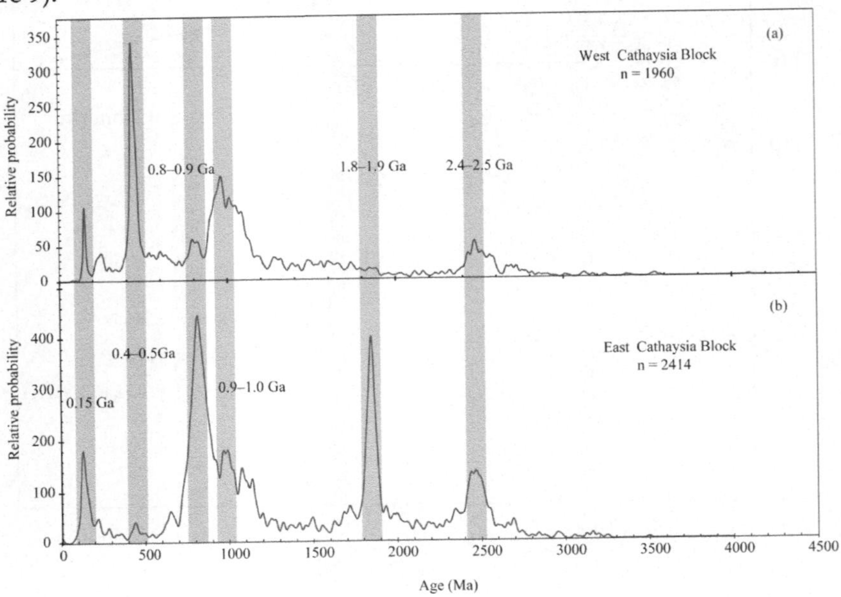

Figure 10. Relative probability density plots of detrital and inherited zircons from rocks of the WestCathaysia Block (a) and the East Cathaysia Block (b). In addition to the data in this study, the data of the West Cathaysia Block are from [4,11,28,29,81,82,85], respectively. The data of the East Cathaysia Block are from [22,25,31], respectively.

The Paleoarchean crustal evolution of the West Cathaysia Block (4000~3200 Ma) is poorly understood since rare detrital zircon grains were identified in comparison to the Neo- and Meso-Archean. It was proposed that the Mesoarchaean (~3.3 Ga) was one of the major magmatic periods in the Nanling-Yunkai terrane of the West Cathaysia Block [4]. In addition, the Mesoarchaean magmatic event was both involved in the juvenile input and the reworking of older crustal materials [4]. A previous study by Wang et al. (2018) [82] also reported a Paleoarchean detrital zircon grain of 3534 Ma with an $\varepsilon_{Hf}(t)$ value of 0.54 and Hf model age of 3774 Ma, demonstrating that juvenile materials should be involved in the formation of the Paleoarchean crust. In this work, a zircon grain from the Dengfuxian granites (#25) has a $^{207}Pb/^{206}Pb$ age of 3252 Ma with a positive $\varepsilon_{Hf}(t)$ value of 3.7 and a Hf model age of 3357 Ma, which is similar to the zircon grain from the Tanghu granites (#3-13) with a $^{207}Pb/^{206}Pb$ age of 3395 Ma, $\varepsilon_{Hf}(t)$ value of 3.09 and Hf model age of 3505 Ma. These data were nearly plotted along the evolutionary line of the depleted mantle in the age vs. $\varepsilon_{Hf}(t)$ diagram (Figure 9). Additionally, the crustal incubation time for these zircon grains is within the range of 150 Ma, indicating a juvenile crustal growth event (Figure 11). In conclusion, juvenile crustal growth might be predominant in the Archean for the West Cathaysia Block, however, limited crustal reworking events occurred in the Neo- and Meso-Archean.

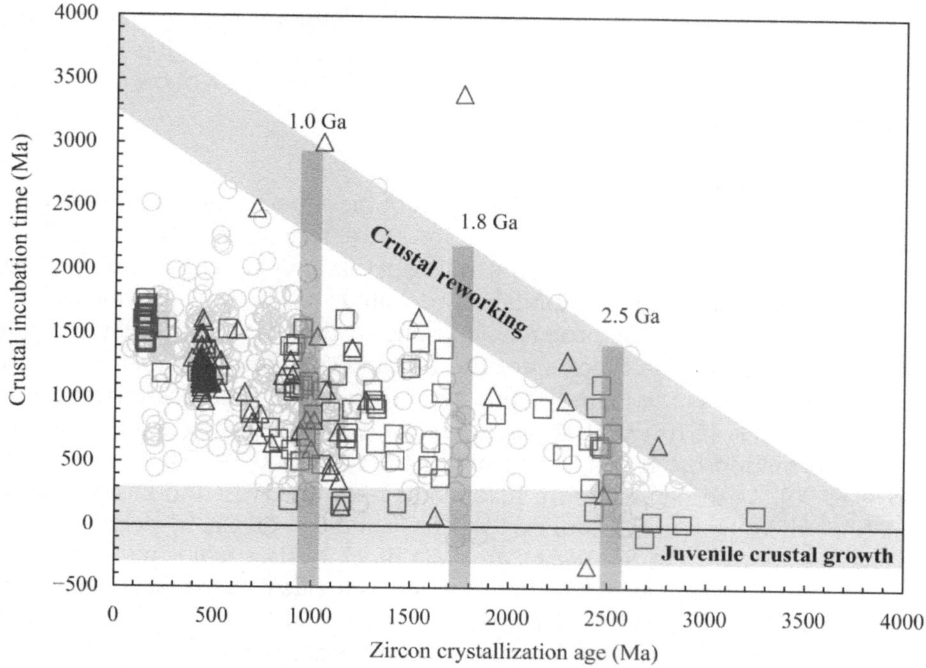

Figure 11. The plot of crustal incubation time versus crystallization ages of the detrital and inherited zircons from the (meta)sedimentary rocks in the Cathaysia Block. Crustal incubation time is defined by the difference between the zircon Hf model age and U–Pb crystallization age [83,84]. The data of the detrital zircons in the West Cathaysia Block are from [4,29,81,82], respectively. Green square, blue triangle and buff circle in this figure represent the inherited zircons from the Dengfuxian Jurassic granites, Tanghu Ordovician granites, and meta-sediments of the West Cathaysia Block, respectively.

Paleoproterozoic (2500~1600 Ma)

It is well accepted that the 1.8–1.9 Ga magmatism is related to the assembly of the Columbia supercontinent [10,86,87]. Based on these exposed 1.89–1.86 Ga S-type granitoids, 1.85~1.83 Ga A-type granites, and ~1.77 Ga amphibolites in the Cathaysia Block, it was regarded that these rock units recorded a complete geodynamic process from collisional orogeny to intracontinental rifting during the Late Paleoproterozoic time, coinciding with the timing of final assembly and initial rifting of the Columbia supercontinent [22]. However, it is noteworthy that these Paleoproterozoic rocks are mainly hosted in the Wuyishan region of the East Cathaysia Block. Unlike the East Cathaysia Block, rocks with Paleoproterozoic ages have not been identified in the West Cathaysia Block so far. In addition, rare detrital zircon grains with Paleoproterozoic ages were discovered in the metamorphic rocks and fluvial sediment [4,22,28,82]. A remarkable difference with the East Cathaysia Block is that the peak ages of 1.8-1.9 Ga are absent in the West Cathaysia Block (Figure 10). It was proposed that the Nanling-Yunkai terrane of the West Cathaysia Block and the Wuyishan terrane of the East Cathaysia Block were separated from each other before the Neoproterozoic and finally united during the Grenvillian orogeny [4]. Therefore, the West Cathaysia

Block might be unaffected by the assembly of the Columbia supercontinent in the Paleoproterozoic. Additionally, the crust generation and reworking of the West Cathaysia Block in this period is likely stagnant, due to the lack of coeval igneous/magmatic activities.

Mesoproterozoic-Neoproterozoic (1600~540 Ma)

A consensus is that the SCB became a part of the supercontinent Rodinia in the Neoproterozoic, caused by the amalgamation of the Yangtze and Cathaysia Blocks during the Sibao orogeny, however, the timing of this amalgamation and the paleo-position of the SCB in Rodinia are still hotly debated [10,12,15,20]. Previous studies have indicated that the Neoproterozoic is one of the important crust accretion periods for the Cathaysia Block since the Precambrian detrital zircon ages from several (meta)sedimentary samples and river sands yield significant peaks at ca. 1.0 Ga and 0.8 Ga [4,10,19,82]. At the same time, these peak ages are consistent with the formation of igneous rocks of 1.0–0.9 Ga and 0.8–0.7 Ga, respectively [4,85,88,89,90]. As shown in Figure 10, zircon ages in the West and East Cathaysia Block both yield the two peaks at ca. 0.9–1.0 Ga and 0.8–0.9 Ga, respectively.

The age spectrum of the West Cathaysia Block hosts a major peak at 0.9–1.0 Ga and a subordinate peak at 0.8–0.9 Ga, whereas the East Cathaysia Block shows the opposite data (Figure 10). This phenomenon might record the subducted–collisional process between the Yangtze and Cathaysia Blocks from northwest to southeast since the Wuyi-Yunkai arc-back-arc system was proposed to be finally closed at ~920 Ma in the West Cathaysia Block, which was earlier than the closure age of 830–820 Ma for the Shuangxiwu arc or back-arc system in the Jiangnan Orogen Belt [89,91,92,93,94].

The Hf isotopes of these Neoproterozoic zircons contain both positive and negative $\varepsilon_{Hf}(t)$ values, indicating that ancient and/or juvenile materials might be involved (Figure 9). However, the vast majority of corresponding Hf models' ages for these zircons are significantly older than the U-Pb ages, resulting in high values of crust incubation time (Figure 11). Consequently, it is inferred that the reworking of the juvenile and ancient crust might be predominant in the Mesoproterozoic-Neoproterozoic for the West Cathaysia Block.

Implications for the Regional Tungsten Mineralization

Tungsten is one of the strongly incompatible lithophile elements and tends to be enriched in crustal rocks. Therefore, the melting of abundant crustal rocks is necessary to form economically significant tungsten deposits [95,96,97]. In most cases, the formation of these tungsten deposits is commonly related to the granitic rocks which were involved in the ancient crustal materials, since tungsten has a high solubility in silicate melts [18,96,98,99]. In addition, tungsten becomes enriched in the late stage of these melts through intense fractional crystallization, supported by the fact that tungsten mineralization is commonly hosted in these highly fractionated granites [17,34,100,101,102].

However, tungsten mineralization is not exclusive to these highly fractionated melts. Lowly differentiated melts can also generate the formation of tungsten deposits, if the source of these melts is particularly rich in tungsten [96]. As discussed above, the crust evolution of the West Cathaysia Block is dominated by crustal reworking since the Proterozoic. Therefore, the long-term crustal reworking of the West Cathaysia Block enhances the enrichment of the economically valuable lithophile elements (e.g., W, Sn, U, Nb, and Ta) in the crust of the West Cathaysia Block.

The large-scale tungsten mineralization in the Nanling Range, explosively occurring in the Middle to Late Jurassic, was caused by the coeval granitic magmatism [103,104,105,106]. The formation of these granites and tungsten deposits in the Nanling Range is one of the hot issues in the last three decades, reaching a consensus that these granites are mostly highly fractionated granites and generated by the partial melting of the Cambrian basement rocks [17]. The Dengfuxian Jurassic granites related to the large-scale Xiangdong tungsten deposit host a series of ancient zircons which are mainly Proterozoic to Archean, indicating that complicated ancient sources were involved in the formation of the Dengfuxian Jurassic granites.

In addition, almost all the Jurassic zircons have negative $\varepsilon_{Hf}(t)$ values and old Hf model ages, suggesting that the Dengfuxian Jurassic granites were likely derived from the Proterozoic basement rocks of the West Cathaysia Block. In addition, this property is one of the favorable factors for the generation of economically valuable metal-enriched magma and related deposits.

CONCLUSIONS

Inherited zircons from the Mesozoic Dengfuxian and Paleozoic Tanghu plutons, the Nanling Range of the West Cathaysia Block, have U-Pb ages from the Neoproterozoic to the Archean. Together with the previously reported data, we conclude as follows: (1) The Hadean crust fragment might exist in the West Cathaysia Block and might have experienced the reworking event since at least 4100 Ma; (2) Archean zircons mainly originated from the depleted mantle and the juvenile crust generation was predominant in the Archean for the West Cathaysia Block; (3) The Proterozoic zircons mostly originated from the crustal source, and the crust evolution of the West Cathaysia Block has been dominated by the crust reworking since the Proterozoic.

The Jurassic granites in the Nanling Range, formed by the partial melting of these Proterozoic rocks, are enriched in the metalliferous lithophile elements W, Sn, U, and Nb, causing the intense regional tungsten polymetallic mineralization.

AUTHOR CONTRIBUTIONS

J.C., Y.L. and L.L. conceived and designed the experiments; J.F., S.Y. and Q.W. took part in the discussion; G.X., X.Q. and Z.Z. took part in the field campaigns; J.C., Y.L. and L.L. wrote the paper. All authors have read and agreed to the published version of the manuscript.

ACKNOWLEDGMENTS

We appreciate the valuable and insightful suggestions and comments from the anonymous reviewers. We also thank the Ding He from the Department of Ocean Science of the Hong Kong University of Science & Technology for this excellent English editing work. We also thank Meirong Zheng for her kind support.

REFERENCES

1. Thieblemont, D.; Delor, C.; Cocherie, A.; Lafon, J.M.; Goujou, J.C.; Balde, A.; Bah, M.; Sane, H.; Fanning, C.M. A 3.5 Ga granite–gneiss basement in Guinea: Further evidence for early archean accretion within the West African Craton. Precambrian Res. 2001, 108, 179–194.
2. Marschik, R.; Mathur, R.; Ruiz, J.; Leveille, R.; de Almeida, A.J. Late Archean Cu–Au–Mo mineralization at Gameleira and Serra Verde, Carajas mineral Province, Brazil: Constraints from Re–Os molybdenite ages. Miner. Depos. 2005, 39, 983–991.
3. Zheng, Y.F.; Zhang, S.B. Formation and evolution of Precambrian continental crust in South China. Chin. Sci. Bull. 2007, 52, 1–12, (In Chinese with English Abstract).
4. Yu, J.H.; O'Reilly, Y.S.; Wang, L.J.; Griffin, W.L.; Zhou, M.F.; Zhang, M.; Shu, L.S. Components and episodic growth of Precambrian crust in the Cathaysia Block, South China: Evidence from U–Pb ages and Hf isotopes of zircons in Neoproterozoic sediments. Precambrian Res. 2010, 181, 97–114.
5. Payne, J.L.; Hand, M.; Pearson, N.J.; Barovich, K.M.; McInerney, D.J. Crustal thickening and clay: Controls on O isotope variation in global magmatism and siliciclastic sedimentary rocks. Earth Planet. Sci. Lett. 2015, 412, 70–76.
6. Tang, M.; Chen, K.; Rudnick, R.L. Archean upper crust transition from mafic to felsic marks the onset of plate tectonics. Science 2016, 351, 372–375.
7. Bucholz, C.E.; Biasi, J.A.; Beaudry, P.; Ono, S. Sulfur isotope behavior during metamorphism and anatexis of Archean sedimentary rocks: A case study from the Ghost Lake batholith, Ontario, Canada. Earth Planet. Sci. Lett. 2020, 549, 116494.
8. Li, R.Y.; Ke, S.; Li, S.G.; Song, S.G.; Wang, C. Origins of two types of Archean potassic granite constrained by Mg isotopes and statistical geochemistry: Implications for continental crustal evolution. Lithos 2020, 368, 105570.
9. Xu, X.S.; O'Reilly, Y.S.; Griffin, W.L.; Deng, P.; Pearson, N.J. Relict proterozoic basement in the Nanling Mountains (SE China) and its tectonothermal overprinting. Tectonics 2005, 24, 1–16.
10. Li, X.H.; Li, Z.X.; Li, W.X. Detrital zircon U–Pb age and Hf isotope constrains on the generation and reworking of Precambrian continental crust in the Cathaysia Block, South China: A synthesis. Gondwana Res. 2014, 25, 1202–1215.
11. Xu, X.; O'Reilly, S.Y.; Griffin, W.L.; Wang, X.; Pearson, N.J.; He, Z. The crust of Cathaysia: Age, assembly and reworking of two terranes. Precambrian Res. 2007, 158, 51–78.
12. Zhao, J.H.; Zhou, M.F.; Yan, D.P.; Zheng, J.P.; Li, J.W. Reappraisal of the ages of Neoproterozoic strata in South China: No connection with the Grenvillian orogeny. Geology 2011, 39, 299–302.
13. Li, X.H.; Li, Z.X.; He, B.; Li, W.X.; Li, Q.L.; Gao, Y.Y.; Wang, X.C. The Early Permian active continental margin and crustal growth of the Cathaysia Block: In situ U–Pb, Lu–Hf and O isotope analyses of detrital zircons. Chem. Geol. 2012, 328, 195–207.
14. Charvet, J. The Neoproterozoic–Early Paleozoic tectonic evolution of the South China Block: An overview. J. Asian Earth Sci. 2013, 74, 198–209.
15. Zhang, S.B.; Zheng, Y.F. Formation and evolution of Precambrian continental lithosphere in South China. Gondwana Res. 2013, 23, 1241–1260.
16. Mao, J.R.; Li, Z.L.; Ye, H.M. Mesozoic tectono–magmatic activities in South China: Retrospect and prospect. Sci. China–Earth Sci. 2014, 57, 2853–2877.

17. Cao, J.Y.; Yang, X.Y.; Du, J.G.; Wu, Q.H.; Kong, H.; Li, H.; Wan, Q.; Xi, X.S.; Gong, Y.S.; Zhao, H.R. Formation and geodynamic implication of the Early Yanshanian granites associated with W–Sn mineralization in the Nanling Range, South China: An overview. Int. Geol. Rev. 2018, 60, 1744–1771.

18. Cao, J.Y.; Li, H.; Algeo, T.J.; Yang, L.; Tamehe, L.S. Two–stage magmatism and tungsten mineralization in the Nanling Range, South China: Evidence from the Jurassic Helukou deposit. Am. Mineral. 2021, 106, 1488–1502.

19. Cawood, P.A.; Zhao, G.C.; Yao, J.L.; Wang, W.; Xu, Y.J.; Wang, Y.J. Reconstructing South China in Phanerozoic and Precambrian supercontinents. Earth–Sci. Rev. 2018, 186, 173–194.

20. Yan, C.L.; Shu, L.S.; Faure, M.; Chen, Y.; Huang, R.B. Time constraints on the closure of the Paleo–South China Ocean and the Neoproterozoic assembly of the Yangtze and Cathaysia blocks: Insight from new detrital zircon analyses. Gondwana Res. 2019, 73, 175–189.

21. Yu, J.H.; O'Reilly, Y.S.; Zhou, M.F.; Griffin, W.L.; Wang, L.J. U–Pb geochronology and Hf–Nd isotopic geochemistry of the Badu Complex, Southeastern China: Implications for the Precambrian crustal evolution and paleogeography of the Cathaysia Block. Precambrian Res. 2012, 222–223, 424–449.

22. Yu, J.H.; Wang, L.J.; Griffin, W.L.; O'Reilly, S.Y.; Zhang, M.; Li, C.Z.; Shu, L.S. A Paleoproterozoic orogeny recorded in a long–lived cratonic remnant (Wuyishan terrane), eastern Cathaysia Block, China. Precambrian Res. 2009, 174, 347–363.

23. Xia, Y.; Xu, X.S.; Zhu, K.Y. Paleoproterozoic S– and A–type granites in southwestern Zhejiang: Magmatism, metamorphism and implications for the crustal evolution of the Cathaysia basement. Precambrian Res. 2012, 216–219, 177–207.

24. Zhao, L.; Zhou, X.W.; Zhai, M.G.; Santosh, M.; Geng, Y.S. Zircon U–Th–Pb–Hf isotopes of the basement rocks in northeastern Cathaysia block, South China: Implications for Phanerozoic multiple metamorphic reworking of a Paleoproterozoic terrane. Gondwana Res. 2015, 28, 246–261.

25. Yu, J.H.; Zhang, C.H.; O'Reilly, Y.S.; Griffin, W.L.; Ling, H.F.; Sun, T.; Zhou, X.Y. Basement components of the Xiangshan–Yuhuashan area, South China: Defining the boundary between the Yangtze and Cathaysia blocks. Precambrian Res. 2018, 309, 102–122.

26. Xu, Y.J.; Du, Y.S.; Huang, H.W.; Huang, Z.Q.; Hu, L.S.; Zhu, Y.H.; Yu, W.C. Detrital zircon of 4.1 Ga in South China. China Sci. Bull. 2012, 57, 4356–4362.

27. Xing, G.F.; Wang, X.L.; Wan, Y.S.; Chen, Z.H.; Jiang, Y.; Kitajima, L.; Ushikubo, T.; Gppon, P. Diversity in early crustal evolution: 4100 Ma zircons in the Cathaysia Block of southern China. Sci. Rep. 2014, 4, 5134.

28. Li, X.Y.; Zhang, J.P.; Xiong, Q.; Zhou, X.; Lu, X. Triassic rejuvenation of unexposed Archean–Paleoproterozoic deep crust beneath the western Cathaysia block, South China. Tectonophysics 2018, 724–725, 65–79.

29. Xiong, C.; Chen, H.D.; Niu, Y.L.; Chen, A.Q.; Zhang, C.G.; Li, F.; Xu, S.L.; Yang, S. Provenance, depositional setting, and crustal evolution of the Cathaysia Block, South China: Insights from detrital zircon U–Pb geochronology and geochemistry of clastic rocks. Geol. J. 2019, 54, 897–912.

30. Zheng, J.P.; Griffin, W.L.; Tang, H.Y.; Zhang, Z.H.; Su, Y.P.; Liu, G.L. Archean basement similar to the North China and Yangtze continents may be existed beneath the western Cathaysia. Geol. J. China Univ. 2008, 14, 549–557, (In Chinese with English Abstract).

31. Yang, Z.Y.; Jiang, S.Y. Detrital zircons in metasedimentary rocks of Mayuan and Mamianshan Group from Cathaysia Block in northwestern Fujian Province, South China: New constraints on their formation ages and paleogeographic implication. Precambrian Res. 2019, 320, 13–30.

32. Brown, M. Crustal melting and granite magmatism: Key issues. Phys. Chem. Earth Part A Solid Earth Geod. 2001, 26, 201–212.

33. Lopez, S.; Fernandez, C.; Castro, A. Evolution of the Archaean continental crust: Insights from the experimental study of Archaean granitoids. Curr. Sci. 2006, 91, 607–621.

34. Wu, F.Y.; Liu, X.C.; Ji, W.Q.; Wang, J.M.; Yang, L. Highly fractionated granites: Recognition and research. Sci. China–Earth Sci. 2017, 60, 1201–1219.

35. Ni, P.; Wang, G.G.; Li, W.S.; Chi, Z.; Li, S.N.; Gao, Y. A review of the Yanshanian ore–related felsic magmatism and tectonic settings in the Nanling W–Sn and Wuyi Au–Cu metallogenic belts, Cathaysia Block, South China. Ore Geol. Rev. 2021, 133, 104088.

36. Pidgeon, R.T.; Bosch, D.; Bruguier, O. Inherited zircon and titanite U–Pb systems in an archaean syenite from southwestern Australia: Implications for U–Pb stability of titanite. Earth Planet. Sci. Lett. 1996, 141, 187–198.

37. Zeck, H.P.; Wingate, M.T.D.; Pooley, G.D.; Ugidos, J.M. A sequence of Pan–African and Hercynian events recorded in zircons from an orthogneiss from the Hercynian belt of western central Iberia —An ion microprobe U–Pb study. J. Petrol. 2004, 45, 1613–1629.

38. Flowerdew, M.J.; Millar, I.L.; Vaughan, A.P.M.; Horstwood, M.S.A.; Fanning, C.M. The source of granitic gneisses and migmatites in the Antarctic Peninsula: A combined U–Pb SHRIMP and laser ablation Hf isotope study of complex zircons. Contrib. Mineral. Petrol. 2006, 151, 751–768.

39. Qiao, L.; Wang, Q.F.; Li, C.S. The western segment of the suture between the Yangtze and Cathaysia blocks: Constraints from inherited and co–magmatic zircons from Permian S–type granitoids in Guangxi, South China. Terra Nova 2015, 27, 392–398.

40. Gao, S.; Ling, W.L.; Qiu, Y.M.; Liang, Z.; Hartmann, G.; Simon, K. Contrasting geochemical and Sm–Nd isotopic compositions of Archean metasediments from the Kongling high–grade terrain of the Yangtze craton: Evidence for cratonic evolution and redistribution of REE during crustal anatexis. Geochim. Et Cosmochim. Acta 1999, 63, 2071–2088.

41. Gao, S.; Yang, J.; Zhou, L.; Li, M.; Hu, Z.C.; Guo, J.L.; Yuan, H.L.; Gong, H.J.; Xiao, G.Q.; Wei, J.Q. Age and growth of the Archean Kongling terrain, South China, with emphasis on 3.3 Ga granitoid gneisses. Am. J. Sci. 2011, 311, 153–182.

42. Zhang, S.B.; Zheng, Y.F.; Wu, Y.B.; Zhao, Z.F.; Gao, S.; Wu, F.Y. Zircon isotope evidence for ≥3.5 Ga continental crust in the Yangtze craton of China. Precambrian Res. 2006, 146, 16–34.

43. Jiao, W.F.; Wu, Y.B.; Yang, S.H.; Peng, M.; Wang, J. The oldest basement rock in the Yangtze Craton revealed by zircon U–Pb age and Hf isotope composition. Sci. China Ser. D Earth Sci. 2009, 52, 1393–1399.

44. Chen, K.; Gao, S.; Wu, Y.B.; Guo, J.L.; Hu, Z.C.; Liu, Y.S.; Zong, K.Q.; Liang, Z.W.; Geng, X.L. 2.6–2.7 Ga crustal growth in Yangtze craton, South China. Precambrian Res. 2013, 224, 472–490.

45. Cui, X.Z.; Wang, J.; Wang, X.C.; Wilde, S.A.; Ren, G.M.; Li, S.J.; Deng, Q.; Ren, F.; Liu, J.P. Early crustal evolution of the Yangtze Block: Constraints from zircon U–

Pb–Hf isotope systematics of 3.1–1.9 Ga granitoids in the Cuoke Complex, SW China. Precambrian Res. 2021, 357, 106155.

46. Zhang, Y.Z.; Wang, Y.J.; Geng, H.Y.; Zhang, Y.H.; Fan, W.M.; Zhong, H. Early Neoproterozoic (similar to 850 Ma) back–arc basin in the Central Jiangnan Orogen (Eastern South China): Geochronological and petrogenetic constraints from meta–basalts. Precambrian Res. 2013, 231, 325–342.
47. Yao, J.L.; Shu, L.S.; Santosh, M.; Zhao, G.C. Neoproterozoic arc–related mafic–ultramafic rocks and syn–collision granite from the western segment of the Jiangnan Orogen, South China: Constraints on the Neoproterozoic assembly of the Yangtze and Cathaysia Blocks. Precambrian Res. 2014, 243, 39–62.
48. Deng, T.; Xu, D.R.; Chi, G.X.; Zhu, Y.H.; Wang, Z.L.; Chen, G.W.; Li, Z.H.; Zhang, J.L.; Ye, T.W.; Yu, D.S. Revisiting the ca. 845–820–Ma S–type granitic magmatism in the Jiangnan Orogen: New insights on the Neoproterozoic tectono–magmatic evolution of South China. Int. Geol. Rev. 2019, 61, 383–403, (In Chinese with English Abstract).
49. Wan, L.; Kusky, T.M.; Jin, W.; Yang, J.; Zeng, Z.X. Neoproterozoic tectonics of the Jiangnan orogen: The magmatic record of continental growth by arc and slab–failure magmatism from 1000 to 780 Ma. Precambrian Res. 2021, 362, 106319.
50. Li, X.H. U–Pb zircon ages of granites from the southern margin of the Yangtze Block: Timing of Neoproterozoic Jinning: Orogeny in SE China and implications for Rodinia Assembly. Precambrian Res. 1999, 97, 43–57.
51. Wang, R.R.; Xu, Z.Q.; Santosh, M.; Zeng, B. Mid–Neoproterozoic magmatism in the northern margin of the Yangtze Block, South China: Implications for transition from subduction to post–collision. Precambrian Res. 2021, 354, 106073.
52. Xie, J.C.; Yang, X.Y.; Sun, W.D.; Du, J.G. Early Cretaceous dioritic rocks in the Tongling region, eastern China: Implications for the tectonic settings. Lithos 2012, 150, 49–61.
53. Mao, Z.H.; Liu, J.J.; Mao, J.W.; Deng, J.; Zhang, F.; Meng, X.Y.; Xiong, B.K.; Xiang, X.K.; Luo, X.H. Geochronology and geochemistry of granitoids related to the giant Dahutang tungsten deposit, middle Yangtze River region, China: Implications for petrogenesis, geodynamic setting, and mineralization. Gondwana Res. 2015, 28, 816–836.
54. Zhou, T.F.; Wang, S.W.; Fan, Y.; Yuan, F.; Zhang, D.Y.; White, N.C. A review of the intracontinental porphyry deposits in the Middle–Lower Yangtze River Valley metallogenic belt, Eastern China. Ore Geol. Rev. 2015, 65, 433–456.
55. Song, S.W.; Mao, J.W.; Xie, G.Q.; Lehmann, B.; Jian, W.; Wang, X.G. The world–class mid–Mesozoic Jiangnan tungsten belt, South China: Coeval large reduced and small oxidized tungsten systems controlled by different magmatic petrogenesis. Ore Geol. Rev. 2021, 139, 104543.
56. Xiao, Q.L.; Zhou, T.F.; Holling, P.; Wang, S.W.; Liu, J.; Yang, Q.G.; Yuan, F. Mineral and whole–rock chemistry of the Chating porphyry Cu–Au deposit related intrusions in the Middle–Lower Yangtze River Belt, Eastern China: Implications for magma evolution and mineralization. Lithos 2021, 380, 105881.
57. Cao, J.Y.; Li, H.; Yang, X.Y.; Tamehe, L.S.; Esmaeili, R. Multi–stage magma evolution recorded by apatite and zircon of adakite–like rocks: A case study from the Shatanjiao intrusion, Tongling region, Eastern China. Am. Mineral. 2022, 107, 178–189.
58. Zhou, X.M.; Sun, T.; Shen, W.Z.; Shu, L.S.; Niu, Y.L. Petrogenesis of Mesozoic granitoids and volcanic rocks in South China: A response to tectonic evolution. Episodes 2006, 29, 26–33.

59. Hu, X.J.; Xu, J.K.; Tong, Z.X.; Chen, C.H. The Precambrian Geology of Southwestern Zhejiang Province; Geological Publishing House: Beijing, China, 1991; pp. 1–278, (In Chinese with English Abstract).

60. Hu, X.J.; Xu, J.K.; Tong, Z.X.; Chen, C.H. Geochronology of the Middle Proterozoic Longquan Group in Southwestern Zhejiang. Geol. Rev. 1992, 38, 271–278, (In Chinese with English Abstract).

61. Wang, W.; Zhou, M.F.; Yan, D.P.; Li, J.W. Depositional age, provenance, and tectonic setting of the Neoproterozoic Sibao group, Southeastern Yangtze Block, South China. Precambrian Res. 2012, 192–195, 107–124.

62. Guo, C.L.; Liu, Z.K. Caledonian granites in South China: The geological and geochemical characteristics on their petrogenesis and mineralization. J. Earth Sci. Environ. 2021, 43, 927–961, (In Chinese with English Abstract).

63. Cai, Y.; Lu, J.J.; Ma, D.S.; Huang, H.; Zhang, H.F. Chronology and geochemical characteristics of Late Indosinian Dengfuxian two–mica granite in eastern Hunan Province, China, and its significance. Acta Petrol. Sin. 2013, 29, 4215–4231, (In Chinese with English Abstract).

64. He, M.; Liu, Q.; Hou, Q.L.; Sun, J.F.; Zhang, J.H.; Wu, S.C.; Zhu, H.F. Petrogenesis of the Dengfuxian granite, eastern Hunan Province and constraints on mineralization: Evidences from zircon and cassiterite U–Pb geochronology, zircon Hf–O isotopes and whole–rock geochemistry. Acta Petrol. Sin. 2018, 34, 637–655.

65. Li, H.; Cao, J.Y.; Algeo, T.J.; Jiang, W.C.; Liu, B.; Wu, Q.H. Zircons reveal multi–stage genesis of the Xiangdong (Dengfuxian) tungsten deposit, South China. Ore Geol. Rev. 2019, 111, 102979.

66. Cao, J.Y.; Wu, Q.H.; Yang, X.Y.; Deng, X.T.; Li, H.; Kong, H.; Xi, X.S. Geochemical factors revealing the differences between the Xitian and Dengfuxian composite plutons, middle Qin–Hang Belt: Implications to the W–Sn mineralization. Ore Geol. Rev. 2020, 118, 103353.

67. Deng, X.T.; Cao, J.Y.; Wu, Q.H.; Kong, H.; Xi, X.S. Difference of sources of the Yanshanian Xitian and Dengfuxian granites in Hunan Province and their implication. J. Cent. South Univ. Sci. Technol. 2017, 48, 212–222.

68. Xia, J.L.; Huang, G.C.; Ding, L.; Ding, L.X.; Chen, X.Q.; Ji, W.B. Chronological Framework of the Zhuguangshan Composite Batholith in the Nanling Area. South China Geol. 2021, 37, 280–297, (In Chinese with English Abstract).

69. Wiedenbeck, M.; Alle, P.; Corfu, F.; Griffin, W.L.; Meier, M.; Ober, F.; Von Quadt, A.; Roddick, J.C.; Speigel, W. Three natural zircon standards for U–Th–Pb, Lu–Hf, trace–element and REE analyses. Geostand. Newsl. 1995, 19, 1–23.

70. Andersen, T. Correction of common lead in U–Pb analyses that do not report 204Pb. Chem. Geol. 2002, 192, 59–79.

71. Liu, Y.; Gao, S.; Hu, Z.; Gao, C.; Zong, K.; Wang, D. Continental and oceanic crust recycling–induced melt–peridotite interactions in the trans–north china orogen: U–Pb dating, Hf isotopes and trace elements in zircons from mantle xenoliths. J. Petrol. 2010, 51, 537–571.

72. Ludwig, K.R. ISOPLOT 3.00: A Geochronological Toolkit for Microsoft Excel; Berkeley Geochronology Center: Berkeley, CA, USA, 2003; p. 39.

73. Thirlwall, M.F.; Anczkiewicz, R. Multidynamic isotope ratio analysis using MC–ICP–MS and the causes of secular drift in Hf, Nd and Pb isotope ratios. Int. J. Mass Spectrom. 2004, 235, 59–81.

74. Blichert-Toft, J.; Chauvel, C.; Albarède, F. Separation of Hf and Lu for high–precision isotope analysis of rock samples by magnetic sector–multiple collector ICP–MS. Contrib. Mineral. Petrol. 1997, 127, 248–260.

75. Wu, F.; Yang, Y.; Xie, L.; Yang, J.; Xu, P. Hf isotopic compositions of the standard zircons and baddeleyites used in U–Pb geochronology. Chem. Geol. 2006, 234, 105–126.

76. Sláma, J.; Košler, J.; Condon, D.J.; Crowley, J.L.; Gerdes, A.; Hanchar, J.M.; Horstwood, M.S.A.; Morris, G.A.; Nasdala, L.; Norberg, N.; et al. Plešovice zircon—A new natural reference material for U–Pb and Hf isotopic microanalysis. Chem. Geol. 2008, 249, 1–35.

77. Söderlund, U.; Patchett, P.J.; Vervoort, J.D.; Isachsen, C.E. The 176Lu decay constant determined by Lu–Hf and U–Pb isotope systematics of Precambrian mafic intrusions. Earth Planet. Sci. Lett. 2004, 219, 311–324.

78. Bouvier, A.; Vervoort, J.D.; Patchett, P.J. The Lu–Hf and Sm–Nd isotopic composition of CHUR: Constraints from unequilibrated chondrites and implications for the bulk composition of terrestrial planets. Earth Planet. Sci. Lett. 2008, 273, 48–57.

79. Griffin, W.L.; Wang, X.; Jackson, S.E.; Pearson, N.J.; O'Reilly, S.Y.; Xu, X.; Zhou, X. Zircon chemistry and magma mixing, SE China: In–situ analysis of Hf isotopes, Tonglu and Pingtan igneous complexes. Lithos 2002, 61, 237–269.

80. Hoskin, P.W.O.; Schaltegger, U. The composition of zircon and igneous and metamorphic petrogenesis. Rev. Mineral. Geochem. 2003, 53, 27–62.

81. Li, X.C.; Wang, A.D.; Wan, J.J.; Lin, L.F. The ctustal growth and evolution of Cathaysia block: Constraints from detrital zircon Lu–Hf isotopes of the ganjiang river sediments. Sci. Technol. Eng. 2017, 17, 145–153.

82. Wang, K.X.; Sun, L.Q.; Sun, T.; Huang, H.; Qin, L.S. Provenance, weathering conditions, and tectonic evolution history of the Cambrian meta–sediments in the Zhuguangshan area, Cathaysia Block. Precambrian Res. 2018, 311, 195–210.

83. Wang, C.Y.; Campbell, I.H.; Allen, C.M.; Williams, I.S.; Eggins, S.M. Rate of growth of the preserved North American continental crust: Evidence from Hf and O isotopes in Mississippi detrital zircons. Geochim. Et Cosmochim. Acta 2009, 73, 712–728.

84. Wang, C.Y.; Campbell, I.H.; Stepanov, A.S.; Allen, C.M.; Burtsev, I.N. Growth rate of the preserved continental crust: II. Constraints from Hf and O isotopes in detrital zircons from Greater russian rivers. Geochim. Et Cosmochim. Acta 2011, 75, 1308–1345.

85. Zhang, A.M.; Wang, Y.J.; Fan, W.M.; Zhang, Y.Z.; Yang, J. Earliest Neoproterozoic (ca. 1.0 Ga) arc–back–arc basin nature along the northern Yunkai Domain of the Cathaysia Block: Geochronological and geochemical evidence from the metabasite. Precambrian Res. 2012, 220–221, 217–233.

86. Meert, J.G.; Santosh, M. The Columbia supercontinent revisited. Gondwana Res. 2017, 50, 67–83.

87. Chaves, A.D. Columbia (Nuna) supercontinent with external subduction girdle and concentric accretionary, collisional and intracontinental orogens permeated by large igneous provinces and rifts. Precambrian Res. 2021, 352, 106017.

88. Shu, L.S.; Faure, M.; Yu, J.H.; Jahn, B.M. Geochronological and geochemical features of the Cathaysia block (South China): New evidence for the Neoproterozoic breakup of Rodinia. Precambrian Res. 2011, 187, 263–276.

89. Wang, Y.J.; Zhang, Y.Z.; Fan, W.M.; Geng, H.Y.; Zou, H.P.; Bi, X.W. Early Neoproterozoic accretionary assemblage in the Cathaysia Block: Geochronological, Lu–Hf isotopic and geochemical evidence from granitoid gneisses. Precambrian Res. 2014, 249, 144–161.
90. Sun, J.J.; Shu, L.S.; Santosh, M.; Wang, L.S. Precambrian crustal evolution of the central Jiangnan Orogen (South China): Evidence from detrital zircon U–Pb ages and Hf isotopic compositions of Neoproterozoic metasedimentary rocks. Precambrian Res. 2018, 318, 1–24.
91. Li, W.X.; Li, X.H.; Li, Z.X. Neoproterozoic bimodal magmatism in the Cathaysia Block of South China and its tectonic significance. Precambrian Res. 2005, 136, 51–66.
92. Li, W.X.; Li, X.H.; Li, Z.X. Ca. 850 Ma bimodal volcanic rocks in northeastern Jiangxi South China: Initial extension during the breakup of Rodinia. Am. J. Sci. 2010, 310, 951–980.
93. Wang, Y.J.; Fan, W.M.; Zhang, G.W.; Zhang, Y.H. Phanerozoic tectonics of the South China Block: Key observations and controversies. Gondwana Res. 2013, 23, 1273–1305.
94. Wang, Y.J.; Zhang, A.M.; Cawood, P.A.; Fan, W.M.; Xu, J.F.; Zhang, G.W.; Zhang, Y.Z. Geochronological, geochemical and Nd–Hf–Os isotopic fingerprinting of an early Neoproterozoic arc–back–arc system in South China and its accretionary assembly along the margin of Rodinia. Precambrian Res. 2013, 231, 343–371.
95. Rasmussen, K.L.; Lentz, D.R.; Falck, H.; Pattison, D.R.M. Felsic magmatic phases and the role of late–stage aplitic dykes in the formation of the world–class Cantung Tungsten skarn deposit, Northwest Territories, Canada. Ore Geol. Rev. 2011, 41, 75–111.
96. Romer, R.L.; Kroner, U. Phanerozoic tin and tungsten mineralization–Tectonic controls on the distribution of enriched protoliths and heat sources for crustal melting. Gondwana Res. 2016, 31, 60–95.
97. Korges, M.; Weis, P.; Lueders, V.; Laurent, O. Depressurization and boiling of a single magmatic fluid as a mechanism for tin–tungsten deposit formation. Geology 2018, 46, 75–78.
98. Walter, M.J.; Thibault, T. Partitioning of tungsten and molybdenum between metallic liquid and silicate melt. Science 1995, 270, 1186–1189.
99. Schmidt, C.; Romer, R.L.; Wohlgemuth–Ueberwasser, C.C.; Appelt, O. Partitioning of Sn and W between granitic melt and aqueous fluid. Ore Geol. Rev. 2020, 117, 103261.
100. Shu, X.J.; Wang, X.L.; Sun, T.; Xu, X.S.; Dai, M.N. Trace elements, U–Pb ages and Hf isotopes of zircons from Mesozoic granites in the western Nanling Range, South China: Implications for petrogenesis and W–Sn mineralization. Lithos 2011, 127, 468–482.
101. Guo, C.L.; Chen, Y.C.; Zeng, Z.L.; Lou, F.S. Petrogenesis of the Xihuashan granites in southeastern China: Constraints from geochemistry and in–situ analyses of zircon U–Pb–Hf–O isotopes. Lithos 2012, 148, 209–227.
102. Chen, Y.; Li, H.; Sun, W.; Ireland, T.; Tian, X.; Hu, Y.; Yang, W.; Chen, C.; Xu, D. Generation of Late Mesozoic Qianlishan A2–type granite in Nanling Range, South China: Implications for Shizhuyuan W–Sn mineralization and tectonic evolution. Lithos 2016, 266, 435–452.
103. Hua, R.M.; Chen, P.R.; Zhang, W.L.; Yao, J.M.; Lin, J.F.; Zhang, Z.S.; Gu, S.Y. Metallogeneses and their geodynamic settings related to Mesozoic granitoids in the Nanling Range. Geol. J. China Univ. 2005, 11, 291–304, (In Chinese with English Abstract).

104. Mao, J.W.; Cheng, Y.B.; Chen, M.H.; Pirajno, F. Major types and time–space distribution of Mesozoic ore deposits in South China and their geodynamic settings. Miner. Depos. 2013, 48, 267–294.

105. Mao, J.W.; Xie, G.Q.; Guo, C.L.; Chen, Y.C. Largescale tungsten–tin mineralization in the Nanling region, South China: Metallogenic ages and corresponding geodynamic processes. Acta Petrol. Sin. 2007, 23, 2329–2338, (In Chinese with English Abstract).

106. Chen, J.; Wang, R.C.; Zhu, J.C.; Lu, J.J.; Ma, D.S. Multiple–aged granitoids and related tungsten–tin mineralization in the Nanling Range, South China. Sci. China Earth Sci. 2013, 56, 2045–2055.

5

MINERALIZATION OF THE TUWU PORPHYRY CU DEPOSIT IN EASTERN TIANSHAN, NW CHINA: INSIGHTS FROM IN SITU TRACE ELEMENTS OF CHLORITE AND PYRITE

Wei Tan[1,2], Qigui Mao[1], Mingjie Yu[1], Yan Sun[1] and Xiaoqiang Lv[1,2]

[1]Beijing Institute of Geology for Mineral Resources Co., Ltd., Beijing, China
[2]Sino-Zijin Resources Ltd., Beijing, China

ABSTRACT

The Tuwu porphyry copper deposit is located on the Dananhu-Haerlik island arc in eastern Tianshan, NW China. Based on geology, petrology, and in situ trace element studies of pyrite and chlorite, we redefined the characteristics of hydrothermal fluids and the following three mineralization stages: premineralization stage (stage Ⅰ), porphyry metallogenic stage (stage Ⅱ), and superimposed transformation stage (stage Ⅲ). Pyrite stage Ⅰ (Py-I) has the highest Co/Ni ratios, and the precipitation crystallization of chlorite (Chl-I_2) has the similar rare earth element distribution patterns with those of volcanic rocks Carboniferous Qieshan (CQ), indicating intense volcanic hydrothermal activity. The Co/Ni ratios of Py-II_1 and Py-II_2 (stage Ⅱ) tend to decrease over time. Moreover, the rare earth element (REE) distribution patterns of Chl-II have similar LREE enrichment, and the Eu anomalies in Chl-II_1, Chl-II_2, and Chl-II_3 range from positive to negative. The initial ore-forming fluid was mainly magmatic hydrothermal fluid, and with the late-stage addition of meteoric water and continuous sulfide precipitation, the trace element composition of the fluid changed, and the whole system became more oxidizing. Py-III (stage Ⅲ) has the lowest Co/Ni ratios, and the REE distribution pattern of Chl-III is characterized by LREE enrichment. Moreover, the Chl-III shows obvious shear deformation characteristics. The results indicate that the host rocks experienced intensely superimposed reformation.

By combining and integrating our results with the regional evolution processes in the eastern Tianshan, we propose that the Tuwu porphyry deposit has undergone magmatic hydrothermal and metamorphic hydrothermal processes. Volcanism (stage Ⅰ) provided the space and initial conditions for the emplacement of the metallogenic body. With the emplacement of the plagiogranite porphyry (stage Ⅱ), the main copper mineralization occurred in the porphyry and surrounding rocks. After porphyry mineralization (stage Ⅲ), regional ductile shearing and collisional compression led to a copper reaction, and its accumulation along the faults formed an ore shoot.

Keywords: *Porphyry deposits, Mineralization, Pyrite, Chlorite, LA-ICP-MS*

INTRODUCTION

The eastern Tianshan orogen, which represents an important part of the tectonic evolution of the Central Asian orogenic belt, has undergone a complicated evolutionary process of stretching, multiblock merging, subduction accretion, collisional orogenesis, and postcollision processes (Figure 1A) (Chen et al., 2007; 2012; Pirajno, 2009, 2013; Zhai et al., 2011). Along with the multiple processes of continental accretion, crust–mantle interactions, and transformation in the eastern Tianshan orogen, numerous polygenetic mineral deposits have been formed, and most have been superimposed by late complex ore-forming processes. These deposits include the Honghai (Huangtupo) Cu-Zn deposit (Deng et al., 2016; Mao et al., 2019), Xiaorequanzi Cu-Zn deposit (Long et al., 2019; Mao et al., 2020), Yamaisu Cu-Fe deposit, Hongyuntan Cu-Fe deposit (Sun et al., 2017; Han et al., 2018; Zhang et al., 2018), and Tuwu-Yandong Cu deposit (Wang Y. F. et al., 2016; Xiao et al., 2017). Research on superimposed metallogenic systems is very important for recognizing the regional metallogenic history and guiding mineral exploration activities in the eastern Tianshan.

The Tuwu porphyry copper deposit is located at the southern part of the Dananhu-Haerlik island arc belt (Figure 1B). Since the Tuwu copper deposit was discovered by the No. 1 geological party of the Xinjiang Bureau of Geological Exploration in 1994, the Yandong, Fuxing, Yanxi, Linglong, and Chihu copper deposits were consecutively discovered afterward. All of these deposits formed the Tuwu-Yandong porphyry copper metallogenic belt, which has attracted extensive attention. Studies have focused on its geological characteristics (Wang et al., 2001; Rui et al., 2001, 2002; Han et al., 2006; Shen et al., 2014b), metallogenic background, metallogenic characteristics, zoning classification (Shen et al., 2012; Pan et al., 2013; Xiao et al., 2018), geochemistry, geochronology (Rui et al., 2002; Zhang et al., 2006; Han et al., 2014; Shen et al., 2014a; Wang et al., 2014), metallogenic chronology (Rui et al., 2002; Wang et al., 2014; Xiao et al., 2017), and compositions and characteristics of ore-forming fluids (Han et al., 2006; Gao et al., 2015; Wang et al., 2015, 2017). We have found that there are many differences between the porphyry metallogenic system of the Tuwu-Yandong deposit and the classical porphyry metallogenic model. For example, 1) the boundaries of the potassic zone are not obvious, 2) the morphology of the ore body is tabular rather than equiaxial, and 3) the temperature of the ore-forming fluid is obviously lower than that of a typical porphyry copper deposit. These findings indicate that the Cenozoic porphyry mineralization model may not be strictly suitable for Paleozoic porphyry deposits which experienced a multistage structural evolution.

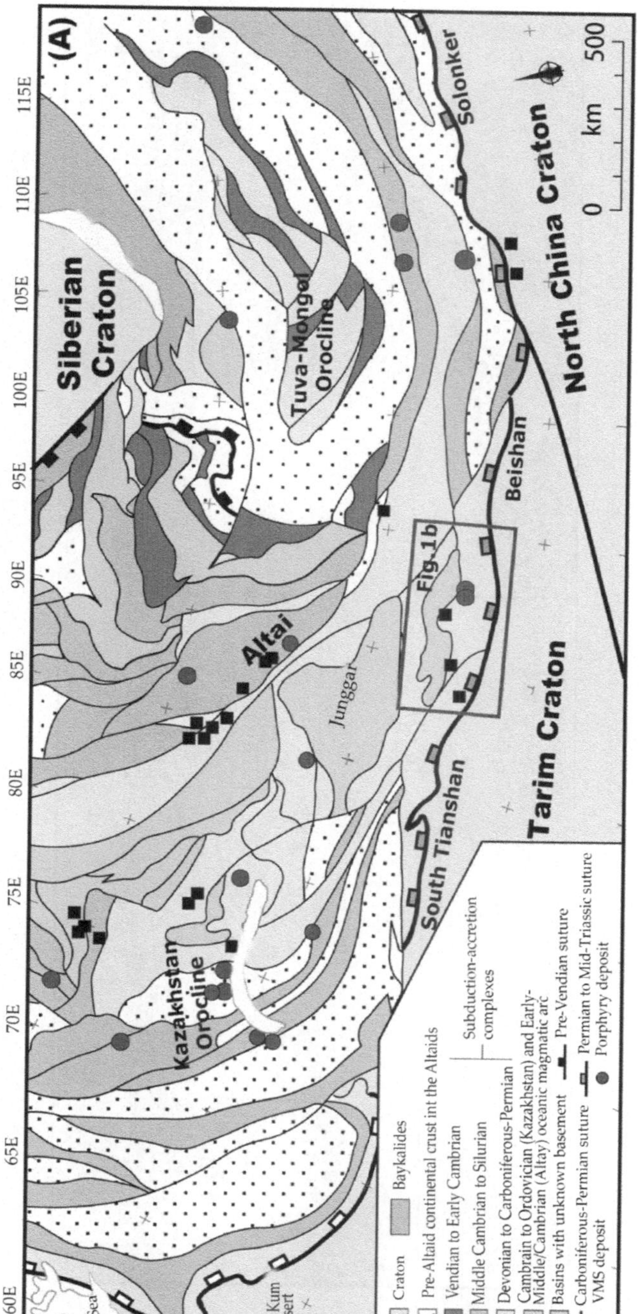
(A)
Siberian Craton
Tuva-Mongol Orocline
Altai
Junggar
Fig. 1b
Beishan
Solonker
North China Craton
Tarim Craton
South Tianshan
Kazakhstan Orocline
Aral Sea
Kizil Kum desert
0 km 500
45N
40N
60E
65E
70E
75E
80E
85E
90E
95E
100E
105E
110E
115E
Craton
Baykalides
Pre-Altaid continental crust int the Altaids
Vendian to Early Cambrian
Middle Cambrian to Silurian
Devonian to Carboniferous-Permian
Subduction-accretion complexes
Cambrain to Ordovician (Kazakhstan) and Early-Middle/Cambrian (Altay) oceanic magmatic arc
Basins with unknown basement
Pre-Vendian suture
Carboniferous-Permian suture
Permian to Mid-Triassic suture
VMS deposit
Porphyry deposit

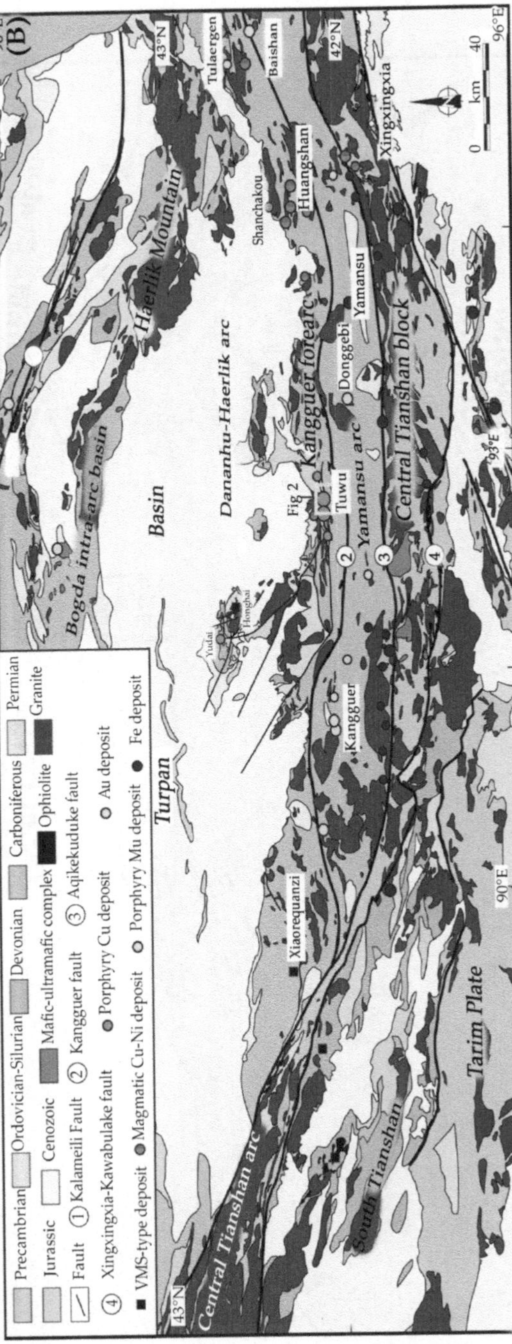

Figure 1. (A) Schematic tectonic map of Central Asia (Şengör et al., 1993; Xiao et al., 2015) showing the position of the eastern Tianshan. (B) Schematic geological map of the eastern Tianshan [modified after Mao et al. (2017) and Xiao et al. (2004)] showing the location of the Tuwu area. The major faults separating the southern Tianshan, Central Tianshan arc, Yamansu arc, and Dananhu-Haerlik arc.

Mineral microanalysis is an important method for defining ore-forming fluid characteristics, metallogenic physicochemical conditions, and ore-forming element migration and enrichment processes (Large et al., 2009; Zhou, et al., 2010; Ye et al., 2011; Zhang, 2011; Reich et al., 2013; Franchini et al., 2015; Zhang et al., 2020). The development of laser ablation inductively coupled plasma mass spectrometry (LA-ICP-MS) analysis technology has allowed for the measurement of most trace elements in sulfide and oxides. Pyrite is an important sulfide in many metallogenic systems, and its chemical composition is often used to trace the genesis of ore deposits and the metallogenic environment and conditions (Cook et al., 2009; Large et al., 2009, 2011; Reich et al., 2013; Franchini et al., 2015). Chlorite is a mineral that commonly forms during diagenesis, medium- to low-temperature metamorphism and hydrothermal alterations. Due to the variability in its structure and the nonstoichiometry of its composition, chlorite and its paragenetic mineral assemblages are often used to constrain the physicochemical conditions and determine the diagenetic and metallogenic environment (Davidson et al., 2001; Bortnikov et al., 2008; Inoue et al., 2010; Fadda et al., 2012; Wilkinson et al., 2015; David et al., 2016). This study used LA-ICP-MS to analyze the composition of trace elements in pyrite and chlorite in the main alteration zone of the Tuwu copper deposit, and the combination of detailed geological surveys and petrographical studies was used to study the genesis of ore minerals and the characteristics of trace elements and further reveal characteristics of hydrothermal fluids and mineralization processes of the Tuwu porphyry deposit.

REGIONAL GEOLOGICAL BACKGROUND

A review of the regional geology and the tectonic evolution processes of the accretionary orogenic belt from the Precambrian to Triassic are given in Xiao et al. (2004). From north to south, eastern Tianshan consists of the Dananhu-Haerlik arc, Kangguer Forearc, Yamansu arc, Central Tianshan arc, and the South Tianshan (Figure 1B).

The Dananhu-Haerlik Ordovician-Carboniferous island arc, as an important Ordovician to Permian metallogenic belt located between the Kalameili and the Kangguer faults (Figure 1B), includes the Paleozoic Dananhu-Haerlik arc, the Bogda Late Paleozoic intra-arc basin, and the Kanggur forearc (Xiao et al., 2004, 2015). The Dananhu-Haerlik arc consists of Ordovician to Permian tholeiite to calc-alkaline mafic-felsic volcanic rock, volcanoclastics, tuff, and flysch sediments (Ma et al., 1997; Li et al., 2004; Xiao et al., 2004; Hou et al., 2005; Tang et al., 2006; Mao et al., 2014b, Mao et al., 2016; Deng et al., 2016; Li et al., 2016). The Ordovician to Silurian marine volcanic and volcaniclastic rocks consist of tholeiite to calc-alkaline basalts, andesite-basalts, andesite, high-Mg andesite, and volcaniclastic rocks (Ma et al., 1997; Mao et al., 2014b, 2016; Deng et al., 2016; Li et al., 2016). The Devonian to Carboniferous rocks mainly consists of tholeiite to calc-alkaline basalts, andesite, dacite rhyolite, volcaniclastic rocks, and sedimentary (Ma et al., 1997; Li et al., 2004; Xiao et al., 2004; Hou et al., 2005; Tang et al., 2006; Mao et al., 2014b). The arc-related granitic intrusions are formed from Ordovician to Permian (Song et al., 2002; Li et al., 2004; Chen et al., 2005; Hou et al., 2005; Sun et al., 2005; Wu et al., 2006; Chao et al., 2006; Guo et al., 2006; Mao et al., 2010, 2019; Long et al., 2016; Zhang et al., 2017).

The Bogda intra-arc basin, which formed within the Dananhu-Haerlik arc from the Carboniferous to Early Permian, is composed of basalt, dacite, rhyolite, volcanoclastics, minor interbedded andesite, and sedimentary rocks. The mafic to granitic intrusions are mainly Early Permian (Gu et al., 2001a, Gu et al., 2001b; Li et al., 2004; Xiao et al., 2004; Li et al., 2006; Wang et al., 2010; Shu et al., 2011; Chen et al., 2013; Gao et al., 2014).

The Kanggur forearc located at the south margin of the Dananhu-Haerlik arc is made up of structurally dismembered Devonian-Carboniferous volcanic-sedimentary rocks and ophiolitic slices (Yang and Scott, 1996; Ma et al., 1997; Zhou et al., 2001; Xiao et al., 2004). The Permian granitoids are extensive, and a zone of earlier Permian mafic-ultramafic complexes is located along the Kangguer forearc terrane and stretches several hundred kilometers (Ma et al., 1997; Mao et al., 2002; Han C. M. et al., 2004; Xiao et al., 2004; Zhou et al., 2004; Qin et al., 2011). Multiple types of mineral deposits developed in the Dananhu-Haerlik arc, including porphyry Cu (Au/Mo) deposits (Rui et al., 2002; Chen et al., 2005; Wu et al., 2006; Zhang et al., 2006; Shen et al., 2014a; Wang et al., 2015; Mao et al., 2018), hydrothermal vein Cu (Au) deposits (Fang et al., 2002; Mao et al., 2014a; Yu et al., 2019), Cu-Ni sulfide deposits (Mao et al., 2002; Han B. F. et al., 2004; Xiao et al., 2004; Zhou et al., 2004; Mao et al., 2006; Han et al., 2010; Qin et al., 2011), VMS Cu-Zn-(Au/Ag) polymetallic deposits (Mao et al., 2010, 2019; Deng et al., 2016; Li et al., 2002), and Au deposits (Wang et al., 2006).

The Dananhu-Haerlik island arc, situated north of the Kanggur fault, comprises Paleozoic volcano-sedimentary rocks and Mesozoic to Cenozoic sedimentary rocks, intercalated with volcanic and pyroclastic rocks (Xiao et al., 2004; Hou et al., 2006; Qin et al., 2011). Paleozoic mafic–ultramafic complexes and granites intruded the sedimentary and volcanic rocks and are associated with some Cu and Au mineralization (Deng et al., 2016; Gao et al., 2015). The Carboniferous rocks are mostly foliated and mylonitized and influenced by ductile shearing processes. Carboniferous acidic igneous rocks and a Permian mafic–ultramafic complex are well developed in this belt and mainly occur in the form of stocks (Han et al., 2006; Pan et al., 2013; Shen et al., 2014b). Late Paleozoic mineral deposits are developed at the southern zone of the Dananhu-Haerlik island arc. These deposits include shear zone-type gold deposits, such as the Tianmu Au deposit and Hongtan Au deposit (Yao, 2006), Cu-Ni sulfide deposits, which are associated with Permian mafic-ultramafic rocks (Wang Y. L. et al., 2016), and Carboniferous-Permian porphyry Cu (Mo) deposits, such as the Tuwu-Yandong Cu deposits (Pirajno, 2013; Shen et al., 2014b) and Sanchakou Cu deposit (Qin et al., 2009).

GEOLOGY OF THE TUWU DEPOSITS

The Tuwu copper deposit is hosted in the Carboniferous Qieshan (CQ) group, diorite porphyrite and plagiogranite porphyry at the northeastern part of the Qieshan anticline (Figure 2A). According to previous studies (Han et al., 2006; Shen et al., 2012; Pan et al., 2013; Shen et al., 2014b), the CQ group can be divided into five lithologic members (Figure 2A). The first lithologic member (CQ^1) is volcanic effusive facies distributed symmetrically around the volcanic crater. The lithology is mainly basalt, followed by andesite, tuff, dacite, breccia lava, volcanic breccia, and other volcanics. Additionally, dioritic porphyrite, which is the main ore-hosting rock, is located at the contact between the basalt in CQ^1 and the pebble-rich lithic sandstone in CQ^3. The second lithologic member (CQ^2), which is an explosive volcanic facies, is located in the north-central part of the mining area near the center of volcanism.

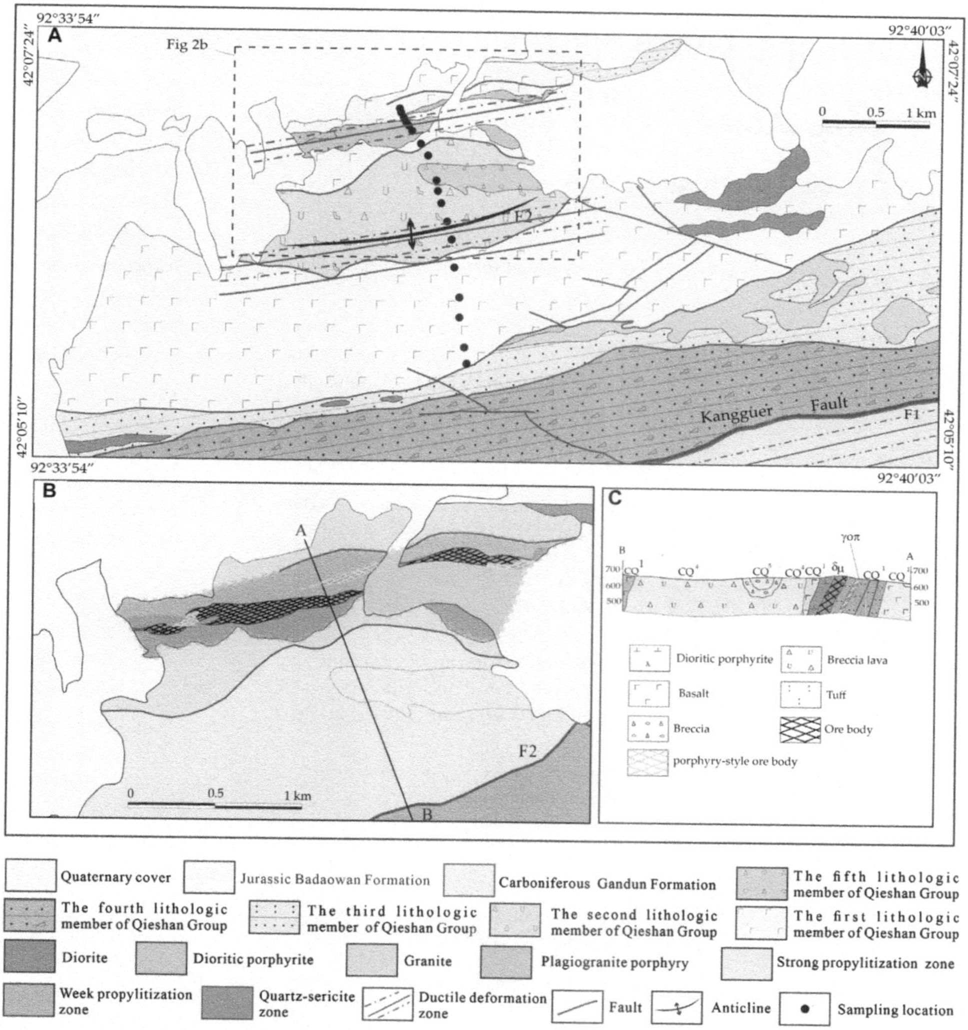

Figure 2. (A) Geological map of the porphyry Cu deposit in eastern Tianshan, Xinjiang, NW China (revision according to Ma et al., 2003; Pan et al., 2013). (B) Alteration zonation map of the Tuwu copper ore district. (C) Profile map of the Tuwu copper ore district.

The lithology is agglomerate breccia lava and minor andesite and basalt. The second lithologic group can be divided into three lithologic members. The third lithologic member (CQ^3) is distributed in a nearly east–west direction in the northern part of the mining area. The lithology is mainly volcanic-sedimentary facies. CQ^3 is a pebble-rich lithic sandstone that contains basalt and andesite layers. The fourth lithologic member (CQ^4) is distributed in the southern part of the mining area in a nearly east–west direction along the northern side of the

Kangguer fault. The lithology is mainly pebble-rich sandstone and minor basalt and volcanic breccia at the top. The fifth lithologic member (CQ^5) is in angular unconformable contact with the underlying breccia lava of CQ^2.

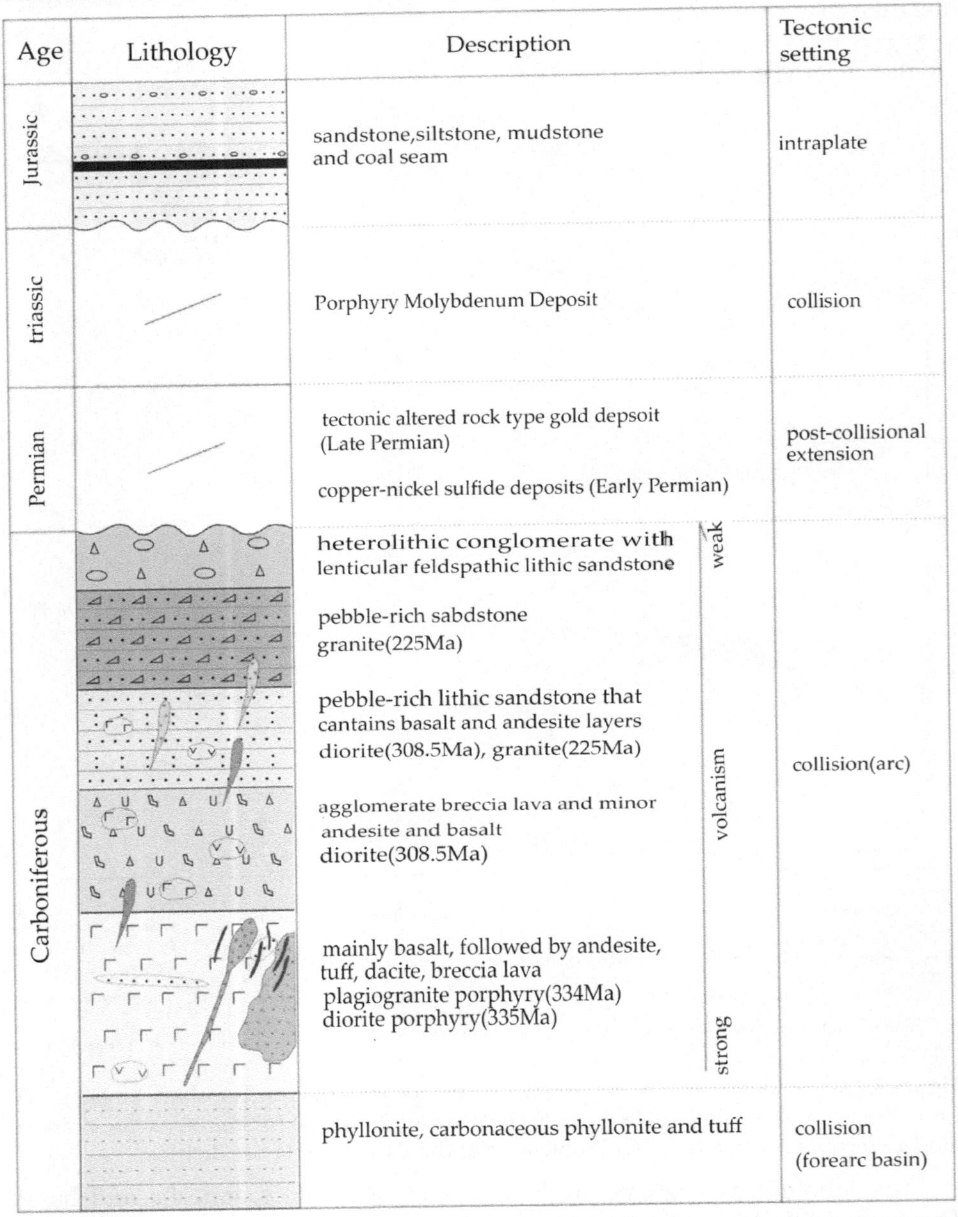

Figure 3. The chronostratigraphic chart of geological events in Tuwu copper ore district.

The lithology is mainly heterolithic conglomerate with lenticular feldspathic lithic sandstone. The volcanic rocks of the CQ Group constitute a relatively complete volcanic eruption cycle and resulted from Carboniferous magmatism in this area. The CQ group strikes nearly east–west and dips to the south at angles of approximately 40–65°. The upper Carboniferous Gandun formation (C_2g) is distributed in the southern part of the mining area in a nearly east–west direction along the southern side of the Kangguer fault. The lithology is phyllonite, carbonaceous phyllonite and tuff, which have been altered by regional ductile shear deformation metamorphism. The Jurassic Xishanyao formation (Jx) is covered in cap rock and exposed in the northern part of the mining area. The lithology is mainly sandstone, siltstone, mudstone, and conglomerate.

The magmatism was developed quite intensely in the Tuwu-Yandong ore district. The intrusive rocks mainly include granite, plagiogranite porphyry, diorite and diorite porphyry. The Carboniferous metallogenic plagiogranite porphyry of the Tuwu deposit occurs as a small stock (Cheng et al., 2005; Wang et al., 2014; Xiao et al., 2018). The plagiogranite porphyry cut through the dioritic porphyrite and intrude into the contemporaneous volcanism-related fault with an E-W trend (Pan et al., 2013). The dioritic porphyrite is the main ore-hosting rock and belongs to subvolcanic rocks (Hou et al., 2005, 2006; Zhao et al., 2014). It cuts through the basalt in CQ^1 and the pebble-rich lithic sandstone in CQ^3, thus forming an east–west–trending stock or dike (Shen et al., 2012) (Figure 3).

The Tuwu deposit is located in the northeastern part of the Qieshan anticlinorium (Figure 2A). The direction of fracture development is mainly east–west, followed by northwest–southeast. The copper ore bodies mainly occur in the dilated portion of the east–west fault. The intensity of the ductile deformation zone in the Tuwu-Yandong mining area is stronger in the south and weaker in the north. The southern part of the ore field features a continuous strong ductile zone, whereas the northern part is a weakly ductile deformation zone composed of several discretely distributed small lenticular and zonal ductile deformation zones. The copper ore body is distributed in an en echelon shape both in cross section and in map view. In general, the ore-controlling brittle–ductile shear zone is a ENE-trending tectonic belt that dips southward (Ma et al., 2003). The exploration data show that the schistosity obviously cuts the strata and ore body, indicating that the ore body suffered late brittle–ductile deformation. However, distinguishing radial and annular contemporaneous faults from the effect of brittle-ductile deformation is difficult.

The ore mainly has medium- to fine-grained hypidiomorphic–xenomorphic granular textures (Figure 4; Figure 5). The ore structure mainly includes veins and dissemination. The metallic minerals are mainly chalcopyrite, followed by pyrite and bornite, with minor chalcocite, covellite, magnetite, hematite, and molybdenite. Pyrite is mainly developed at the top and bottom of the ore body and it is negatively related to copper minerals. The gangue minerals are mainly quartz, sericite and chlorite, followed by biotite, epidote, actinolite, gypsum, feldspar, and calcite. The supergene minerals are malachite, atacamite, and limonite.

Propylitization is extensively developed in the volcanic-pyroclastic rocks of the CQ Group. According to systematic alteration mapping and a comprehensive laboratory study, we find that weak propylitization related to volcanic hydrothermal activity in the late volcanic activity is concentrated in the range of 1.2–3 km from the ore body while strong propylitization obviously superimposed by late hydrothermal activity is concentrated in the range of 0–1.2 km. The alteration intensity differs significantly between these zones, with the F2 fault as the dividing line. The altered minerals associated with weak propylitization are chlorite, epidote, quartz, zoisite, calcite, and pyrite, and the rock color is dark green in general (Figures 4A,F; Figure 5A). Chlorite is mainly in the form of interstitial mafic minerals developed between feldspars. The formation of chlorite is related to the late hydrothermal alteration of dark minerals, such as amphibole. Epidote mainly exists discontinuously in the form of veins in faulted structures. Pyrite grains with particle sizes of 0.01–0.05 nm exhibit a stellate distribution.

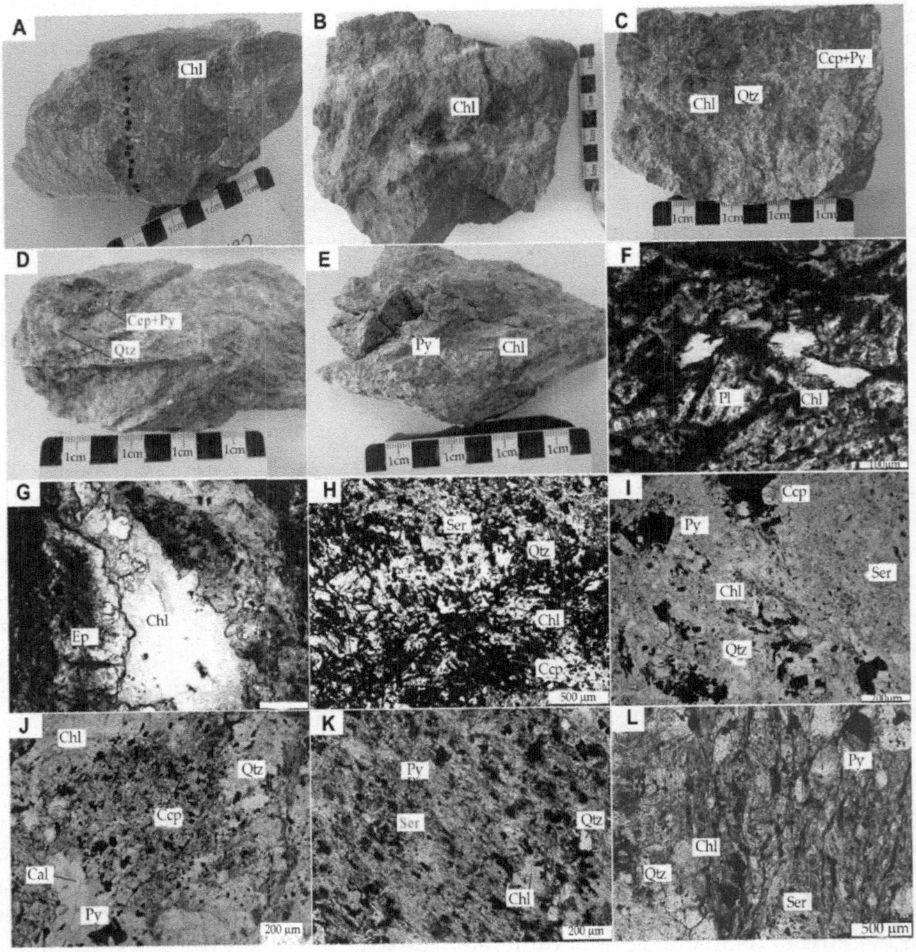

Figure 4. Photomicrographs of the chlorite and specimens of typical ore samples from different mineralization stages of Tuwu porphyry Cu deposit in eastern Tianshan, Xinjiang, NW China. (A) Weakly propylitized basalt; (B) strongly propylitized basalt; (C) pyrite + chalcopyrite + chlorite + sericite in diorite porphyrite; and (D) pyrite + chalcopyrite + silicification development in plagiogranite porphyry; (E) mylonitized diorite porphyrite; (f) weak chlorite development in basalt, Chl-I; (G) strong epidote and chlorite development in basalt, Chl-II_1; (H) pyritization, sericitization, silicification, and chloritization of plagiogranite porphyry, Chl-II_2; (I) sericite, silicification, and chlorite development in plagiogranite porphyry, Chl-II_2; (J) chlorite, epidote, pyrite, and calcite development in plagiogranite porphyry,

Chl-II_3; (K) chlorite, chalcopyrite, and sericite in diorite porphyry (chlorite, sericite, and pyrite grains have an oriented arrangement) Chl-III; and (L) chlorite, chalcopyrite, and sericite in diorite porphyry rock with an augen structure, Chl-III. Ccp, chalcopyrite; Py, pyrite; Pl, plagioclase; Chl, chlorite; Ep, epidote; Qtz, quartz; Cal, calcite; Ser, sericite.

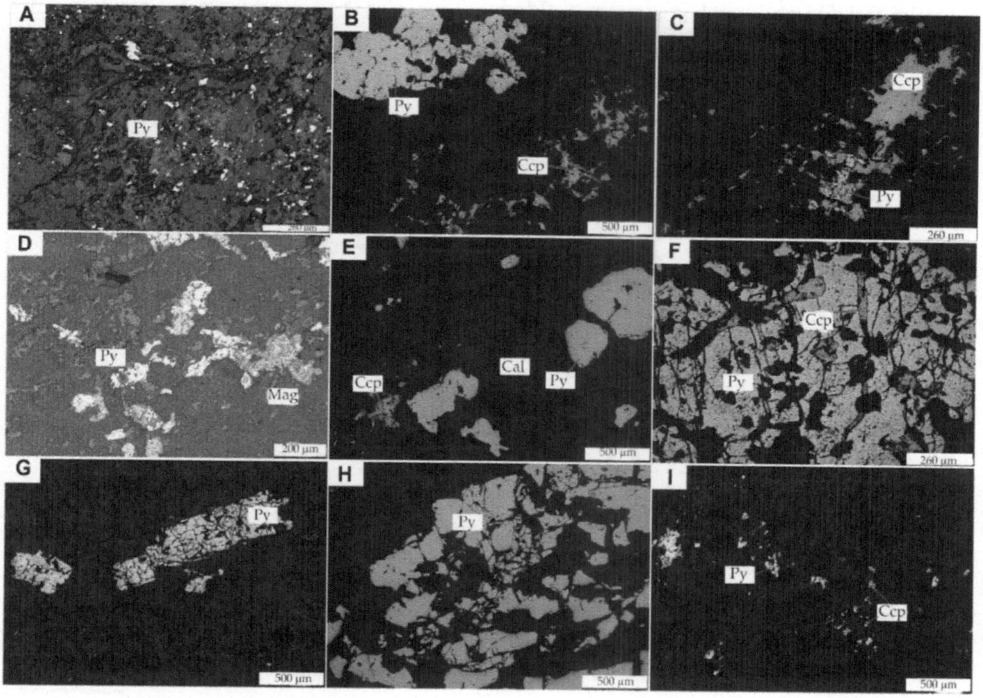

Figure 5. Photomicrographs of pyrite from different mineralization stages in the Tuwu porphyry Cu deposit in eastern Tianshan, Xinjiang, NW China. (A) Disseminated pyrite in basalt with weak propylitization, Py-I; (B) Pyrite associated with chalcopyrite in plagiogranite porphyry, Py-II_1; (C) pyrite growing along chalcopyrite fissure in plagiogranite porphyry, Py-II_1; (D) pyrite associated with magnetite in plagiogranite porphyry, Py-II_2; (E) veinlets of pyrite, chalcopyrite, calcite, and silicification, in plagiogranite porphyry, Py-II_2; (F) chalcopyrite growing along pyrite fissure in plagiogranite porphyry in diorite porphyry, Py-III; (G) chlorite, sericite, and pyrite in diorite porphyrite, pyrite with crushed structure, Py-III; (H) sericite and pyrite in plagiogranite porphyry, pyrite with crushed structure, Py-III; and (I) chlorite and sericite in diorite porphyrite with pyrite and chalcopyrite vein. Ccp, chalcopyrite; Py, pyrite; Mag, magnetite; Qtz, quartz; Cal, calcite; Ser, sericite.

The concentric alteration zones from the interior outward in succession are the potassic zone, the quartz–sericite zone, and the propylitic zone.

The potassic zone is mainly distributed in plagiogranite porphyry and dioritic porphyrite. The alteration mineral assemblage of disseminated and vein types includes potassium feldspar, biotite, and magnetite. The fine-grained and scaly biotite originated from biotitization of hornblende and recrystallization of primary biotite. The hydrothermal biotite is poorly preserved and often altered into chlorite, ilmenite, and rutile. Identifying the potassic zone is difficult because it has been altered by later hydrothermal activity. Consequently, the boundary of the potassic zone is not clear. Pyritization, sericitization, and silicification are well developed in plagiogranite porphyry (Figures 4C,H). Plagioclase has been almost completely altered to sericite and fine-grained quartz. Pyrite is also widespread (Figures 5B,C). The alteration assemblage is mainly sericite, pyrite, quartz, chalcopyrite, and calcite with net veins and disseminated forms. The dioritic porphyrite is locally developed and sericitized. Plagioclase has been altered to sericite and minor epidote and albite. The alteration assemblage is sericite, chlorite, pyrite, epidote, and quartz (Figures 4D,I,J; Figures 5D–F).

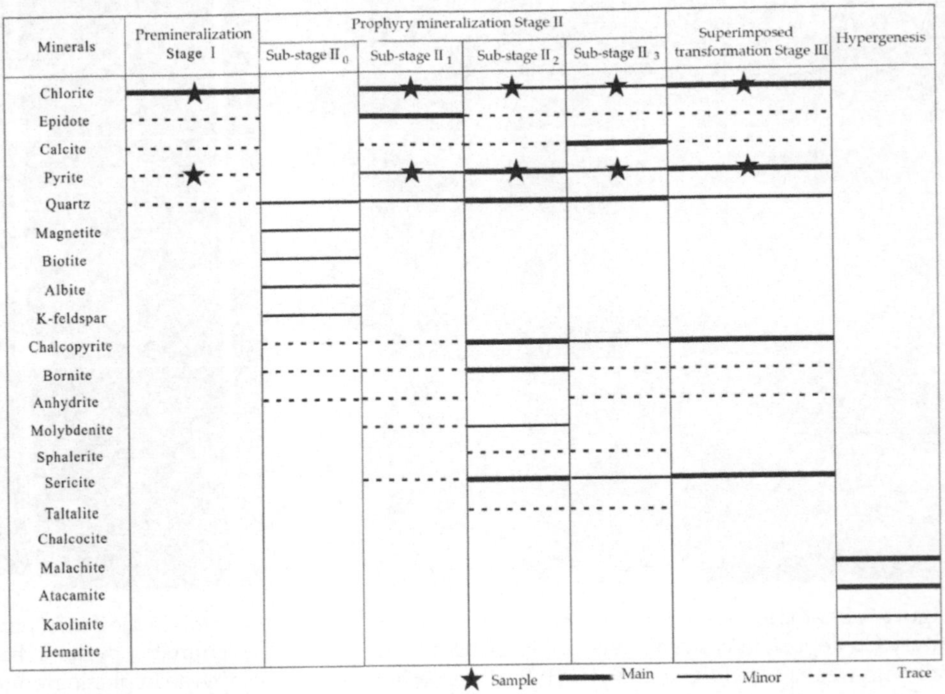

Figure 6. Alteration and mineral time–space diagram for the paragenetic history of the Tuwu porphyry Cu deposit in eastern Tianshan, Xinjiang, NW China.

The strongly propylitized rocks are bright green in color and exhibit strong silicification. The alteration assemblage is mainly epidote, chlorite, quartz, calcite, and pyrite. Quartz, epidote, and chlorite are mainly distributed in dense disseminated and vein forms. The disseminated quartz is mainly associated with propylitic alteration, while the vein quartz that appears to have a hydrothermal filling genesis is an obviously younger overprinting feature. Radial and granular epidote aggregates can be seen in some veins. The basalt near the ore is altered and contains minor pyrite, epidote, chlorite, quartz, calcite, and chalcopyrite veins (Figures 4B,G).

Alterations related to ductile shear deformation are developed in the local area and characterized by intense sericitization and chloritization. The altered minerals show obvious crumpled and directional arrangement. The rocks have augen and cataclastic structures. The plagioclase in the dioritic porphyrite has been altered to scaly sericite and fine-grained quartz. Chlorite is present in the form of blocks and stockwork and well developed in the local ductile shear zone (Figures 4E,K,L). The altered mineral assemblage is sericite, chlorite, and quartz. The metallic minerals are mainly chalcopyrite and pyrite, which are distributed in disseminated and vein types (Figures 5G,H,I). The alteration zone is related to the rich ore body in this area.

The main supergene alteration mineral assemblage is kaolinite, atacamite, malachite, and hematite.

Previous researchers have conducted a large amount of research on the alteration zonation and metallogenic timing of the Tuwu copper deposit based on the association of alteration minerals, mineral relationships, and crosscutting relationships (Shen et al., 2012; Pan et al., 2013; Wang Y. F. et al., 2016). After extensive research on the petrography of the main rock types exposed in the Tuwu mining area (Figures 2B,C), this study identified four stages of hydrothermal activity, namely, alterations related to volcanic activity (stage Ⅰ), alterations related to porphyry mineralization (stage Ⅱ), alterations related to ductile shear deformation (stage Ⅲ), and alterations related to supergenesis (stage Ⅳ) (Figure 6).

SAMPLING AND ANALYSIS METHODS

This study carried out detailed geological observations and profile surveys of outcrops, troughs, open mining pits and drilling samples from the Tuwu mining area. Typical samples of various alteration zones were collected from the area, and thin sections with a thickness of 0.8 mm were prepared. Two to three typical samples were selected from each alteration zone, and 19 samples were selected for testing and analysis. The mineral types of the tested samples are mainly pyrite Py-I, chlorite Chl-I_1 (metasomatic crystallization), and Chl-I_2 (precipitation crystallization) related to volcanic activity, pyrite Py-II, and chlorite Chl-II related to porphyry mineralization, pyrite Py-III, and chlorite Chl-III related to ductile shear deformation.

The in situ analysis of chlorite composition was completed in the LA-ICP-MS laboratory of the State Key Laboratory of Ore Deposit Geochemistry, Institute of Geochemistry, Chinese Academy of Sciences. The LA-ICP-MS instrument is composed of an Agilent 7700cs quadrupole ICP-MS instrument connected to a GeoLasPro 193 nm LA system. The operating conditions of the instrument were as follows: the laser beam width was 44 nm, the frequency of the laser pulse was 6 Hz, and the pulse energy was 0.032–0.105. The analysis time was 90 s (30 s for the blank test and 60 s for the sample analysis). During testing, the LA spot coincided with the electron microprobe analysis spot. The internal standards NIST610 and NIST612 were analyzed once every nine chlorite samples to monitor the instrument deviation. The international standard samples ML3B-G and BCR-2G were used as unknown samples to monitor the data quality. Data processing was calibrated by Igor-pro software. The Si content was analyzed by electron microprobe as the internal standard to calibrate the Si content determined by LA-ICP-MS. The analysis results of GOR128-G, BIR-1G, BHVO-2G, ML3B-G, and BCR-2G were in good agreement with their reference values. The analytical error of most elements was less than 10%, although Cr, Y, and Zr had errors of 10–15%.

The analysis of pyrite was carried out in the State Key Laboratory of Ore Deposit Geochemistry, Institute of Geochemistry, Chinese Academy of Sciences. The standard sample of the LA system was NWR-213, and the experimental analysis instrument was an Agilent 7700x type quadrupole mass spectrometer. Helium (480 ml/min) was used as a transport medium for the ablated material, and it was mixed evenly with argon (900 ml/min) after leaving the ablation compartment. The mixture then entered the ICP-MS instrument, where the element contents were measured. The analysis time was 90 s (30 s for blank test and 60 s for sample analysis). The laser beam width was 44 nm, the frequency of the laser pulse was 10 Hz, and the pulse energy was 5 J/cm^2. The analysis method was a multiple external standard–no internal standard method, and the external standards were GSE-1G, MASS, and Py. Data processing was performed with ICPMSDataCal software.

TEST RESULTS

In Situ Trace Element Composition of Pyrite

The results of the in situ trace element composition of pyrite in the Tuwu copper mining area are shown in Table 1. Overall, the contents of chalcophile elements (Cu, Zn, Pb, As, Se, Te, Ge, and Ag), siderophile elements (Cr, Co, and Ni), V, and Ba are all above the limit of detection. The contents of Sn, Cd, Sb, Au, Bi, and other elements were mostly lower than or near the detection limit (Figures 7, 8).

The results showed that Py-I has the highest total amount of trace elements (average 4,351 ppm) and highest Co/Ni ratio (average 33). Py-I is rich in Co, Ni, and As and poor in Cu and Pb, and the Zn, Au, and Ag contents are below or near the detection limit. The total amount of trace elements of Py-II$_1$ (average 4,049 ppm) and Py-II$_2$ (average 544 ppm) displayed an obvious decrease with the occurrence of porphyry mineralization. Moreover, the Co/Ni ratio of Py-II$_1$ (average 3) and Py-II$_2$ (average 0.53) also showed a certain decrease. Py-II$_1$ and Py-II$_2$ are rich in Co, Ni, and Cu, and poor in As, Pb, Zn, Au, and Ag.

Table 1. In situ trace element compositions of pyrite for different metallogenic stages of the Tuwu porphyry Cu deposit in eastern Tianshan, Xinjiang, NW China (ug/g)

Trace element (ppm)	Py-I	Py-II$_1$	Py-II$_2$	Py-III
	n = 3 (total)	*n* = 11 (total)	*n* = 10 (total)	*n* = 6 (total)
V	0.68–1.11; Av = 0.87 (*n* = 3)	0.06–32.17; Av = 7.34 (n = 11)	0.03–0.29; Av = 0.09 (*n* = 10)	Mostly bdl; 0.53–0.53; Av = 0.53
Cr	3.02–9.07; Av = 5.65 (*n* = 3)	1.21–27.41; Av = 7.28 (*n* = 11)	1.08–18.45; Av = 6.35 (*n* = 10)	2.25–20.96; Av = 7.56
Co	288.5–6,417.17; Av = 3,027.09 (*n* = 3)	115.7–6,339.2; Av = 2,260.87 (*n* = 11)	5.73–318.06; Av = 91.40 (*n* = 10)	0.65–415.9; Av = 75.21
Ni	25.38–105.89; Av = 71.9 (*n* = 3)	27.58–3,396.17; Av = 1,635.83 (*n* = 11)	11.19–1,623.02; Av = 325.23 (*n* = 10)	47.38–1,298.36; Av = 509.45
Cu	8.95–50.68; Av = 23.16 (*n* = 3)	0.66–336.25; Av = 86.28 (*n* = 10)	2.96–43.6; Av = 16.38 (*n* = 10)	5.94–4,133.8; Av = 950.33
Zn	0.1–0.7; Av = 0.4 (*n* = 2)	0.13–10.42; Av = 3.13 (*n* = 6)	0.29–2.61; Av = 1.17 (*n* = 8)	Mostly bdl; 8.84–8.84; Av = 8.84
Ge	7.87–11.69; Av = 9.15 (*n* = 3)	6.75–12.88; Av = 9.28 (*n* = 11)	7.23–13.85; Av = 9.24 (*n* = 10)	6.17–10.85; Av = 8.38
As	50.73–2,933.98; Av = 1,072.03 (*n* = 3)	0.27–21.14; Av = 5.03 (*n* = 10)	0.01–58.85; Av = 16.95 (*n* = 8)	10.1–1817.01; Av = 913.56
Se	87.3–167.07; Av = 133.16 (*n* = 3)	21.15–82.59; Av = 39.62 (*n* = 11)	41.95–174.06; Av = 71.49 (*n* = 10)	19.65–137.42; Av = 68.10
Ag	0.02–0.16; Av = 0.11 (*n* = 3)	0.02–0.57; Av = 0.13 (*n* = 9)	0.02–1.4; Av = 0.37 (*n* = 9)	0.02–20.22; Av = 5.21
Cd	0.22–0.33; Av = 0.27 (*n* = 3)	0.1–0.44; Av = 0.24 (*n* = 6)	0.01–0.49; Av = 0.17 (*n* = 7)	0.39–1.41; Av = 0.69
Sn	bdl	0.01–3.63; Av = 0.75 (*n* = 8)	0.01–0.49; Av = 0.26 (*n* = 6)	0.12–1.07; Av = 0.46
Sb	0.22–0.22; Av = 0.22 (*n* = 2)	0.01–0.37; Av = 0.09 (*n* = 6)	0.06–4.8; Av = 0.73 (*n* = 8)	0.08–14.34; Av = 3.66
Te	0.43–8.24; Av = 3.42 (*n* = 3)	0.02–0.74; Av = 0.42 (*n* = 8)	0.01–3.05; Av = 0.68 (*n* = 8)	0.61–12.15; Av = 5.57
Ba	0.05–0.42; Av = 0.19 (*n* = 3)	0.03–8.12; Av = 2.18 (*n* = 8)	0.06–0.91; Av = 0.25 (*n* = 9)	0.15–0.3; Av = 0.21
Au	0.03–0.03; Av = 0.03 (*n* = 2)	0.01–0.04; Av = 0.02 (*n* = 7)	0.01–0.1; Av = 0.03 (*n* = 7)	0.01–4.91; Av = 1.24
Pb	1.93–6.86; Av = 3.93 (*n* = 3)	0.02–3.41; Av = 1.16 (*n* = 11)	0.04–27.01; Av = 5.99 (*n* = 10)	0.66–164.32; Av = 48.61
Bi	0.09–0.19; Av = 0.12 (*n* = 3)	0.02–0.79; Av = 0.34 (*n* = 8)	0.01–6.2; Av = 1.28 (*n* = 8)	0.01–13.82; Av = 3.39
Co/Ni	11.37–60.60; Av = 33.37 (*n* = 3)	0.09–9.94; Av = 2.71 (*n* = 11)	0.12–2.02; Av = 0.53 (*n* = 10)	0.003–0.57; Av = 0.16
Total content	478.09–9,692.23; Av = 4,351.34 (*n* = 3)	594–7,839.08; Av = 4,049.19 (*n* = 11)	125.64–2028.18; Av = 543.68 (*n* = 10)	132.52–6,401.02; Av = 1982.41

Note: Av stands for average and bdl stands for below the detection limit.

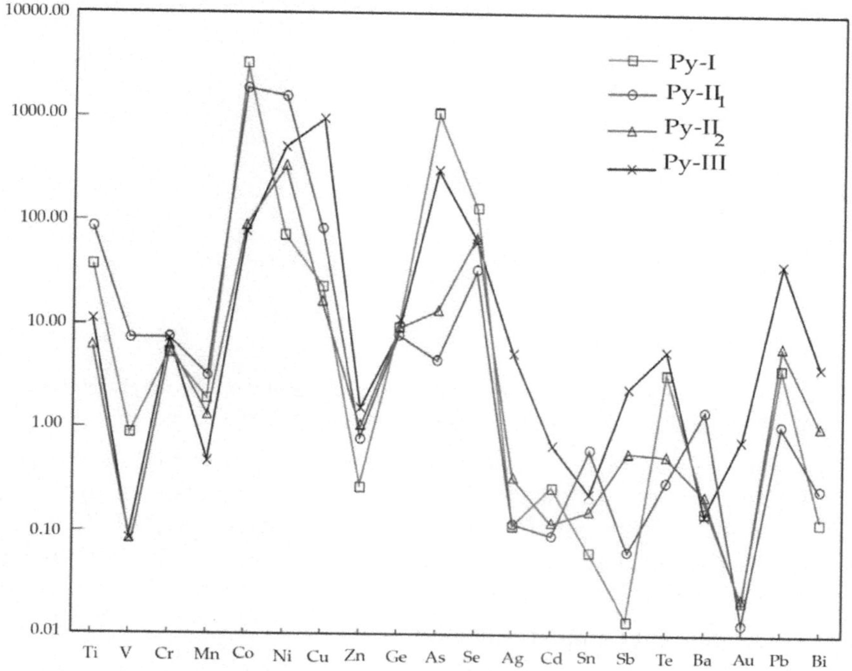

Figure 7. The average trace element composition of pyrite in different mineralization stages in the Tuwu porphyry Cu deposit in eastern Tianshan, Xinjiang, NW China.

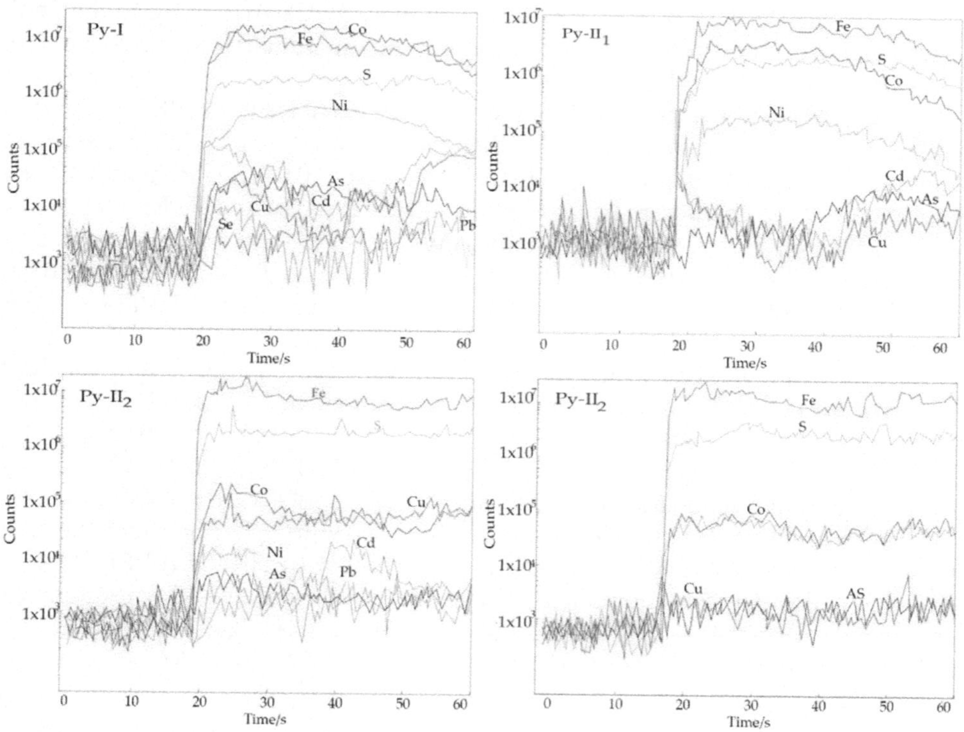

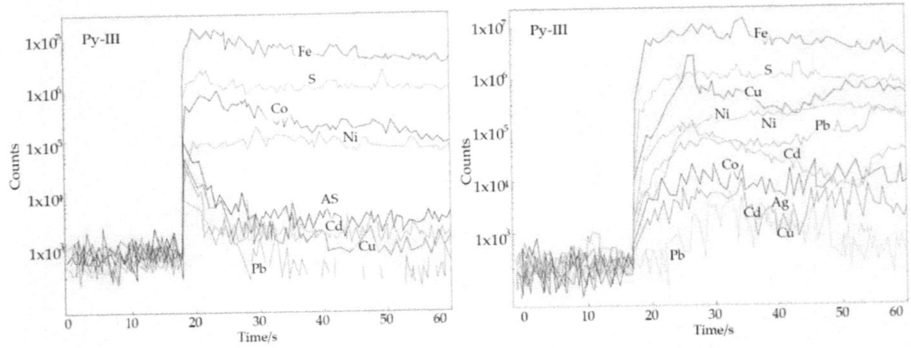

Figure 8. LA-ICP-MS ablation curves of pyrite in different mineralization stages of the Tuwu porphyry Cu deposit in eastern Tianshan, Xinjiang, NW China.

The total amount of trace elements in Py-III ranges from 133 to 6,401 ppm (average 1982 ppm). Py-III has the lowest Co/Ni ratio (average 0.16). This phase features obviously higher mineralization element contents than Py-I, Py-II_1, and Py-II_2, and the contents of elements, such as Co, Ni, Cu, Pb, As, Pb, Zn, Au, and Ag, have extremely uneven distributions (Table 1).

In Situ Trace Element Composition of Chlorite

The results of the in situ trace element composition of chlorite in the Tuwu copper mining area are shown in Table 2. To ensure that the mineral in the experiment is chlorite, the chlorite are filtered based on the standard of ω (Na_2O + K_2O + CaO) < 0.5% (Foster, 1962; Zang and Fyfe, 1995). According to the Fe-Si classification of chlorite, the chlorite samples are mainly clinochlore and pyrochlore (unpublished data).

In this study, the data show that the total rare earth element (REE) content of chlorite from different stages is above the detection limit, thus, REEs are selected for research. The content of REEs in Chl-I_1 ranges from 0.03 to 17.62 ppm, and the REE patterns are characterized by obvious enrichment of light REEs over heavy REEs (LREE/HREE = 0.12–5.95), negative Eu anomalies (δEu = 0.223–0.866), and no Ce anomalies (δCe = 0.53–2.11). The results of (La/Sm)n (0.14–1.33) and (Gd/Yb) n (0.95–4.70) indicate that the internal fractionation of LREEs and HREEs is weak (Figure 9A).

The content of REEs in Chl-I_2 ranges from 0.02 to 0.71 ppm, and the REE patterns are mainly characterized by the enrichment of LREEs (LREE/HREE = 0.18–3.91), positive Eu anomalies (δEu = 1.0–3.78), and weak Ce anomalies (δCe = 0.54–2.01). The results of (La/Sm)n (0.07–1.81) and (Gd/Yb)n (0.22–5.61) indicate that the internal fractionation of LREEs and HREEs is relatively weak (Figure 9B).

The content of REEs in Chl-II_1 ranges from 0.056 to 0.77 ppm, and the REE patterns are characterized by the relative enrichment of HREEs (LREE/HREE = 0.07–0.59), positive Eu anomalies (δEu = 1.08–1.62), and weakly negative Ce anomalies (δCe = 0.43–1.52). The results of (La/Sm) n (0.11–0.497) and (Gd/Yb) n (1.1–12.02) indicate that the internal fractionation of LREEs and HREEs is stronger (Figure 9C).

Table 2. In situ trace element composition and characteristic REE values of chlorite for different metallogenic stages of the Tuwu porphyry Cu deposit in eastern Tianshan, Xinjiang, NW China (μg/g)

Trace element	Chl-I$_1$	Chl-I$_2$	Chl-II$_1$	Chl-II$_2$	Chl-II$_3$	Chl-III
	n = 17 (total)	*n* = 15 (total)	*n* = 18 (total)	*n* = 10 (total)	*n* = 13 (total)	*n* = 6 (total)
Na	76.46–1,050.02; Av = 498.1 (*n* = 17)	76.75–1,413.61; Av = 562.06 (*n* = 15)	269.87–1,274.41; Av = 637.36 (*n* = 18)	24.09–1,346.2; Av = 512.28 (*n* = 10)	20.07–864.1; Av = 202.91 (*n* = 13)	42–96.48; Av = 66.36 (*n* = 6)
K	66.64–660.13; Av = 252.8 (*n* = 17)	79.72–596.30; Av = 274.13 (*n* = 15)	27.8–398.35; Av = 180.18 (*n* = 18)	12.58–804.58; Av = 198.13 (*n* = 10)	11.51–2,164.91; Av = 337.54 (*n* = 13)	63.82–483.25; Av = 144.47 (*n* = 6)
Ca	101.30–2,525.45; Av = 828.3 (*n* = 17)	186.82–2,104.71; Av = 581.74 (*n* = 15)	219.25–1,550.54; Av = 455.25 (*n* = 18)	68.92–569.66; Av = 246.97 (*n* = 8)	10.25–849.34; Av = 203.19 (*n* = 13)	140–553.39; Av = 310.82 (*n* = 6)
Ti	12.22–159.33; Av = 29.64 (*n* = 17)	10.96–573.20; Av = 71.45 (*n* = 15)	11.07–176.09; Av = 37.54 (*n* = 18)	19–486.8; Av = 165.92 (*n* = 10)	21.64–1,055.86; Av = 334.43 (*n* = 13)	177.9–572.4; Av = 299.68 (*n* = 6)
V	44.63–210.72; Av = 115.75 (*n* = 17)	35.38–188.80; Av = 99.53 (*n* = 15)	38.78–181.6; Av = 103.62 (*n* = 18)	28.67–933.29; Av = 229.91 (*n* = 10)	127.41–780.29; Av = 363.64 (*n* = 13)	153.74–254.63; Av = 214.70 (*n* = 6)
Mn	1946.41–3,466.17; Av = 2,275.41 (*n* = 17)	1,376.12–2,973.11; Av = 2099.84 (*n* = 15)	2,368.23–3,662.86; Av = 2,711.42 (*n* = 18)	380.31–2048.76; Av = 1,505.32 (*n* = 10)	348.55–2,715.22; Av = 1,139.89 (*n* = 13)	1951.67–2,853.68; Av = 2,247.81 (*n* = 6)
Ni	153.95–387.96; Av = 309.24 (*n* = 17)	147.21–397.03; Av = 301.59 (*n* = 15)	251.76–371.36; Av = 329.01 (*n* = 18)	80.73–477.43; Av = 325.56 (*n* = 10)	117.74–382.95; Av = 249.15 (*n* = 13)	262.68–313.44; Av = 284.64 (*n* = 6)
Cu	1.81–345.91; Av = 98.07 (*n* = 17)	3.11–215.25; Av = 106.21 (*n* = 15)	0.02–388.38; Av = 68.54 (*n* = 18)	0.05–102.99; Av = 37.96 (*n* = 7)	0.01–1.03; Av = 0.288 (*n* = 10)	0.04–1.2; Av = 0.64 (*n* = 6)
Zn	190.46–411.24; Av = 257.89 (*n* = 17)	185.61–293.06; Av = 256.57 (*n* = 15)	157.33–597.87; Av = 251.98 (*n* = 18)	40.31–395.41; Av = 192.57 (*n* = 10)	57.63–154.24; Av = 106.17 (*n* = 13)	154.04–188.16; Av = 168.21 (*n* = 6)
Ge	0.18–0.76; Av = 0.413 (*n* = 6)	0.02–1.11; Av = 0.506 (*n* = 7)	0.02–1.05; Av = 0.576 (*n* = 15)	2.98–3.66; Av = 3.33 (*n* = 3)	2.02–4.15; Av = 3 (*n* = 5)	bdl
Sr	1.79–29.16; Av = 10.94 (*n* = 17)	0.86–7.49; Av = 4.79 (*n* = 15)	1.83–7.1; Av = 4.45 (*n* = 18)	0.51–58.69; Av = 12.79 (*n* = 10)	0.37–121.19; Av = 17.18 (*n* = 13)	0.71–2.02; Av = 1.42 (*n* = 6)
Mo	0.01–0.159; Av = 0.088 (*n* = 16)	0.13–0.023; Av = 0.071 (*n* = 15)	0.024–0.13; Av = 0.082 (*n* = 18)	0.01–0.09; Av = 0.056 (*n* = 9)	0.012–0.099; Av = 0.045 (*n* = 12)	0.04–0.34; Av = 0.12 (*n* = 6)
Cd	0.053–0.107; Av = 0.062 (*n* = 6)	0.02–0.09; Av = 0.052 (*n* = 5)	0.0064–0.065; Av = 0.037 (*n* = 14)	0.07–0.11; Av = 0.09 (*n* = 2)	0.022–0.15; Av = 0.075 (*n* = 5)	bdl
In	0.002–0.025; Av = 0.012 (*n* = 6)	0.005–0.02; Av = 0.011 (*n* = 6)	0.0012–0.056; Av = 0.022 (*n* = 15)	0.007–0.013; Av = 0.01 (*n* = 3)	0.004–0.027; Av = 0.013 (*n* = 6)	bdl
Sb	0.002–0.456; Av = 0.085 (*n* = 14)	0.0004–0.62; Av = 0.193 (*n* = 12)	0.001–0.0186; Av = 0.067 (*n* = 14)	0.001–0.20; Av = 0.068 (*n* = 10)	0.003–0.0302; Av = 0.072 (*n* = 10)	0.11–0.25; Av = 0.2 (*n* = 3)
La	0.001–2.132; Av = 0.208 (*n* = 16)	0.0009–0.043; Av = 0.012 (*n* = 15)	0.0018–0.014; Av = 0.008 (*n* = 16)	0.001–0.01; Av = 0.008 (*n* = 10)	0.002–0.045; Av = 0.019 (*n* = 11)	0.0004–0.0758; Av = 0.03 (*n* = 6)
Ce	0.003–6.37; Av = 0.566 (*n* = 17)	0.0008–0.18; Av = 0.037 (*n* = 12)	0.0038–0.067; Av = 0.027 (*n* = 18)	0.002–0.1; Av = 0.03 (*n* = 10)	0.006–0.22; Av = 0.056 (*n* = 12)	0.011–0.244; Av = 0.099 (*n* = 6)

Pr	0.001–0.93; Av = 0.101 (*n* = 14)	0.0007–0.02; Av = 0.059 (*n* = 15)	0.0008–0.016; Av = 0.007 (*n* = 15)	0.001–0.01; Av = 0.005 (*n* = 10)	0.005–0.03; Av = 0.011 (*n* = 10)	0.001–0.052; Av = 0.022 (*n* = 5)
Nd	0.008–4.56; Av = 0.506 (*n* = 14)	0.0056–0.048; Av = 0.026 (*n* = 12)	0.0086–0.11; Av = 0.038 (*n* = 16)	0.008–0.12; Av = 0.047 (*n* = 8)	0.008–0.14; Av = 0.07 (*n* = 10)	0.029–0.351; Av = 0.144 (*n* = 5)
Sm	0.007–1.06; Av = 0.155 (*n* = 12)	0.0005–0.026; Av = 0.013 (*n* = 11)	0.0039–0.04; Av = 0.024 (*n* = 16)	0.006–0.05; Av = 0.02 (*n* = 8)	0.006–0.1; Av = 0.034 (*n* = 11)	0.007–0.106; Av = 0.053 (*n* = 5)
Eu	0.002–0.085; Av = 0.017 (*n* = 13)	0.002–0.021; Av = 0.01 (*n* = 12)	0.0033–0.02; Av = 0.011 (*n* = 17)	0.004–0.042; Av = 0.017 (*n* = 8)	0.002–0.04; Av = 0.015 (*n* = 11)	0.011–0.047; Av = 0.019 (*n* = 5)
Gd	0.007–0.847; Av = 0.153 (*n* = 12)	0.0022–0.064; Av = 0.026 (*n* = 15)	0.0093–0.084; Av = 0.048 (*n* = 18)	0.0001–0.125; Av = 0.045 (*n* = 7)	0.005–0.18; Av = 0.065 (*n* = 11)	0.008–0.1071; Av = 0.048 (*n* = 5)
Tb	0.0009–0.12; Av = 0.017 (*n* = 16)	0.0009–0.014; Av = 0.006 (*n* = 11)	0.0028–0.024; Av = 0.01 (*n* = 16)	0.0008–0.045; Av = 0.016 (*n* = 9)	0.0003–0.04; Av = 0.015 (*n* = 11)	0.0001–0.025; Av = 0.0068 (*n* = 6)
Dy	0.0007–0.672; Av = 0.115 (*n* = 14)	0.0008–0.088; Av = 0.044 (*n* = 12)	0.039–0.11; Av = 0.078 (*n* = 17)	0.003–0.454; Av = 0.11 (*n* = 7)	0.001–0.43; Av = 0.116 (*n* = 13)	0.016–0.109; Av = 0.054 (*n* = 50
Ho	0.002–0.139; Av = 0.028 (*n* = 15)	0.0001–0.035; Av = 0.01 (*n* = 14)	0.0085–0.042; Av = 0.021 (*n* = 16)	0.001–0.134; Av = 0.027 (*n* = 8)	0.009–0.12; Av = 0.031 (*n* = 13)	0.0076–0.033; Av = 0.013 (*n* = 5)
Er	0.002–0.43; Av = 0.085 (*n* = 16)	0.0028–0.10; Av = 0.037 (*n* = 13)	0.0058–0.145; Av = 0.071 (*n* = 18)	0.005–0.796; Av = 0.191 (*n* = 10)	0.005–0.69; Av = 0.17 (*n* = 11)	0.006–0.069; Av = 0.03 (*n* = 6)
Tm	0.0003–0.045; Av = 0.012 (*n* = 16)	0.0008–0.017; Av = 0.007 (*n* = 9)	0.0018–0.015; Av = 0.01 (*n* = 18)	0.0008–0.13; Av = 0.027 (*n* = 7)	0.0003–0.124; Av = 0.029 (*n* = 11)	0.001–0.009; Av = 0.004 (*n* = 6)
Yb	0.005–0.25; Av = 0.073 (*n* = 17)	0.0059–0.21; Av = 0.054 (*n* = 12)	0.0236–0.269; Av = 0.087 (*n* = 18)	0.0001–1.175; Av = 0.206 (*n* = 9)	0.009–1.094; Av = 0.217 (=13)	0.006–0.027; Av = 0.018 (*n* = 5)
Lu	0.001–0.034; Av = 0.011 (*n* = 14)	0.0008–0.029; Av = 0.01 (*n* = 10)	0.0007–0.037; Av = 0.013 (*n* = 18)	0.002–0.23; Av = 0.048 (*n* = 7)	0.0009–0.198; Av = 0.042 (*n* = 12)	0.0003–0.01; Av = 0.0004 (*n* = 6)
Tl	0.0008–0.011; Av = 0.006 (*n* = 5)	0.0009–0.01; Av = 0.006 (*n* = 6)	0.0003–0.0089; Av = 0.0045 (*n* = 10)	0.005–0.05; Av = 0.02 (*n* = 3)	0.004–0.125; Av = 0.029 (*n* = 6)	bdl
ΣREE	0.03–17.62; Av = 1.8	0.02–0.71; Av = 0.248	0.056–0.77; Av = 0.425	0.02–3.42; Av = 0.66	0.05–3.34; Av = 0.808	0.04–1.27; Av = 0.49
LREE/HREE	0.12–5.95; Av = 1.32	0.18–3.91; Av = 0.96	0.07–0.59; Av = 0.296	0.11–1.19; Av = 0.64	0.08–0.85; Av = 0.55	1.26–3.17; Av = 1.75
(La/Sm) n	0.14–1.33; Av = 0.70	0.07–1.81; Av = 0.64	0.11–0.497; Av = 0.268	0.07–0.71; Av = 0.298	0.124–0.78; Av = 0.37	0.2–0.88; Av = 0.52
(Gd/Yb) n	0.95–4.70; Av = 1.90	0.22–5.61; Av = 2.28	1.1–12.02; Av = 3.33	0.502–11.60; Av = 3.28	0.51–7.38; Av = 2.99	1.8–8.08; Av = 4.84
(La/Yb) n	0.066–6.16; Av = 1.09	0.09–1.09; Av = 0.31	0.01–0.09; Av = 0.06	0.01–8.11; Av = 1.05	0.02–1.93; Av = 0.26	0.57–12.08; Av = 3.89
δEu	0.223–0.866; Av = 0.55	1–3.78; Av = 1.83	1.08–1.62; Av = 1.15	1.02–2.52; Av = 1.36	0.385–1.49; Av = 0.87	0.8–4.27; Av = 1.61
δCe	0.53–2.11; Av = 0.915	0.54–2.01; Av = 0.99	0.43–1.52; Av = 1.03	0.52–1.95; Av = 1.24	0.64–1.5; Av = 1.06	0.73–3.92; Av = 1.53

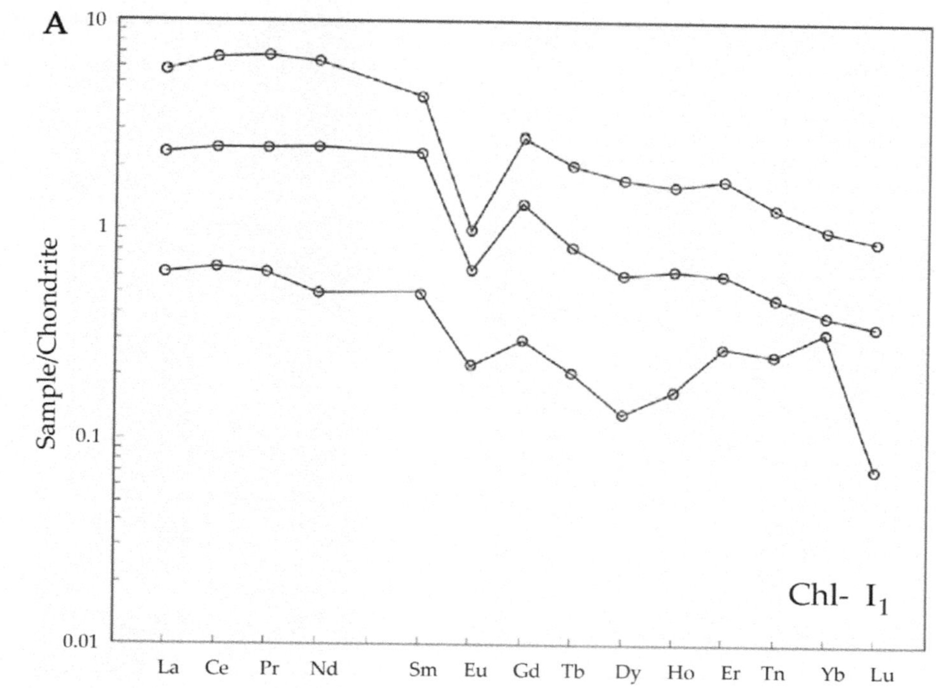
A
10
1
0.1
0.01
Sample/Chondrite
Chl- I_1
La Ce Pr Nd Sm Eu Gd Tb Dy Ho Er Tn Yb Lu

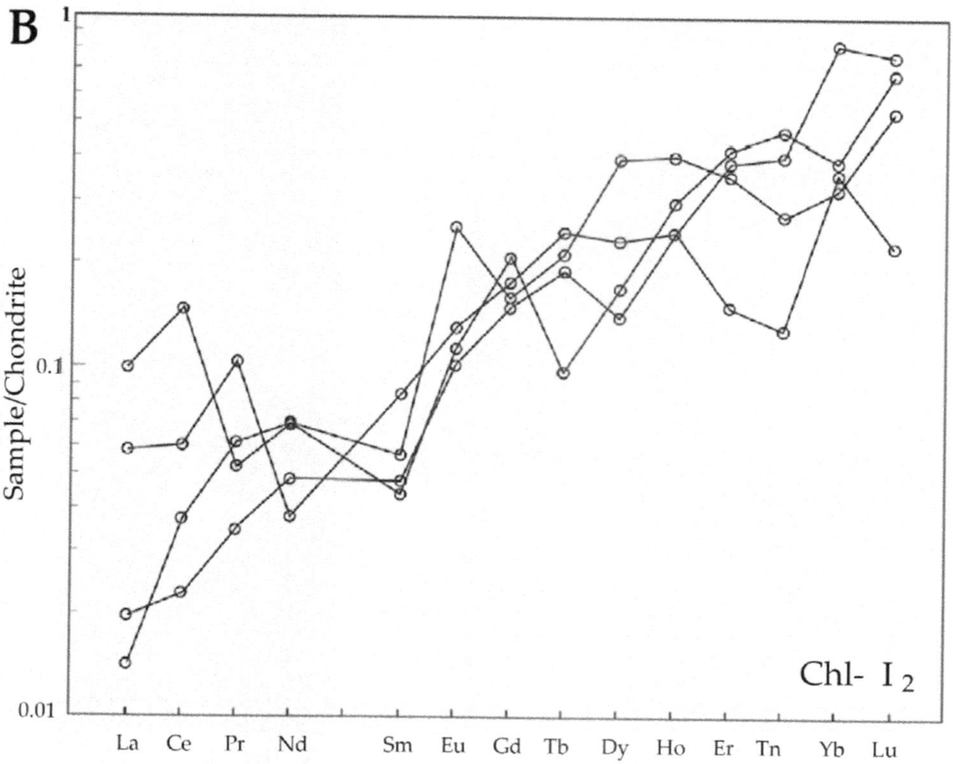
B
1
0.1
0.01
Sample/Chondrite
Chl- I_2
La Ce Pr Nd Sm Eu Gd Tb Dy Ho Er Tn Yb Lu

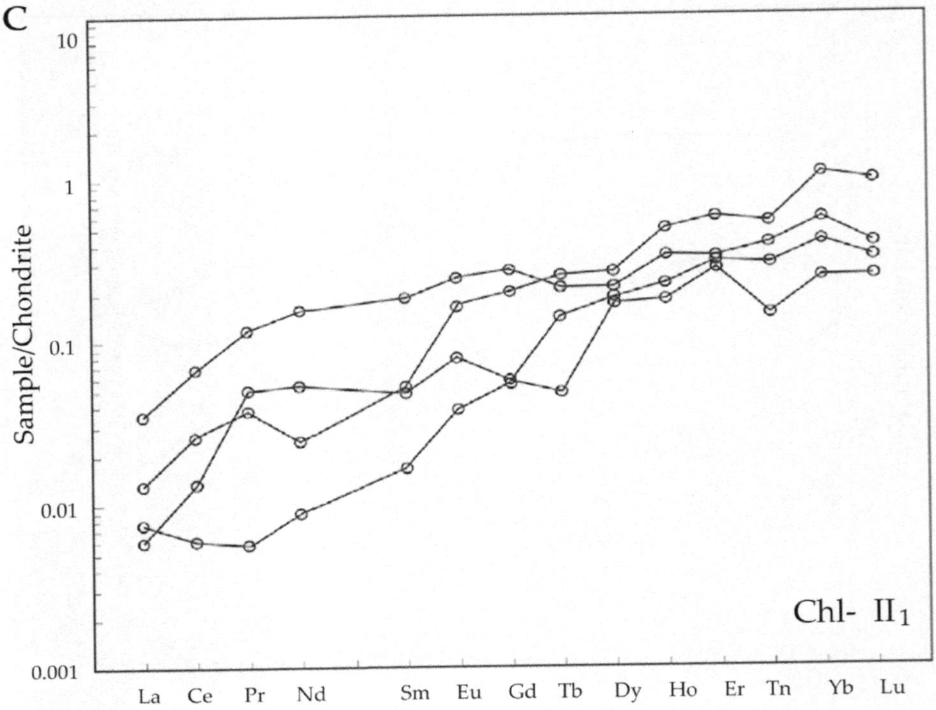
C
10
1
0.1
0.01
0.001
Sample/Chondrite
Chl- II_1
La Ce Pr Nd Sm Eu Gd Tb Dy Ho Er Tn Yb Lu

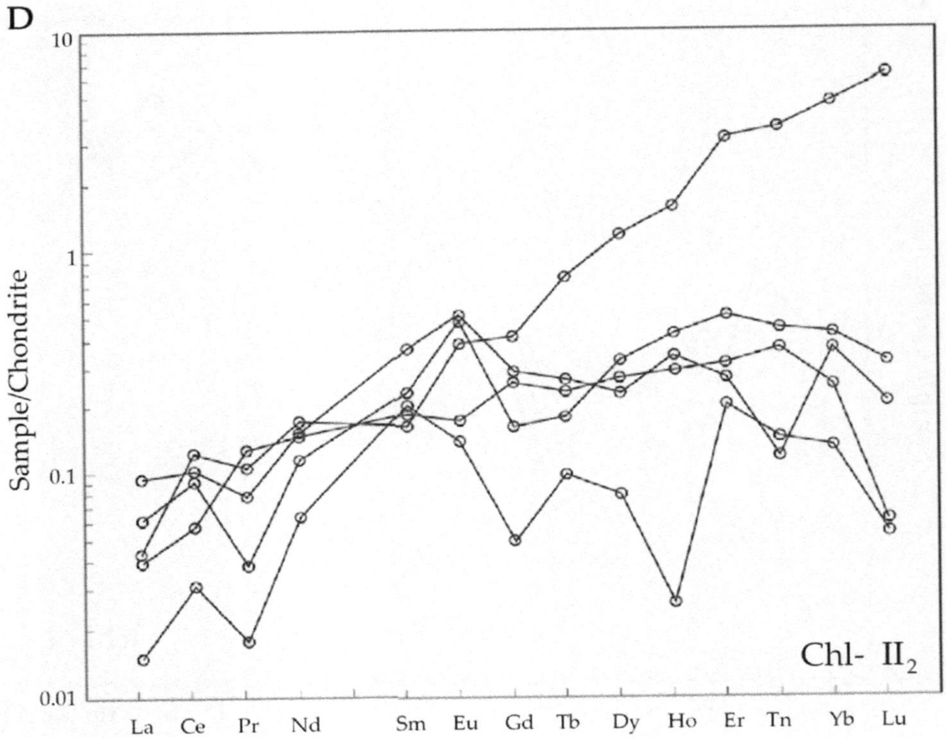
D
10
1
0.1
0.01
Sample/Chondrite
Chl- II_2
La Ce Pr Nd Sm Eu Gd Tb Dy Ho Er Tn Yb Lu

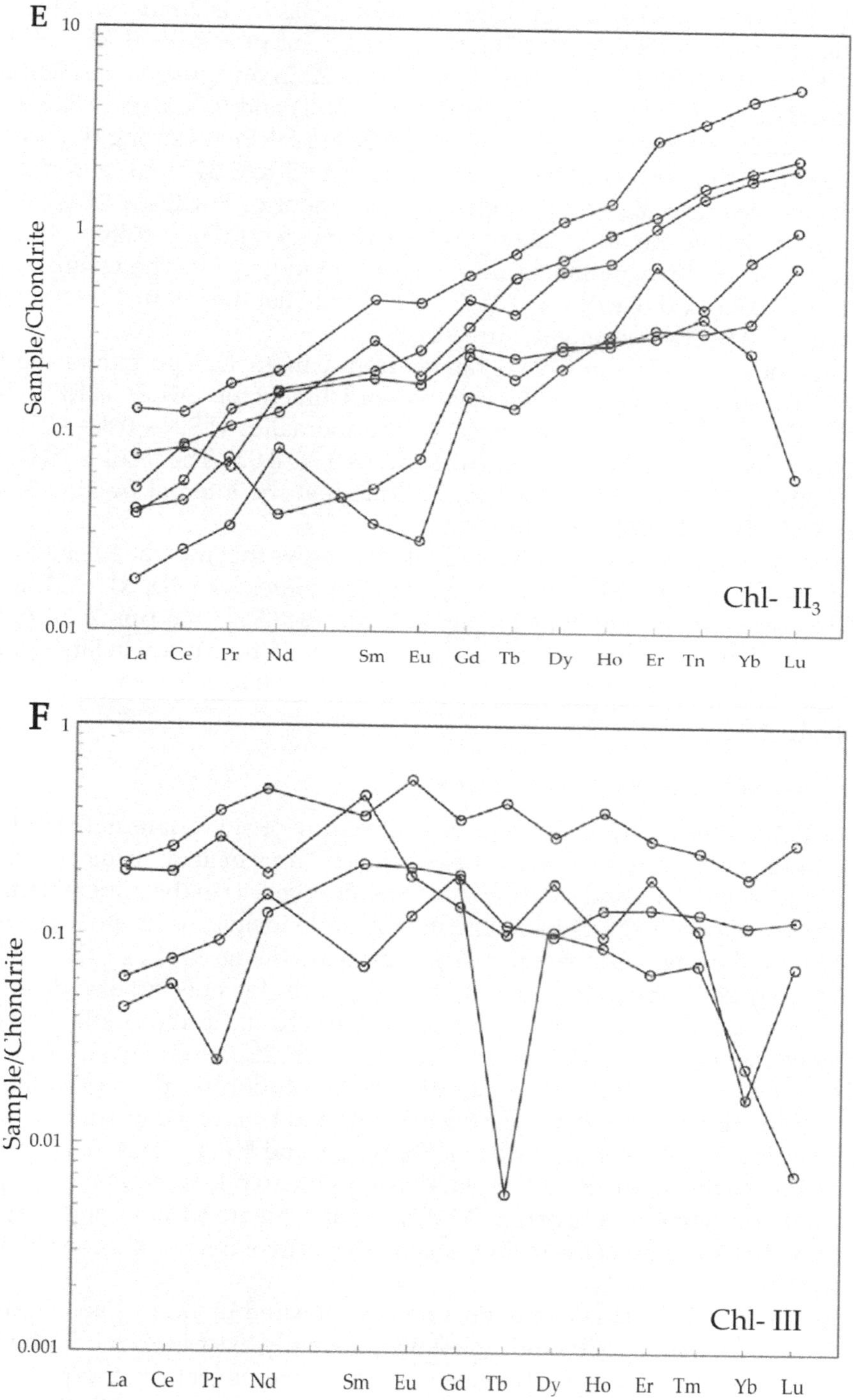

Figure 9. Chondrite-normalized rare earth element patterns of chlorite from different mineralization stages of the Tuwu porphyry Cu deposit in eastern Tianshan, Xinjiang, NW China (normalization values from Talor and McLennan, 1985).

The content of REEs in Chl-II_2 ranges from 0.02 to 3.42ppm, and the REE patterns are characterized by the relative enrichment of HREEs (LREE/HREE = 0.11–1.19), weak positive Eu anomalies (δEu = 1.02–2.52), and positive Ce anomalies (δCe = 0.52–1.95). The results of (La/Sm)n (0.07–0.71) and (Gd/Yb)n (0.502–11.60) indicate that the internal fractionation of LREEs and HREEs is strong (Figure 9D).

The content of REEs in Chl-II_3 ranges from 0.05 to 3.34 ppm, and the REE patterns are characterized by the relative enrichment of HREEs (LREE/HREE = 0.08–0.85), negative Eu to weakly positive anomalies (δEu = 0.385–1.49), and negative Ce to positive anomalies (δCe = 0.64–1.5) in general. The results of (La/Sm)n (0.124–0.78) and (Gd/Yb)n (0.51–7.38) indicate that the internal fractionation of LREEs and HREEs is strong (Figure 9E).

The content of REEs in Chl-III ranges from 0.04 to 1.27 ppm, and the REE patterns are characterized by the relative enrichment of LREEs (LREE/HREE = 1.26–3.17), positive to weakly negative Eu anomalies (δEu = 0.80–4.27), and positive to weakly negative Ce anomalies (δCe = 0.73–3.92). The results of (La/Sm)n (0.20–0.88) and (Gd/Yb)n (1.80–8.08) indicate that the internal fractionation of LREEs and HREEs is strong (Figure 9F).

The in situ trace element analysis of chlorite shows that the trace element with the highest content in Chl-I ~ Chl-III is Mn. The contents of Ge, Mo, Cd, In, Sb, and REEs are usually less than 1 ppm, and Mn (348.55–3,466 ppm), Ni (80.73–477.43 ppm), Cu (0.01–388.38 ppm), and Sr (0.37–121.19 ppm) vary in large ranges. The contents of V, Ni, Ti, and Zn mostly range from 100 to 500 ppm.

DISCUSSION

Occurrence of Trace Elements in Pyrite

The distribution of trace elements of pyrite is obviously various in the different mineralization stages in the Tuwu copper deposit. Three main occurrence forms of trace elements in pyrite and other sulfides are observed: 1) in the mineral lattice in the form of isomorphism, 2) in visible micron-scale inclusions of other minerals, and 3) as invisible nanometer-scale mineral particles (Leng, 2017). LA-ICP-MS can not only accurately obtain the content of in situ trace elements but also determine the occurrence forms of elements in minerals according to the spatial variation trend of elements with depth during LA (Zhou et al., 2010; Keith et al., 2016). The trend of each element in the LA-ICP-MS time resolution profile shows that the signals of Sb, Se, Co, and Ni are generally stable and basically consistent with the signal distribution of Fe, indicating that Sb, Se, Co, and Ni may exist mainly in the form of isomorphism. Co and Ni replace Fe in pyrite, while Se and Sb can replace S. The ablation signal curves of Cu, Pb, Zn, Cu, and Au are variable and irregular, indicating that these elements mainly exist in the form of microinclusions in pyrite (Figure 8).

Moreover, most curves of Cu, Pb, and Ag appeared in local peaks (Figure 8), which indicates that Cu, Pb, and Ag elements mainly exist in pyrite in the form of microinclusions (chalcopyrite, galena, and native silver). In the correlation diagram (Figure 10), Cu shows a significant positive correlation with Pb and Ag. The degree of correlation was better in Py-II_1 and Py-II_2 than in Py-I and Py-III because Pb and Ag replacements of Fe are difficult by isomorphism, while Cu is known to occur as a substitution for Fe in many Fe-bearing sulfides (i.e., pyrite and pyrrhotite) and silicates (i.e., biotite and hornblende), indicating that more Cu exists in the form of isomorphism in Py-I and Py-III than in Py-II_1 and Py-II_2. A weak positive correlation occurs between Cu and As in pyrite. The signal distribution of As is consistent with that of Fe.

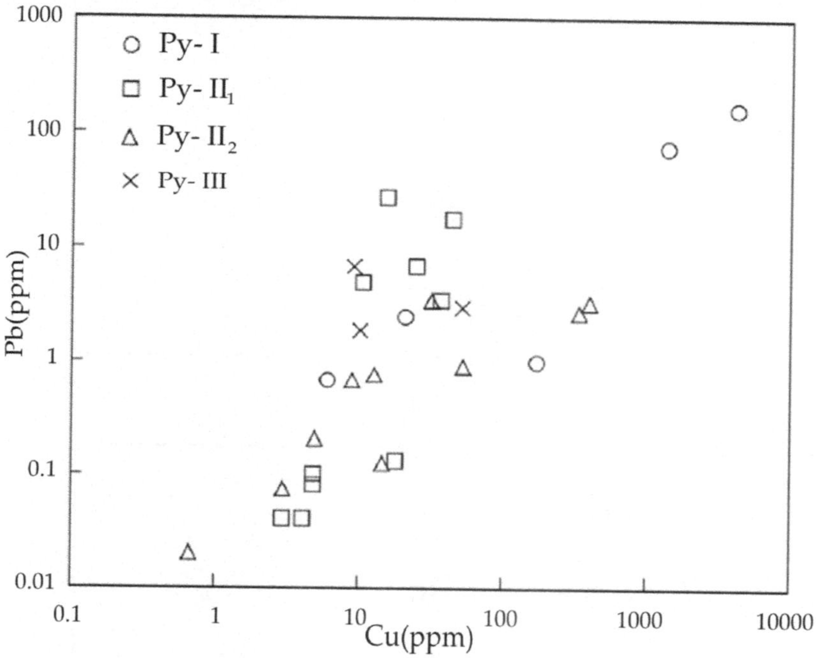

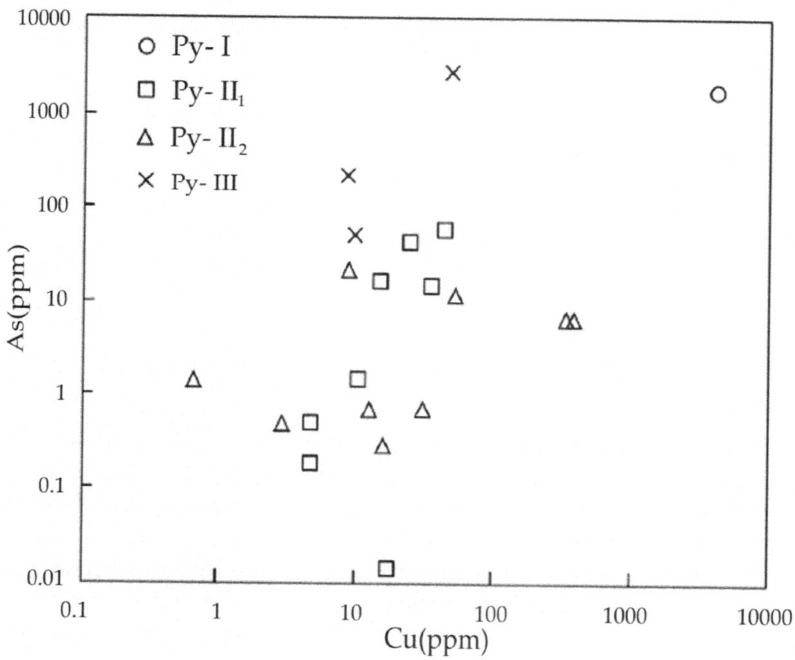

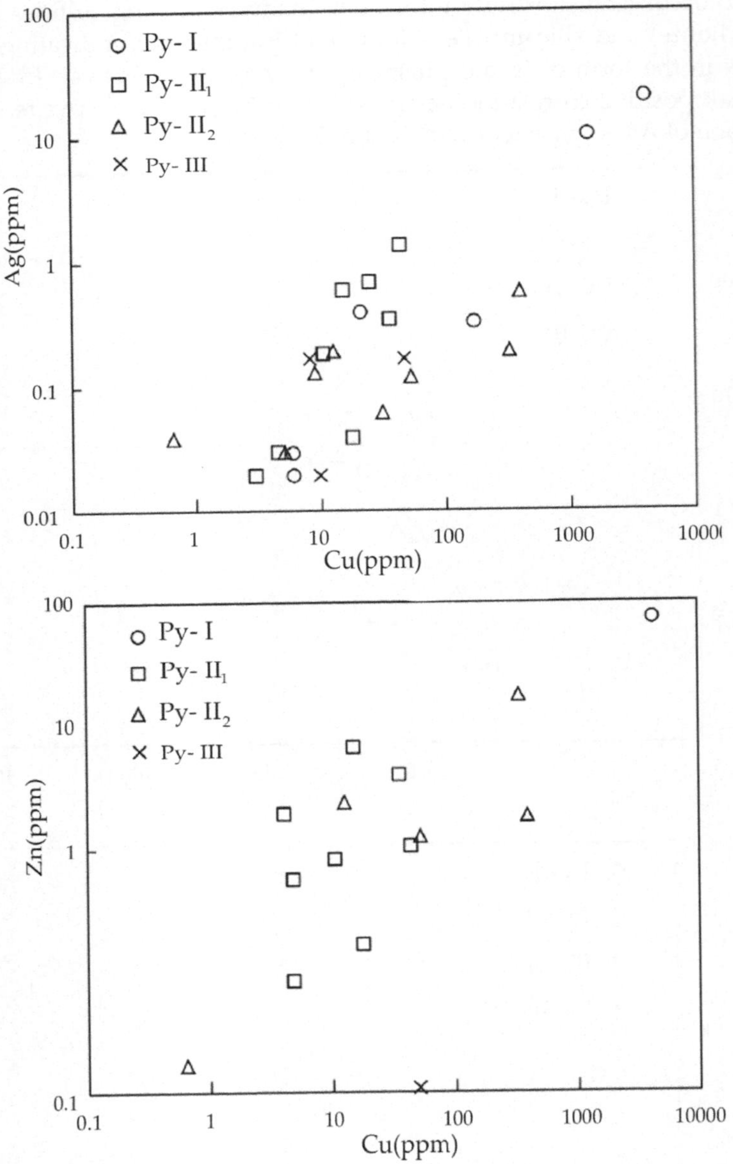

Figure 10. Correlation diagrams for the trace elements of pyrite from the Tuwu porphyry Cu deposit in eastern Tianshan, Xinjiang, NW China.

A local peak of As appears in a small number of samples and corresponds to the high value point of Cu, which indicates that As not only exists in arsenopyrite but also replaces S in pyrite via isomorphism. The mass fraction of As fluctuates greatly, with obvious high values in Py-I and Py-III. Above all, these results indicate that the content of As was greatly affected by fluid in later stages, and its content may have a certain effect on the precipitation and occurrence of Cu. Zinc appeared in local peaks (Figure 8), which indicates that Zn may exist in the form of sphalerite microinclusions in pyrite, and the content of Zn appears to be highly variable in Py-III1 and Py-III. Moreover, considering that the ionic radius of Zn^{2+} is 0.74 Å and Fe^{2+} is 0.76 Å, Zn may replace Fe in the pyrite lattice via isomorphism.

Genesis of Pyrite

Trace elements entering into the pyrite lattice in solid solution form are restricted by the chemical composition of the surrounding rock, mineral assemblage, and fluid properties. Because different hydrothermal fluids usually form under different physicochemical conditions, the contents of trace elements in pyrite vary significantly. Co and Ni are the most common trace elements in pyrite, and their contents and the Co/Ni ratio are often used to trace the genesis, deposition environment, and formation temperature of pyrite (Bralia et al., 1979; Campbell and Ethier, 1984; Bajwah et al., 1987; Cook, 1996). Different genetic types of pyrite usually have different Co/Ni ratios (Bralia et al., 1979; Brill, 1989; Large et al., 2011). The Co/Ni ratio of synsedimentary pyrite is usually less than 1, whereas the Co/Ni ratio of hydrothermal pyrite varies widely but is generally greater than 1; in particular, the Co/Ni ratio of volcanic hydrothermal pyrite is generally between 5 and 100 (Bralia et al., 1979; Craig et al., 1998; Franchini et al., 2015; Keith et al., 2016) because of the difference in the octahedral site preference energies of Co (7.4 cal/mol) and Ni (20.6 cal/mol) (Bralia et al., 1979). With the process of magma crystallization differentiation, Ni prefers to enter minerals with a high octahedral ratio (such as olivine and orthopyroxene) during early crystallization, while Co becomes relatively enriched in minerals forming in later-stage crystallization, thus causing the Co/Ni ratio of magmatic and hydrothermal pyrite to be greater than 1. Therefore, the Co/Ni ratio of pyrite that precipitated in response to magmatic or hydrothermal activity is generally greater than 1 (Song and Zhang, 1986; Zhou et al., 2010; Leng, 2017).

The test results show that the Co and Ni contents of pyrite in the Tuwu mining area vary greatly and tend to decrease from Py-I, Py-II_1, and Py-II_2 to Py-III, which reflects the fluid evolution characteristics of the Tuwu mining area to a certain (Figure 11). The Co/Ni ratio of the pyrite in the weakly propylitized zone (Py-I) ranges from 11.37 to 60.60. The siderophile elements Co and Ni are often enriched in basic or ultrabasic rocks. Thus, the anomalous enrichment of Co and Ni elements and high Co/Ni ratios in Py-I reflect a magmatic origin and are closely related to the formation of the CQ group volcanic rocks.

The Co/Ni ratio of the pyrite in the pyritization, sericitization, and silicification zone (Py-II_1) ranges from 0.09 to 9.94 and is generally consistent with the characteristics of magmatic or hydrothermal pyrite. However, the lower the Co/Ni ratio of Py-II_1 is, the more the fragmentized pyrite is. Some studies have shown that using the Co/Ni ratio as a criterion to distinguish the boundary of different types of ore deposits is not reliable, especially for hydrothermal deposits (Song and Zhang, 1986; Wang, 1989). Geochemistry characteristics of fluid inclusions and H, O, and S isotopes show that the ore-forming fluid mainly evolved from magmatic hydrothermal fluids, meteoric water, and metamorphic fluid (Liu et al., 2009; Wang et al., 2018). Therefore, Py-II_1 may have been affected by the addition of meteoric water or other fluids in the late stage of mineralization, and the Co/Ni ratio of the pyrite shows certain characteristics of sedimentary origin. Py-II_1 is depleted in low-temperature elements such as As, Se, and Te, which is consistent with the geochemical characteristics of high-temperature hydrothermal pyrite.

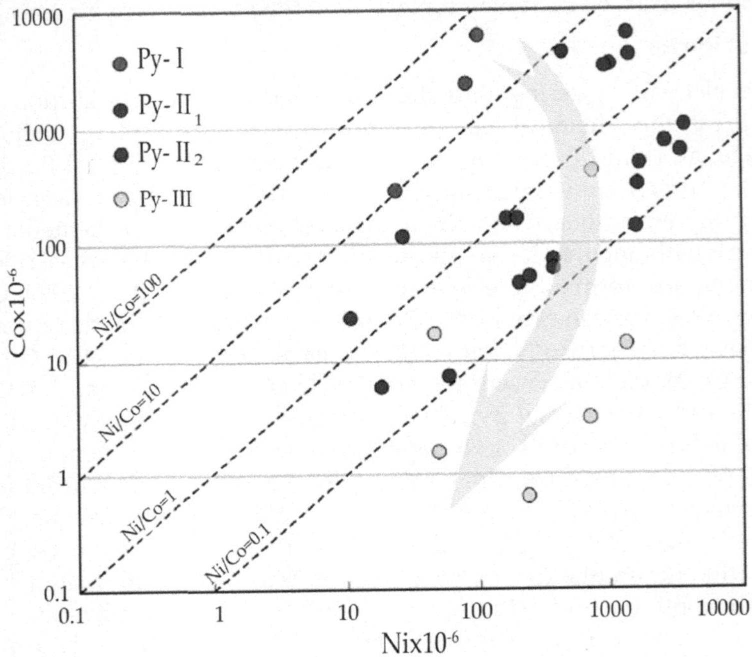

Figure 11. Co-Ni genetic diagram for the different types of pyrite from the Tuwu porphyry Cu deposit in eastern Tianshan, Xinjiang, NW China.

The Co/Ni ratio of the pyrite in the sericitization zone (Py-II_2) ranges from 0.12 to 2.02 (average 0.53), and the corresponding data points mainly plot in the field of sedimentary-origin pyrite. The characteristics of trace elements in Py-II_2 indicate the complexity of pyrite genesis. The formation of this pyrite may have involved the mixing of magmatic hydrothermal fluid and meteoric water. The meteoric water erased most of the magmatic-hydrothermal information. In addition, calcite, a typical low-temperature mineral, is abundant in this stage.

The Co/Ni ratio of the pyrite in the chloritization–sericitization zone (Py-III) ranges from 0.003 to 0.57. Cobalt and Ni in pyrite can form CoS_2 and NiS_2 by replacing Fe via isomorphism. The unit cell parameters of pyrite are lengthening because the bond lengths of Co-S and Ni-S are greater than that of Fe-S. Pyrite can form a continuous solid solution series of FeS_2–CoS_2 and a discontinuous solid solution series of FeS_2–NiS_2 (Bralia et al., 1979; Song and Zhang, 1986; Craig et al., 1998). The content of FeS_2–CoS_2 is 2-3 orders of magnitude lower than that of FeS_2–NiS_2 in Py-III. The rocks in the chloritization–sericitization zone are characterized by shear deformation, which manifests as highly schistose local ore bodies and significant elongation of minerals. Overall, the formation of pyrite in this stage appears to involve superimposed transformation by metamorphic hydrothermal fluids, which is consistent with the geochemistry of fluid inclusions and isotopes (Liu et al., 2009; Wang et al., 2018).

The temperature controls the types and contents of trace elements in pyrite. Generally, a higher temperature of formation corresponds to more trace element types in pyrite and a higher content of trace elements (Craig et al., 1998; Large et al., 2007; Keith et al., 2016). Pyrite with different total trace element contents can be divided into three different stages. Comprehensive research on the contents of different trace elements and the Co/Ni ratio in pyrite shows that the total trace element content is highest in Py-I (4,351.34 ppm), which mainly formed in the later period of volcanic activity. The high copper content may reflect the high copper background value of the volcanic rocks in the CQ Group.

The total trace element contents in $PyII_1$ and $PyII_2$ are 4,049.19 and 543.68 ppm, respectively. Combined with previous studies of fluid inclusions (Liu et al., 2009; Han et al., 2010; Wang et al., 2018), these results reflect the variation characteristics of trace elements in pyrite from the main mineralizing stage to the late stage of the porphyry mineralization period. The total trace element content in Py-III is 1982.41 ppm. The temperature of the metamorphic hydrothermal fluid is between 150°C and 300°C, and the copper is obviously enriched by the superimposed transformation of the metamorphic hydrothermal solution. This finding explains why the temperature of the main metallogenic stage in the Tuwu copper deposit was approximately 200°C (Liu et al., 2009; Han et al., 2010), which is mainly because earlier high-temperature fluid inclusions were superimposed by ductile shearing.

Rare Earth Element Geochemistry of Chlorite

The geochemical similarity of REEs allows them to be used in tracing applications, which is similar to isotopic values for the process of mineralization (Michard and Albarède, 1986; Maclean, 1988; Fulignati et al., 1999). By studying the REE composition characteristics of minerals from different metallogenic stages, we can determine the geochemical behavior of REEs during the processes of ore formation and alteration (Maclean, 1988; Chen and Fu, 1991, 1997). Such work can effectively constrain the ore-forming material source, the metallogenic physicochemical environment, the fluid evolution process, and ore genesis (Chen and Fu, 1991; Bortnikov et al., 2008; Parsapoor et al., 2009; Tan et al., 2016, 2018). Research has shown that the geochemical behavior of REEs in minerals is mainly related to the secondary minerals formed by alteration, the content of REEs in hydrothermal fluids, the formation temperature of minerals, the pH and Eh of hydrothermal systems, the crystal chemical properties of minerals, and the partition coefficient of REEs between minerals and fluids (Michard and Albarède, 1986; Michard, 1989; Bortnikov et al., 2008). Previous studies on the REE characteristics of argillization zones (Karakaya et al., 2012), silicification zones (Terakado and Fujitani, 1998; Khalili and Malekmahmoudi, 2012), sericitization–silicification zones (Parsapoor et al., 2009), propylitization zones (Fulignati et al., 1999), and potassic zones (Taylor and Mclennan, 1985) show that the contents of REEs differ among different alteration zones.

Rare Earth Element Characteristics of Chlorite in the Premineralization Period

Microstructure research has shown that there are two main formation mechanisms of chlorite in the Tuwu copper deposit: 1. metasomatic crystallization, which is chlorite formation via the metasomatism of biotite and hornblende, and in those rocks by late-stage hydrothermal fluids, 2. precipitation crystallization, which is the chlorite precipitated from hydrothermal fluids rich in Fe and Mg in the cracks or gaps between minerals.

Metasomatic crystallization chlorite (Chl-I_1) in the weakly propylitized zone mainly formed by late hydrothermal alteration of hornblende, pyroxene, and biotite in basalt. This chlorite generally inherited the REE distribution pattern of the volcanic rocks in the CQ group in the Tuwu mining area (Hou et al., 2005, 2006; Zhao et al., 2014).

The precipitation crystallization chlorite (Chl-I_2) is mainly associated with quartz, epidote, and minor pyrite. The REE patterns of Chl-I_2 are characterized by the relative enrichment of LREEs and positive Eu anomalies, which may be due to the Eu in minerals that were in equilibrium with the hydrothermal solution entering the fluid in the form of Eu^{2+} during the rock–water interaction and separated from the other REEs. As a result, Chl-I_1 has stronger negative Eu anomalies than the volcanic rocks. In contrast, Eu is enriched in the fluid, and Chl-I_2, which precipitated from the fluid, has positive Eu anomalies.

The enrichment degree of Eu in the fluid is closely related to temperature. Even in relatively oxidized environments, Eu in hydrothermal solutions mainly exists in the form of Eu^{2+} at relatively high temperatures (Bortnikov et al., 2008; Parsapoor et al., 2009; Fadda et al., 2012). Therefore, Chl-I_2 has a high crystallization temperature, and the total content of REEs in Chl-I_1 is generally 1-2 orders of magnitude higher than those in Chl-II and Chl-III, indicating the characteristics of the fluids related to Chl-I were obviously different from those of the fluids in the metallogenic epoch. The fluid likely was a volcanogenic hydrothermal solution that formed during late volcanic activity. The content of Cu in chlorite is generally high. The closer to the ore body, the higher the Cu content. This relationship reflects that the volcanic rocks in the CQ group have a high background value of Cu. In addition, the volcanic syngenetic hydrothermal activity primarily led to the enrichment in copper.

Rare Earth Element Characteristics of Chlorite in Porphyry Metallogenic Stage

Similar LREE enrichment is observed in the REE distribution patterns of Chl-II_1, which is associated with epidote, quartz, calcite, and zoisite in the strong propylitization zone. Chl-II_2 is associated with sericite, pyrite, and chalcopyrite in the pyritization–sericitization–silicification zone, and Chl-II_3 is associated with sericite, calcite, pyrite, and chalcopyrite in the sericitization–silicification zone. In addition, the total REE content in Chl-II is significantly lower than that of Chl-I and the Eu anomalies in Chl-II_1, Chl-II_2, and Chl-II_3 range from positive to negative. The reasons for the above REE characteristics in Chl-II are as follows.

1) Because chloritization occurs in neutral to alkaline pH conditions with relatively low temperatures (<300°C) (Reed, 1997), the REEs in the fluid mainly form complexes with SO_4^{2-} and CO_3^{2-}. While SO_4^{2-} does not preferentially bind with REEs, CO_3^{2-} does preferentially bind with REEs to form complexes (Foster, 1962; Bortnikov et al., 2008; Inoue et al., 2010). The secondary minerals formed from this fluid are relatively rich in heavy REEs.

LREEs have a larger ionic radius and smaller atomic weight than HREEs; therefore, LREEs are more easily leached from minerals (Bortnikov et al., 2008; Ye et al., 2011; Karakaya et al., 2012). The REE patterns of Chl- II are characterized by the relative enrichment of HREEs, indicating that it experienced leaching action of late hydrothermal fluids. Additionally, the similarity in REE distribution patterns indicates that Chl-II$_1$, Chl-II$_2$, and Chl-II$_3$ have the same material source and geochemical behavior.

2) Although accessory minerals exist in only very small quantities in most rocks, the REE distribution coefficients in accessory minerals are very high; therefore, they play a great role in controlling the REE content in rocks. The content of REEs in plagiogranite porphyry (24.87–55.88 ppm) is lower than that in volcanic rock (50.64–117.73 ppm) (Wang et al., 2014, 2015; Zhao et al., 2014), and the accessory minerals in plagiogranite porphyry are more than that in the volcanic rock, which causes the content of REEs in Chl-II to be lower than that in Chl-I.

3) The stability of Eu^{2+} is controlled not only by the temperature, pH, Eh, and oxygen fugacity of the open system but also by the supply of alteration minerals. Previous studies have found that Eu^{3+} tends to occur under low pH and high oxygen fugacity (Sverjensky, 1984; Bau, 1991; Hedenquist, 1995). The valence state of Eu will transform from Eu^{3+} to Eu^{2+} and fractionate from other $REEs^{3+}$ in a reducing environment, and Eu^{2+} will remain in a stable state under a low oxygen fugacity environment. The Chl-II$_1$ in the strong propylitization zone and the Chl-II$_2$ in the sericitization–silicification zone exhibit positive Eu anomalies, indicating that the hydrothermal system featured a higher temperature (>250°C), a low pH, and relatively reducing conditions and the chlorite generally inherited the REE distribution pattern of the ore-forming fluid. The stronger positive Eu anomalies in Chl-II$_2$ may is consistent with the large amount of sulfide that crystallized from the ore-forming fluid. With the development of mineralization and the addition of meteoric water, the metallogenic environment will transform from a relatively reducing environment to a relatively oxidizing environment. The occurrence of a large number of low-temperature minerals, such as calcite, indicates that the mineralization temperature has obviously decreased. Chl-II$_3$ formed at this stage shows obviously negative Eu anomalies. Eu^{2+} has an ionic radius similar to that of Ca^{2+} and often occurs in calcite via isomorphism. Therefore, a large amount of calcite precipitation also reduces the content of Eu in the ore-forming fluid.

Rare Earth Element Characteristics of Chlorite in the Superimposed Transformation Period

The chlorite in the rich ore bodies (Chl-III) is characterized by obvious shear deformation characteristics. The REE distribution pattern exhibits LREE enrichment, and the fractionation of HREEs and LREEs is not obvious. This result suggests that the REE distribution pattern of Chl-III is apparently different from that of Chl-I and Chl-II. The formation depth of the Tuwu copper deposit is approximately 2–3 km (Wang et al., 2018). Combined with the significant deformation characteristics of the minerals in the chloritization–sericitization zone, we can infer that the Tuwu copper deposit was located in the brittle-ductile transitional zone. As the water/rock ratio increased, the diffusion pattern of the fluids shifted from percolation to infilling. The REE distribution pattern of Chl-III was mainly controlled by the fluid properties. In addition, the fluid source is also an important factor in determining the composition of REEs. The REE distribution patterns of the plagiogranite porphyry and dioritic porphyrite are generally consistent with that of Chl-III. However, the fractionation of HREEs and LREEs in Chl-III is obviously low, which indicates that the physical and chemical conditions of the Chl-III formation are apparently different from those of Chl-I and Chl-II. The positive Eu anomalies of Chl-III may indicate that Chl-III formed in a relatively reducing environment at relatively high temperatures.

Spatial Variation Pattern of Trace Elements in Chlorite

In the formation process of porphyry deposits, a large alteration halo usually occurs around the center of the ore body. In this alteration halo, the mineral association and chemical composition may show certain zonation (Lowell and Guilbert, 1970). The classical metallogenic model of porphyry copper provides theoretical support for exploring porphyry deposits on the regional scale. However, ore bodies of economic value tend to be concentrated in the range of 1 km, and there are many alterations that are irrelevant to the ore-forming process (Inoue et al., 2010; Fadda et al., 2012). Therefore, effective identification of metallogenic and nonmetallogenic alteration systems is of great significance for the exploration of porphyry deposits. In recent years, the geochemical characteristics of altered minerals have attracted the widespread attention (Inoue et al., 2010; Cooke et al., 2014; Wilkinson et al., 2015; Xiao et al., 2018). Understanding these characteristics not only helps reveal the genesis of ore deposits but also provides technical guidance for prospecting exploration. Currently, based on electron probe and LA-ICP-MS analyses of altered minerals in porphyry systems, scholars have documented spatial variations in trace elements in altered minerals (Cooke et al., 2014; Wilkinson et al., 2015).

These achievements have been applied and verified many times in the field of exploration in China and elsewhere. They are commonly used to assist exploration. The systematic electron probe and LA-ICP-MS analysis of chlorite in the propylitization zone of the Batu Hijau superlarge porphyry copper (gold) deposit in Indonesia has shown that K, Li, Mg, Ca, Sr, Ba, Ti, V, Mn, Co, Ni, Zn, and Pb can enter the crystal lattice of chlorite. With distance from the ore body, the contents of Ti, V, and Mg decrease exponentially, while the contents of K, Li, Ca, Sr, Ba, Mn, Co, Ni, Zn, and Pb increase exponentially (Cooke et al., 2014; Wilkinson et al., 2015; Xiao et al., 2018).

The spatial variations in the trace element compositions in chlorite in the Tuwu mining area are similar to those in the Batu Hijau superlarge porphyry copper (gold) deposit. For example, the contents of high field strength elements, such as Ti and V, are higher in Chl-II and Chl-III near the ore body than in Chl-I far from the ore body. The degree of isomorphism of Ti in chlorite is mainly controlled by the temperature of mineral crystallization (Wilkinson et al., 2015).

Therefore, the spatial variation in Ti content in chlorite also provides information on the thermal anomaly around the mineralization center. The variation range of Ti contents in Chl-II and Chl-III can reach 2–3 orders of magnitude within 300 m. However, the migration ability of high field strength elements, such as Ti and V, in hydrothermal fluid is quite limited. The significant differences in Ti content in chlorite in the porphyry mineralization stage in the Tuwu mining area indicate that the migration and diffusion of Ti were affected not only by temperature but also by the repeated action of hydrothermal. Moreover, the Ti contents in Chl-I vary little within a range of 1 km, which suggests that Chl-I in the volcanic rocks mainly formed in the volcanism and was scarcely affected by late hydrothermal activity. With increasing distance from the ore body, the contents of K, Ca, Sr, Ba, and other large ion lithophile elements decrease but the degree of spatial variation is not as obvious as in a previous study (Wilkinson et al., 2015).

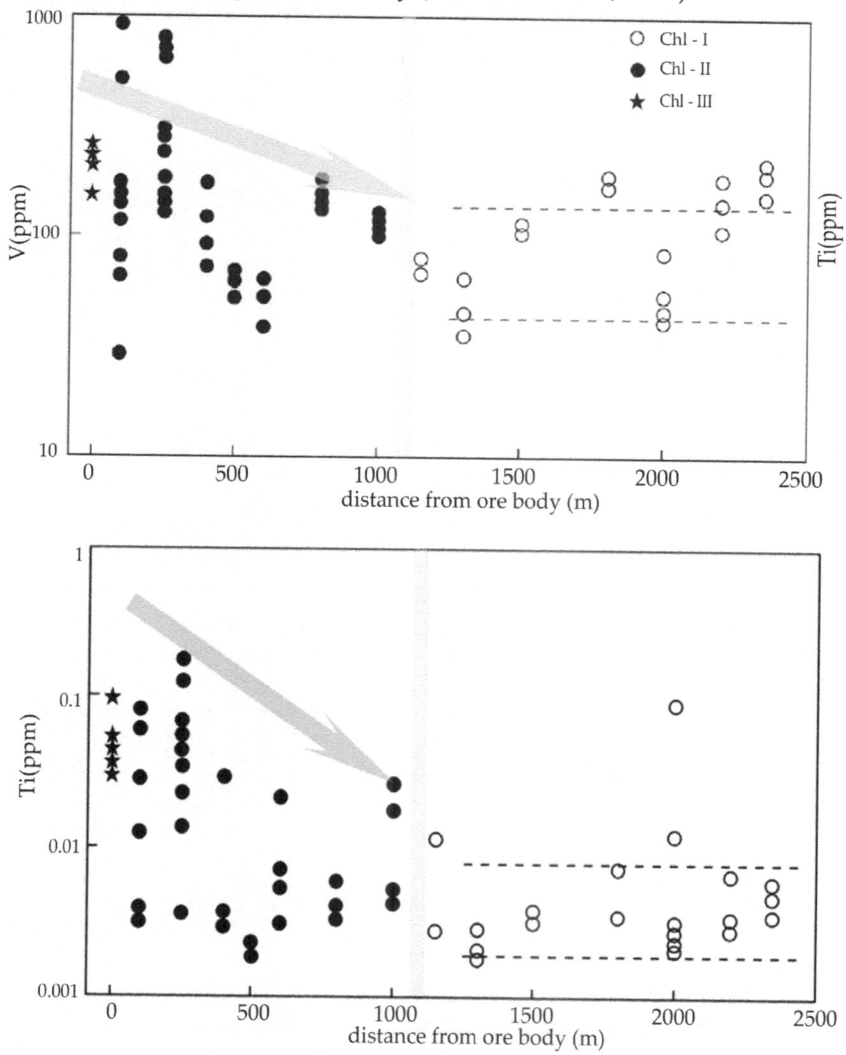

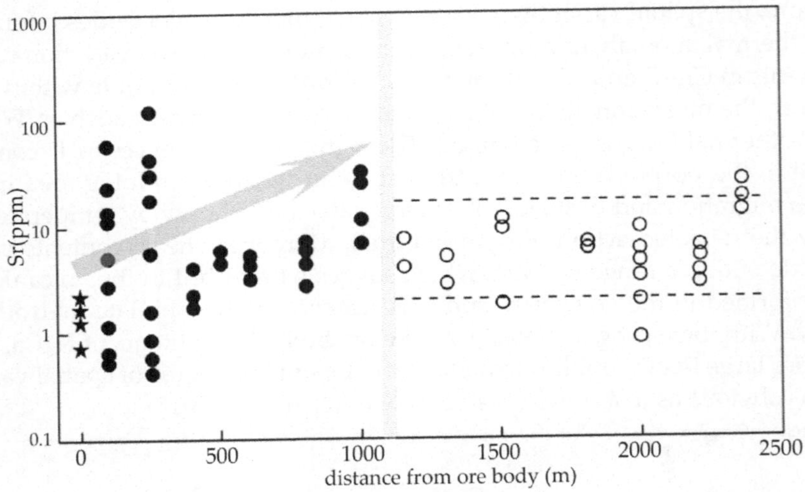
Sr(ppm)
1000
100
10
1
0.1
0
500
1000
1500
2000
2500
distance from ore body (m)

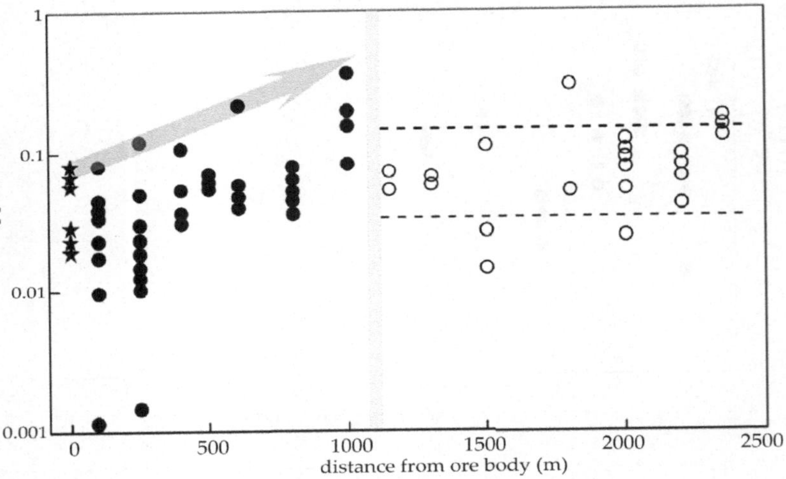
Cr(ppm)
1
0.1
0.01
0.001
0
500
1000
1500
2000
2500
distance from ore body (m)

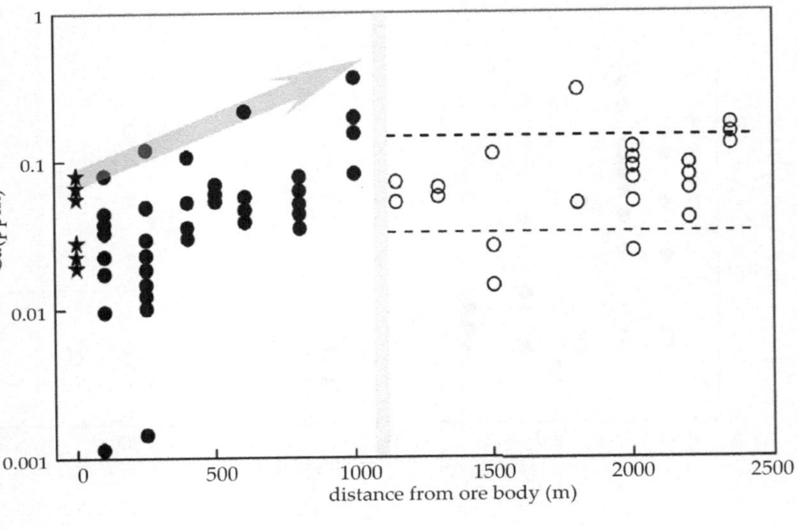
Ca(ppm)
1
0.1
0.01
0.001
0
500
1000
1500
2000
2500
distance from ore body (m)

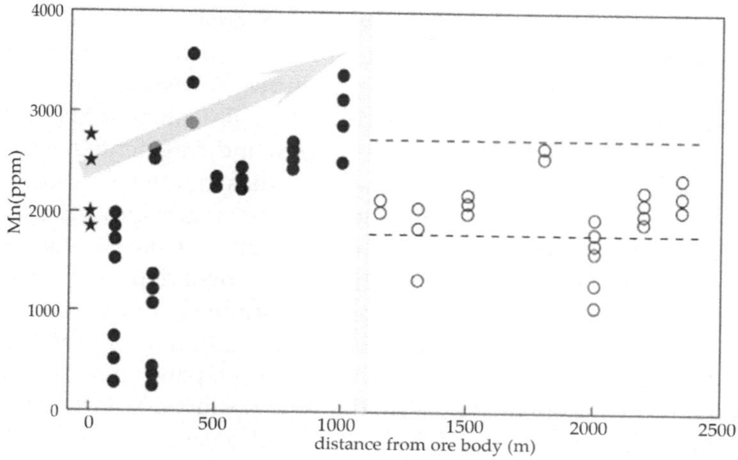

Figure 12. Trace element spatial variations in chlorite for different alteration zones of the Tuwu porphyry Cu deposit in eastern Tianshan, Xinjiang, NW China.

This finding shows that the large ion lithophile elements migrated more easily in the late-stage hydrothermal solution. The content ranges of K, Ca, Sr, and Ba in Chl-I are obviously less than those in Chl-II and Chl-III, which suggests that Chl-II and Chl-III were affected by multiple phases of hydrothermal activity. The contents of Mn and Zn in chlorite gradually increase with distance from the center of the ore body, with a maximum value at a distance of approximately 1.2 km; beyond this distance, they gradually decrease until they stabilize. This pattern reflects the horizontal movement of Mn and Zn with the magmatic hydrothermal fluid in the surrounding rock. The high contents of Mn and Zn in Chl-I reflect the high background values of Mn and Zn in the volcanic rocks in the CQ group (Figure 12).

Influence of Ductile Shearing on the Tuwu Deposit

Although the Tuwu deposit experienced intense superimposed reformation, there are many features and characteristics that match those of a porphyry system. 1) Although most of the Cu in the Tuwu deposit is hosted in the surrounding rock, the ore-bearing plagiogranite porphyries are almost wholly mineralized. 2) Although the shape of the orebodies is mainly tabular, the ore structures are disseminated and veinlet-disseminated and form stockwork. 3) Although the boundaries of the alteration zone are not obvious, the concentric alteration zones from the interior outward in succession are the potassic zone, the quartz–sericite zone, and the propylitic zone in general.

Compared with young porphyry systems, ancient porphyry systems tend to be subjected to later superimposed transformations. However, few studies have focused on the deformation and mineralization enrichment of porphyry deposits. The following evidence is presented in this article to demonstrate that the Tuwu porphyry deposit has undergone strong deformation and transformation. 1) The Tuwu porphyry deposit has experienced regional ductile shear activity after mineralization. The mining area is located in the northern influence zone of the large Qiugemingtashi-Huangshan ductile shear zone. 2) The intensity of the ductile deformation zone in the Tuwu-Yandong mining area is stronger in the south and weaker in the north. The southern part features a continuous strong ductile deformation zone and the northern part features a weak ductile deformation zone composed of several discretely distributed small lenticular and zonal ductile deformation zones. The ore has mylonite, augen, and fragmented structures. The mylonitic foliation developed in the ore body provides the main fine structural space for ore minerals (Ma et al., 2002, 2003). These characteristics indicate that the Tuwu mining area formed in the brittle–ductile transition zone. 3) Considerable research has already been done on fluid inclusions in the Tuwu copper deposit. Previous studies have found that the temperatures of the main ore-forming periods are approximately 200°C (Liu et al., 2009; Han et al., 2010), while others found that the temperatures are approximately 350°C (Wang et al., 2015). This discrepancy may be related to the different fluid properties of ore bodies in different positions. 4) Oriented secondary fluid inclusions are well developed in the deformation zone (Liu et al., 2009; Wang et al., 2018), indicating that the shape of the orebody changes from a disseminated low-grade to a vein high-grade orebody by tectonism. 5) The Tuwu mining area can be divided into three tectonic facies according to the characteristics of deformation. The ductile compression zone belongs to the moderately deep ductile deformation facies; the brittle–ductile shear zone belongs to the moderately shallow transition facies between ductile and brittle deformation, and the brittle fault zone belongs to the shallow structural facies. The fact that the moderately deep, moderately shallow, and shallow structures appear on the same plane indicates that they did not form at the same time but are instead the result of superimposing different structures in ore-bearing geologic bodies that have been uplifted several times and that have undergone multiple phases of structural deformation at different depths (Ma et al., 2003). 6) Closer to the ductile deformation zone, the contents of copper, sericite, chlorite, and quartz increase, and the oriented arrangement characteristics of these mineral become more obvious. There are some chalcopyrite and pyrite distributed along foliations of mylonite in the ductile deformation zone, and the chalcopyrite veins filled the fractures in the pyrite. These characteristics indicate that the ore has experienced strong deformation and the copper remobilized. 7) In situ trace element analysis of chlorite and pyrite in different stages shows distinctly different geochemical compositions Table 3.

The mineralization of the Tuwu and Yandong copper deposits occurred in the early Carboniferous at the southern belt of Dananhu-Haerlik island arc, and these deposits are separated by only 5 km. The ore-forming bodies are all plagiogranite porphyry and have similar mineralization processes (Chen et al., 2005; Hu et al., 2011; Shen et al., 2012; Pan et al., 2013). However, 80% of the Cu in the Tuwu mining area is hosted in the diorite porphyry, while 90% of the Cu in the Yandong mining area is hosted in the plagiogranite porphyry (Hu et al., 2011). The superposition and transformation of late ductile shearing on the two deposits in different spatial positions may be one of the reasons for this difference. The plagiogranite porphyry brought metallogenic materials and formed the initial mineralization of the Tuwu copper deposit, while the late ductile shear process accompanied by strong hydrothermal activity resulted in the activation and enrichment of copper in the mining area. Moreover, the morphology of the ore bodies is mainly tabular, which is quite different from that of typical porphyry deposits.

Table 3. elationship table between the sequence of events and the chemistry of pyrite and chlorite

Age		**Permian**	**Carboniferous**				
Tectonic setting		Post-collisional extension	Collision				
Stage		Superimposed transformation stage	Porphyry metallogenic stage			Premineralization stage	
Mineral structure		Chl-III (disseminated and vein type with shear deformation characteristics)	$Chl-II_1$ (dense disseminated and stockwork type)	$Chl-II_2$ (disseminated and vein type)	$Chl-II_3$ (mainly disseminated type)	$Chl-i_1$ (replacement remnant, pseudomorphic)	$Chl-i_2$ (disseminated type)
		Py-III (disseminated and vein type with fragmentized)	Spotted (fine-grained)	$Py-II_1$ (disseminated and vein type)	$Py-II_2$ (disseminated and vein type)	Py-I (disseminated type)	
Pyrite (ppm)	Co/Ni	0.003–0.57; Av = 0.16	–	0.09–9.94; Av = 2.71	0.12–2.02; Av = 0.53	11.37–60.60; Av = 33.37	
	Co + Ni	313.88–6,523.06; Av = 3,098.99	–	143.28–9,735.37; Av = 3,896.7	16.92–1941.08; Av = 416.63	48.03–1714.26; Av = 584.66	
	Total content	478.09–9,692.23; Av = 4,351.3	–	594–7,839.08; Av = 4,049.19	125.64–2028.18; Av = 543.68	132.52–6,401.02; Av = 1982.41	
	Cu	8.95–50.68; Av = 23.16	–	0.66–336.25; Av = 86.28	2.96–43.6; Av = 16.38	5.94–4,133.8; Av = 950.33	
Chlorite (ppm)	REE	0.04–1.27; Av = 0.49	0.056–0.77; Av = 0.425	0.02–3.42; Av = 0.66	0.05–3.34; Av = 0.808	0.03–17.62; Av = 1.8	0.02–0.71; Av = 0.248
	LREE/HREE	1.26–3.17; Av = 1.75	0.07–0.59; Av = 0.296	0.11–1.19; Av = 0.64	0.08–0.85; Av = 0.55	0.12–5.95; Av = 1.32	0.18–3.91; Av = 0.96
	δEu	0.8–4.27; Av = 1.61	1.08–1.62; Av = 1.15	1.02–2.52; Av = 1.36	0.385–1.49; Av = 0.87	0.223–0.866; Av = 0.55	1–3.78; Av = 1.83
	δCe	0.73–3.92; Av = 1.53	0.43–1.52; Av = 1.03	0.52–1.95; Av = 1.24	0.64–1.5; Av = 1.06	0.53–2.11; Av = 0.915	0.54–2.01; Av = 0.99
	Ti	177.9–572.4; Av = 299.68	11.07–176.09; Av = 37.54	19–486.8; Av = 165.92	21.64–1,055.86; Av = 334.43	12.22–159.33; Av = 29.64	10.96–573.20; Av = 71.45
	Sr	0.71–2.02; Av = 1.42	1.83–7.1; Av = 4.45	0.51–58.69; Av = 12.79	0.37–121.19; Av = 17.18	1.79–29.16; Av = 10.94	0.86–7.49; Av = 4.79

These findings may be related to changes in the tectonic environment of the Tuwu mining area caused by ductile shearing to brittle shearing after metallogenesis, which not only led to the activation and migration of copper but also the enrichment of copper in the narrow brittle fractures. Therefore, the ore bodies are characterized by obvious tectonic control.

Mineralization of the Tuwu Copper Deposit

The Tuwu deposit is located approximately 10 km north of the Kangguer fault. The southern part of the Tuwu mining area experienced complicated subduction, island arc growth, plate collision, and ductile shearing from the Carboniferous to Permian. Finally, a faulted structural zone with complex structures and properties developed (Han et al., 2006; Wang et al., 2006; Cai et al., 2012), and the obviously superimposed mineralization characteristics of the Tuwu copper deposit developed. By combining and integrating our results with previously published data on the eastern Tianshan orogens, we suggest the following new model to interpret the evolution of the Tuwu deposit in eastern Tianshan and then provide a short discussion. Stage I: In the early Carboniferous, the northward subduction of the North Tianshan Ocean formed the CQ group volcanic rocks in the southern belt of the Dananhu-Haerlik island arc (Hou et al., 2005, 2006). A series of circular structures and radial structures around the volcanic crater were formed in the Tuwu mining area. The intersections of these structures formed dilational zones that provided the space and initial conditions for the emplacement of the plagiogranite porphyry. Typical minerals in the weakly propylitized zone, such as chlorite, epidote, dolomite, and calcite in the CQ Group volcanic rocks, were formed by volcanic hydrothermal processes. The higher contents of trace elements in pyrite and chlorite show that the mixing action of meteoric water played only a relatively minor role, while the high copper contents reflect the high copper background values of the volcanic rocks and the initial enrichment of Cu by volcanic hydrothermal processes. The volcanic rocks that were far from the ore bodies mainly suffered weakly volcanic hydrothermal process because of the undeveloped fracture system, and the trace element content of chlorite does not change obviously with the spatial position and generally inherited the REE distribution pattern of dark minerals in volcanic rocks Table 3.

Stage II: In the early Carboniferous, with the subduction and partial melting of the northern Tianshan oceanic plate at depth under certain physical and chemical conditions, adakitic magmas intruded into the shallow crust and formed the plagiogranite porphyry (Wang et al., 2014, Han et al., 2014; Shen et al., 2014b). The contemporaneous faults of the CQ group provided the tectonic environment for porphyry intrusion.

The pyrite that formed in the early stage of porphyry mineralization under high-temperature and high-salinity conditions has an obvious magmatic-hydrothermal origin. With the development of mineralization, the input of meteoric water, and the long-term metasomatism of fluid in the late stage, the Co/Ni ratio of pyrite continuously decreased. From the early to late stages of porphyry mineralization, the REEs in chlorite show similar distribution patterns, indicating that the fluid had a similar source of ore-forming material Table 3. With the addition of meteoric water in the late stage II, athe metallogenic system became more open, and the fluid mixing caused minerogenetic matter precipitation (Liu et al., 2009; Shen et al., 2014b; Wang et al., 2015; Gao et al., 2017; Wang et al., 2018). As a consequence, the REE composition of the fluid changed, and the metallogenic system became a relatively oxidizing environment.

Stage III: In the Late Carboniferous-Triassic, the North Tianshan oceanic basin subducted and closed along the Kangguer ductile shear zone (Xiao et al., 2004; Mao et al., 2014b; Chen et al., 2019). The Tuwu copper deposit, which underwent multiple phases of compression and shear deformation, is located in the influence zone of the Kangguer ductile deformation zone, and the ore body underwent a strong superimposed transformation. The ore body was obviously mylonitized, and the Co/Ni ratio of pyrite decreased continuously under the action of multiphase fluids. The trace element characteristics of Chl-III also show distinct differences from those of Chl-II in the porphyry metallogenic stage Table 3.

CONCLUSION

1) We redefined the altered zones and the following three mineralization stages: premineralization stage (stage Ⅰ), porphyry metallogenic stage (stage Ⅱ), and superimposed transformation stage (stage Ⅲ). Pyrite and chlorite are widely distributed in the three mineralization stages and different altered zones. Elements in pyrite, such as Sb, Se, Co, and Ni, mainly occur in the pyrite crystal lattice via isomorphism. Cu has obvious positive correlations with Pb, Zn, Ag, and Au in different ore-forming stages, and these elements may occur in the form of microinclusions or nanometer-scale particles.
2) The trace element compositions of pyrites from three stages obviously changed, thus reflecting the evolution of fluid properties with the mineralization process. The Co/Ni ratios changed from Py-I (11.37–60.60) to Py-II_1 (0.09–9.94), Py-II_2 (0.12–2.02), and finally Py-III (0.003–0.57), indicating that the fluid evolved from a volcanic genesis to a magmatic hydrothermal genesis. The fluid was affected by meteoric water and other fluids in the late metallogenic period and experienced a superimposed complicated structural hydrothermal fluid.
3) The REE distribution pattern of chlorite has obviously different features in the three stages. Chl-I generally inherited the REE distribution pattern of the volcanic rocks in the CQ Group, which shows that the fluid exsolved from the products of late volcanic activity. The similar REE distribution pattern of Chl-II_1 to Chl-II_3 indicates that the fluid had a similar ore-forming material source. The different features from Chl-II_1 to Chl-II_3 show that the fluid evolved from magmatic hydrothermal to the mixture of magmatic and meteoric fluid. The REE distribution pattern of Chl-III shows that the fluid in ductile deformation zones experienced strong water–rock interactions with the surrounding rocks.
4) The contents of Ti, V, and Mg in the chlorite decreased exponentially with increasing distance, whereas the contents of K, Li, Ca, Sr, Ba, Mn, Co, Ni, Zn, and Pb increased. The maximum value appears at approximately 1.2 km and the values then gradually stabilize.

5) The Tuwu porphyry deposit has undergone volcanic activity, porphyry metallogenesis, and structural deformation from the early Carboniferous to early Triassic.

AUTHOR CONTRIBUTIONS

We the undersigned declare that this manuscript entitled "Mineralization of the Tuwu porphyry Cu deposit in eastern Tianshan, NW China: Insights from in situ trace elements of chlorite and pyrite" is original, has not been published before, and is not currently being considered for publication elsewhere. We would like to draw the attention of the editor to the following publications of one or more of us that refer to aspects of the manuscript presently being submitted, where relevant copies of such publications are attached. We confirm that the manuscript has been read and approved by all named authors and that there are no other persons who satisfied the criteria for authorship but are not listed. We further confirm that the order of authors listed in the manuscript has been approved by all of us. We understand that the corresponding author is the sole contact for the editorial process. He is responsible for communicating with the other authors about the progress, submissions of revisions, and final approval of proofs. Signed by all authors as follows: WT, QM*, MY, YS, and XL.

ACKNOWLEDGMENTS

Han Runsheng of the Kunming University of Science and Technology and Researcher Chen Zhengle of the Institute of Geomechanics provided valuable revision suggestions that improved this paper.

REFERENCES

1. Bau, M. (1991). Rare-earth Element Mobility during Hydrothermal and Metamorphic Fluid-Rock Interaction and the Significance of the Oxidation State of Europium. Chem. Geology. 93, 219–230. doi:10.1016/0009-2541(91)90115-8
2. Bajwah, Z. U., Seccombe, P. K., and Offler, R. (1987). Trace element distribution, Co: Ni ratios and genesis of the Big Cadia iron-copper deposit, New South Wales, Australia. Mineralium Deposita 22, 292–300. doi:10.1007/BF00204522
3. Bortnikov, N. S., Gorelikova, N. V., Korostelev, P. G., and Gonevchuk, V. G. (2008). Rare Earth Elements in Tourmaline and Chlorite from Tin-Bearing Assemblages: Factors Controlling Fractionation of REE in Hydrothermal Systems. Geol. Ore Deposits. 50, 445–461. doi:10.1134/s1075701508060032
4. Bralia, A., Sabatini, G., and Troja, F. (1979). A Revaluation of the Co/Ni Ratio in Pyrite as Geochemical Tool in Ore Genesis Problems. Mineralium Deposita. 14, 353–374. doi:10.1007/bf00206365
5. Brill, B. (1989). Trace-element Contents and Partitioning of Elements in Ore Minerals from the CSA Cu-Pb-Zn deposit, Australia. Can. Mineral. 27, 263–274. doi:10.1007/BF01164492
6. Cai, Z. H., Xu, Z. Q., He, B. Z., and Wang, R. R. (2012). Age and Tectonic Evolution of Ductile Shear Zones in the Eastern Tianshan-Beishan Orogenic belt. Acta Petrologica Sinica. 28, 1875–1895. [in Chinese with English abstract]
7. Campbell, F. A., and Ethier, V. G. (1984). Nickel and Cobalt in Pyrrhotite and Pyrite from the Faro and Sullivan Orebodies. Can. Mineral. 22, 503–506. doi:10.1007/BF01082965

8. Chao, F. G., Tu, Q. J., Zhang, X. M., Ren, Y., Li, G. L., and Dong, F. R. (2006). Preliminary Determination of the Early Paleozoic Magmatic Arc in the Karlik Mountains, East Tianshan, Xinjiang, China-Evidence from Zircon SHRIMP U-Pb Dating of Granite Bodies in the Tashuihe Area. Geol. Bull. China. 25, 923–927. doi:10.3969/j.issn.1671-2552.2006.08.004 [in Chinese with English abstract]
9. Chen, F. W., Li, H. Q., Chen, Y. C., Wang, D. H., Wang, J. L., Liu, D. Q., et al. (2005). Zircon SHRIMP U-Pb Dating and its Geological Significance of Mineralizationin Tuwu- Yandong Porphyry Copper Mine, East Tianshan Mountain. Acta Geol. Sinica. 79, 256–261. doi:10.3321/j.issn:0001-5717.2005.02.011 [in Chinese with English abstract]
10. Chen, X., Shu, L., Santosh, M., and Zhao, X. (2013). Island Arc-type Bimodal Magmatism in the Eastern Tianshan Belt, Northwest China: Geochemistry, Zircon U-Pb Geochronology and Implications for the Paleozoic Crustal Evolution in Central Asia. Lithos. 168-169, 48–66. doi:10.1016/j.lithos.2012.10.006
11. Chen, Y.-J., Chen, H.-Y., Zaw, K., Pirajno, F., and Zhang, Z.-J. (2007). Geodynamic Settings and Tectonic Model of Skarn Gold Deposits in China: An Overview. Ore Geology. Rev. 31, 139–169. doi:10.1016/j.oregeorev.2005.01.001
12. Chen, Y. J., and Fu, S. G. (1991). Variation of REE Patterns in Early Pre-cambrian Sediments: the Oretical Study and Evidence from the Southern Margin of the Northern China Craton. Chin. Sci. Bull. 36, 1100–1104. doi:10.1159/000202829 [in Chinese with English abstract]
13. Chen, Y. J., and Zhao, Y. C. (1997). Geochemical Characteristics and Evolution of REE in the Early Precambrian Sediments: Evidences from the Southern Margin of the North China Craton. Episodes. 20, 109–116.
14. Chen, Z., Xiao, W., Windley, B. F., Schulmann, K., Mao, Q., Zhang, Z., et al. (2019). Composition, Provenance, and Tectonic Setting of the Southern Kangurtag Accretionary Complex in the Eastern Tianshan, NW China: Implications for the Late Paleozoic Evolution of the North Tianshan Ocean. Tectonics. 38, 2779–2802. doi:10.1029/2018tc005385
15. Cook, N. J., Ciobanu, C. L., Pring, A., Skinner, W., Shimizu, M., Danyushevsky, L., et al. (2009). Trace and Minor Elements in Sphalerite: A LA-ICPMS Study. Geochimica Et Cosmochimica Acta. 73, 4761–4791. doi:10.1016/j.gca.2009.05.045
16. Cook, N. J. (1996). Mineralogy of the Sulphide Deposits at Sulitjelma, Northern Norway. Ore Geology. Rev. 11, 303–338. doi:10.1016/s0169-1368(96)00009-1
17. Cooke, D. R., Baker, M., Hollings, P., Sweet, G., Chang, Z., Danyushevsky, L., et al. (2014). "New Advances in Detecting the Distal Geochemical Footprints of Porphyry Systems: Epidote mineral Chemistry as a Tool for Vectoring and Fertility Assessments,". Building Exploration Capability for the 21st Century. Editors KD Kelley, and HC Golden (Colorado, USA: Society of Economic Geologists, Special Publication), 18, 127–152.
18. Craig, J. R., Vokes, F. M., and Solberg, T. N. (1998). Pyrite: Physical and Chemical Textures. Mineralium Deposita. 34, 82–101. doi:10.1007/s001260050187
19. David, R. C., Mike, B., Pete, H., Gabe, S., Zhao, S. G., Zhou, T. F., et al. (2016). New Advances in Detecting the Distal Geochemical Footprints of Porphyry Systems Epidote Mineral Chemistry as a Tool for Vectoring and Fertility Assessments. Society of Economic Geology 18, 127–152. doi:10.5382/SP.18.07
20. Davidson, J., Tepley, F., Palacz, Z., and Meffan-Main, S. (2001). Magma Recharge, Contamination and Residence Times Revealed by In Situ Laser Ablation Isotopic

Analysis of Feldspar in Volcanic Rocks. Earth Planet. Sci. Lett. 184, 427–442. doi:10.1016/s0012-821x(00)00333-2

21. Deng, X.-H., Wang, J.-B., Pirajno, F., Wang, Y.-W., Li, Y.-C., Li, C., et al. (2016). Re-Os Dating of Chalcopyrite from Selected mineral Deposits in the Kalatag District in the Eastern Tianshan Orogen, China. Ore Geology. Rev. 77, 72–81. doi:10.1016/j.oregeorev.2016.01.014
22. Fadda, S., Fiori, M., Grillo, S. M., and Prochaska, W. (2012). REE Mobilization in Complex Hydrothermal-Metasomatic Systems: Fluid Chemistry Evidence of Albitite and Chlorite-Talc Mineralisations in Central Sardinia, Italy. Proc. 10th Int. Congress Appl. Mineralogy, 171–178. doi:10.1007/978-3-642-27682-8_22
23. Fang, T. H., Qin, K. Z., Wang, S. L., Wang, S. L., Jiang, F. Z., Gan, X. P., and Zhou, Z. J. (2002). Preliminary analysis on the geological back ground of Kalatage Cu-Au deposit, Xinjiang. Mineral Deposits 21, 380–384. [in Chinese with English abstract]
24. Foster, M. D. (1962). Interpretation of the Composition and a Classification of the Chlorites. U.s.geological Surv. Prof. Paper. 414, 33. doi:10.3133/pp414A
25. Franchini, M., Mcfarlane, C., Maydagán, L., Reich, M., Lentz, D. R., Meinert, L., et al. (2015). Trace Metals in Pyrite and Marcasite from the Agua Rica Porphyry-High Sulfidation Epithermal deposit, Catamarca, Argentina: Textural Features and Metal Zoning at the Porphyry to Epithermal Transition. Ore Geology. Rev. 66, 366–387. doi:10.1016/j.oregeorev.2014.10.022
26. Fulignati, P., Gioncada, A., and Sbrana, A. (1999). Rare-earth Element (REE) Behaviour in the Alteration Facies of the Active Magmatic-Hydrothermal System of Vulcano (Aeolian Islands, Italy). J. Volcanology Geothermal Res. 88, 325–342. doi:10.1016/s0377-0273(98)00117-6
27. Gao, J.-F., Zhou, M.-F., Qi, L., Chen, W. T., and Huang, X.-W. (2015). Chalcophile Elemental Compositions and Origin of the Tuwu Porphyry Cu deposit, NW China. Ore Geology. Rev. 66, 403–421. doi:10.1016/j.oregeorev.2014.08.009
28. Gao, J. F., Zhou, M. F., Zhang, L. C., Chen, W. T., and Huang, X. W. (2017). Chalcophile elemental compositions and origin of the Tuwu porphyry Cudeposit, NW China. Ore Geology Reviews 66, 403–421. doi:10.1016/j.oregeorev.2014.08.009
29. Gao, J. G., Li, W. Y., Guo, X. C., Zhou, Y., and Fan, T. B. (2014). Geochemistry, Zircon U-Pb Age and Hf Isotopes of Late Carboniferous Rift Volcanics in the Sepikou Region, Eastern Bogda, Xinjiang. Acta Petrologica Sinica. 30, 3539–3552. [in Chinese with English abstract]
30. Gu, L. X., Hu, S. X., Yu, C. S., Wu, C. Z., and Yan, Z. (2001a). Initiation and Evolution of the Bogda Subduction-torn-type Rift. Acta Petrologica Sinica. 17, 585–597. doi:10.3321/j.issn:1000-0569.2001.04.009 [in Chinese with English abstract]
31. Gu, L. X., Hu, S. X., Yu, C. S., Zhao, M., Wu, C. Z., and Li, H. Y. (2001b). Intrusive Activities during Compression-Extension Tectonic Conversion in the Bogda Intracontinental Orogen. Acta Petrologica Sinica. 2, 187–198. doi:10.3321/j.issn:1000-0569.2001.02.002 [in Chinese with English abstract]
32. Guo, H. C., Zhong, L., and Li, L. Q. (2006). Zircon SHRIMP U-Pb Dating of Quartz Diorite in the Koumenzi Area, Karlik Mountains, East Tianshan, Xinjiang, China, and its Geological Significance. Geol. Bull. China 25, 928–931. doi:10.3969/j.issn.1671-2552.2006.08.005 [in Chinese with English abstract]
33. Han, B. F., Song, B., Chen, L. H., and Li, Z. H. (2004). Zircon U-Pb SHRIMP Ages of the Mafic-Ultramafic Complexes from the Halatongke and Huangshandong Cu-Ni Sulfide Deposits in Xinjiang, and Their Tectonic Significance. Chin. Sci. Bull. 49,

2324–2328. [in Chinese with English abstract]. doi:10.1360/04wd0163

34. Han, C. M., Xiao, W. J., Zhao, G. C., Mao, J. W., and Mao, Q. G. (2004). Major Types, Characteristics and Geodynamic Mechanism of Upper Paleozoic Copper Deposits in Northern Xinjiang, Northwestern China. Ore Geology. Rev. 28, 308–328. doi:10.1016/j.oregeorev.2005.04.002

35. Han, C. M., Xiao, W. J., Wan, B., Ao, S. J., Zhang, J. E., Song, D. F., et al. (2018). Late Palaeozoic-Mesozoic Endogenetic Metallogenic Series and Geodynamic Evolution in the East Tianshan Mountains. Acta Petrologica Sinica. 34, 1914–1932. [in Chinese with English abstract]

36. Han, C., Xiao, W., Zhao, G., Ao, S., Zhang, J., Qu, W., et al. (2010). In-situ U-Pb, Hf and Re-os Isotopic Analyses of the Xiangshan Ni-Cu-Co deposit in Eastern Tianshan (Xinjiang), Central Asia Orogenic Belt: Constraints on the Timing and Genesis of the Mineralization. Lithos. 120, 547–562. doi:10.1016/j.lithos.2010.09.019

37. Han, C., Xiao, W., Zhao, G., Mao, J., Yang, J., Wang, Z., et al. (2006). Geological Characteristics and Genesis of the Tuwu Porphyry Copper deposit, Hami, Xinjiang, Central Asia. Ore Geology. Rev. 29, 77–94. doi:10.1016/j.oregeorev.2005.07.032

38. Han, Z. K., Han, C. M., Xiao, W. J., Zhao, G. C., Wang, Z. M, Ao, S. J., et al. (2014). Palaeozoic porphyry Cu-Au and ultramafic Cu-Ni deposits in the eastern Tianshan orogenic belt: Temporal constraints from U-Pb geochronology. Inter. Geol. Review 55, 842–862. doi:10.1080/00206814.2012.745977

39. Hedenquist, J. W. (1995). "The Ascent of Magmatic Fluid, Discharge versus Mineralization,". Magmas, Fluids and Ore Deposits Mineralogical Association of Canada, Short Course. Editor JFH Thompson, 23, 263–289.

40. Hou, G. S., Tang, H. F., and Liu, C. Q. (2006). Geochemical Characteristics of the Late Paleozoic Volcanics in Jueluotage Tectonic belt, Eastern Tianshan and its Implications. Acta Petrologica Sinica. 22, 1167–1177. doi:10.3321/j.issn:1000-0569.2006.05.009 [in Chinese with English abstract]

41. Hou, G. S., Tang, H. F., Liu, C. Q., and Wang, Y. B. (2005). Geochronological and Geochemical Study on the Wallrock of Tuwu-Yandong Porphyry Copper Deposits, Eastern Tianshan Mountains. Acta Petrologica Sinica. 21, 1729–1736. doi:10.3969/j.issn.1000-0569.2005.06.020 [in Chinese with English abstract]

42. Inoue, A., Kurokawa, K., and Hatta, T. (2010). Application of Chlorite Geothermometry to Hydrothermal Alteration in Toyoha Geothermal System, Southwestern Hokkaido, Japan. Resource Geology. 60, 52–70. doi:10.1111/j.1751-3928.2010.00114.x

43. Karakaya, M. Ç., Karakaya, N., Küpeli, Ş., and Yavuz, F. (2012). Mineralogy and Geochemical Behavior of Trace Elements of Hydrothermal Alteration Types in the Volcanogenic Massive Sulfide Deposits, NE Turkey. Ore Geology. Rev. 48, 197–224. doi:10.1016/j.oregeorev.2012.03.007

44. Keith, M., Häckel, F., Haase, K. M., Schwarz-Schampera, U., and Klemd, R. (2016). Trace Element Systematics of Pyrite from Submarine Hydrothermal Vents. Ore Geology. Rev. 72, 728–745. doi:10.1016/j.oregeorev.2015.07.012

45. Khalili, M., and Malekmahmoudi, F. (2012). "Geochemical Variations during Alteration of an Andesite-Basalt to Bentonite, in Khur, East of Isfahan, Iran," in Proceedings of the 10th International Congress for Applied Mineralogy (ICAM), 43, 361–368. doi:10.1007/978-3-642-27682-8_43

46. Large, R. R., Bull, S. W., and Maslennikov, V. V. (2011). A Carbonaceous Sedimentary Source-Rock Model for Carlin-Type and Orogenic Gold Deposits. Econ. Geology. 106, 331–358. doi:10.2113/econgeo.106.3.331

47. Large, R. R., Danyushevsky, L., Hollit, C., Maslennikov, V., Meffre, S., Gilbert, S., et al. (2009). Gold and Trace Element Zonation in Pyrite Using a Laser Imaging Technique: Implications for the Timing of Gold in Orogenic and Carlin-Style Sediment-Hosted Deposits. Econ. Geology. 104, 635–668. doi:10.2113/gsecongeo.104.5.635

48. Large, R. R., Maslennikov, V. V., Robert, F., Danyushevsky, L. V., and Chang, Z. (2007). Multistage Sedimentary and Metamorphic Origin of Pyrite and Gold in the Giant Sukhoi Log Deposit, Lena Gold Province, Russia. Econ. Geology. 102, 1233–1267. doi:10.2113/gsecongeo.102.7.1233

49. Leng, C. B. (2017). Genesis of Hongshan Cu Polymetallic Large deposit in the Zhongdian Area, NW Yunnan, Constraints from LA-IPC-MS of Trace Elements of Pyrite and Pyrrhotite. Earth Sci. Front. 24, 162–175. doi:10.13745/j.esf.yx.2016-11-61 [in Chinese with English abstract]

50. Li, H. Q., Chen, F. W., Lu, Y. F., Yang, H. M., Guo, J., and Mei, Y. P. (2004). Zircon SHRIMP U-Pb Age and Strontium Isotopes of Mineralized Granitoids in the Sanchakou Copper Polymetallic Depoist, East Tianshan Mountains. Acta Geologica Sinica. 22, 191–195. doi:10.3321/j.issn:1006-3021.2004.02.018 [in Chinese with English abstract]

51. Li, J. Y., Song, B., Wang, K. Z., Li, Y. P., Sun, G. H., and Qi, D. Y. (2006). Permian mafic-ultramafic complexes on the southern margin of the Tu-Ha basin, eastern Tianshan Mountians: geological records of vertical crustal growth in central Asia. Acta. Geoscientica Sinica 27, 424–446. doi:10.3321/j.issn:1006-3021.2006.05.006 [in Chinese with English abstract]

52. Li, F. M., Wang, Z. S., and Hou, W. B. (2002). Synthetic prospectiong model developed from Xiaorequanzi copper deposit in east Tianshan, Xinjiang. Xingjiang Geology 20, 38–43. doi:10.3969/j.issn.1000-8845.2002.01.009 [in Chinese with English abstract]

53. Li, W., Chen, J. L., Dong, Y. P., Xu, X. Y., Li, Z. P., Liu, X. M., et al. (2016). Early Paleozoic subduction of the Paleo-Asian Ocean: Zircon U-Pb geochronological and geochemical evidence from the Kalatag high-Mg andesites, East Tianshan. Acta Petrologica. Sinica. 32, 505–521. [in Chinese with English abstract]

54. Liu, M., Wang, Z. L., Zhang, Z. H., Chen, W. S., and Yang, D. (2009). Fluid Inclusion Geochemistry of Tuwu Porphyry Copper deposit, Eastern Tianshan in Xinjiang. Acta Geologica Sinica. 25, 1446–1455. [in Chinese with English abstract]

55. Long, L. L., Wang, J. B., Wang, Y., Deng, X. H., Mao, Q. G., Sun, Y., et al. (2019). Metallogenic Regularity and Metallogenic Model of the Paleo Arc-basin System in Eastern Tianshan. Acta Petrologica Sinica. 35, 3161–3188. [in Chinese with English abstract]

56. Long, L. L., Wang, J. B., Wang, Y. W., Mao, Q. G., Deng, X. H., Zhao, L. T., et al. (2016). Discussion on the Age of Ore-Hosting Volcanic Strata in Kalatage Ore Concentration Area, Eastern Tianshan, Xinjiang. Mineral. Exploration. 7, 31–37. doi:10.3969/j.issn.1674-7801.2016.01.004 [in Chinese with English abstract]

57. Lowell, J. D., and Guilbert, J. M. (1970). Lateral and Vertical Alteration-Mineralization Zoning in Porphyry Ore Deposits. Econ. Geology. 65, 373–408. doi:10.2113/gsecongeo.65.4.373

58. Ma, R. S., Shu, L. S., and Sun, J. (1997). Tectonic Evolution and Metalogeny of Eastern Tianshan Mountains. Beijing: Geological Publishing House

59. Ma, T. L., Wang, L. Q., Sun, L. Q., and Shu, B. (2003). Application of Magnetic Fabric Analysis to the Ductile Deformation Belt. Acta Geoscientia Sinica. 449–452. doi:10.3321/j.issn:1006-3021.2003.05.010 [in Chinese with English abstract]

60. MacLean, W. H. (1988). Rare Earth Element Mobility at Constant Inter-REE Ratios in the Alteration Zone at the Phelps Dodge Massive Sulphide deposit, Matagami, Quebec. Mineral. Deposita. 23, 231–238. doi:10.1007/bf00206399

61. Mao, J. W., Yang, J. M., Qu, W. J., Du, A. D., Wang, Z. L., and Han, C. M. (2002). Re-Os Dating of Cu-Ni Sulfide Ores from Huangshandong deposit in Xinjiang and its Geodynamic Significance. Mineral. Deposits. 21, 323–330. doi:10.1007/s11769-002-0037-5

62. Mao, Q. G., Fang, T. H., Wang, J. B., Wang, S. L., and Wang, N. (2010). The Geochronology Studies of the Early Paleozoic Honghai Massive Sulfide Deposits and its Geological Significance Kalatage belt, in Eastern Tianshan Mountain, Xinjiang, Northwest China. Acta Petrologica Sinica. 26, 3017–3026. [in Chinese with English abstract]

63. Mao, Q. G., Wang, J. B., Fang, T. H., Yu, M. J., and Sun, Y. (2017). Discovery of the Middle Devonian Yudai Porphyric Cu(Au) Deposit in the Kalatage Area of Eastern Tianshan Mountain,Xinjiang and and its Geological Prospecting Significance. Geology Exploration 53, 1–11. [in Chinese with English abstract]

64. Mao, Q. G., Wang, J. B., Fang, T. H., Yu, M. J., Zhu, J. J., Zhuang, R., et al. (2016). Geological Characteristics of Honghai VMS Type deposit of Kalatage Ore belt in the Eastern Tianshan. Mineral. Exploration. 7, 17–30. [in Chinese with English abstract]

65. Mao, Q. G., Wang, J. B., Yu, M. J., Ao, S. J., and Li, Y. C. (2020). Re-Os and U-Pb Geochronology for the Xiaorequanzi VMS deposit in the Eastern Tianshan, NW China: Constraints on the Timing of Mineralization and Stratigraphy. Ore Geology. Rev. 122, 1–14. doi:10.1016/j.oregeorev.2020.103473

66. Mao, Q. G., Xiao, W. J., Han, C. M., Sun, M., Yuan, C., Yan, Z., et al. (2006). Zirocn U-Pb Age and the Geochemistry of the Baishiquan Mafic-Ultramafic Complex in the Eastern Tianshan, Xinjiang: Constraints on the Closure of the Paleo-Asian Ocean. Acta Petrologica Sinica 22, 153–162. doi:10.3321/j.issn:1000-0569.2006.01.016 [in Chinese with English abstract]

67. Mao, Q. G., Xiao, W. J., Windley, B. F., yU, M. J., and Zhang, J. E. (2019). Early Permian Subduction-Related Transtension in the Turpan Basin, East Tianshan (NW China): Implications for Accretionary Tectonics of the Southern Altaids. Geol. Mag., 1–24.

68. Mao, Q., Wang, J., Xiao, W., Fang, T., Yu, M., Ao, S., et al. (2014a). Stratigraphic, U-Pb (Zircon) and Geochemical Constraints on Magmas, Mineralization and Geological Evolution of the Kalatage District, the Central Part of Dananhu Arc in Eastern Tianshan Mountains. Acta Geologica Sinica - English Edition. 88, 885–886. [in Chinese with English abstract]. doi:10.1111/1755-6724.12376_11

69. Mao, Q., Xiao, W., Fang, T., Windley, B. F., Sun, M., Ao, S., et al. (2014b). Geochronology, Geochemistry and Petrogenesis of Early Permian Alkaline Magmatism in the Eastern Tianshan: Implications for Tectonics of the Southern Altaids. Lithos. 190-191, 37–51. doi:10.1016/j.lithos.2013.11.011

70. Mao, Q., Yu, M., Xiao, W., Windley, B. F., Li, Y., Wei, X., et al. (2018). Skarn-mineralized Porphyry Adakites in the Harlik Arc at Kalatage, E. Tianshan (NW China): Slab Melting in the Devonian-Early Carboniferous in the Southern Central Asian Orogenic Belt. J. Asian Earth Sci. 153, 365–378. doi:10.1016/j.jseaes.2017.03.021

71. Michard, A., and Albarède, F. (1986). The REE Content of Some Hydrothermal Fluids. Chem. Geology. 55, 51–60. doi:10.1016/0009-2541(86)90127-0

72. Pan, H. D., Shen, P., Chen, G., Yang, J. T., and Dai, H. W. (2013). Volcanic-plutonic Complex, Ore-Forming Rocks and Their Alterations in Tuwu Porphyry Cu deposit of Xinjiang. Mineral. Deposits. 32, 794–808.

73. Parsapoor, A., Khalili, M., and Mackizadeh, M. A. (2009). The Behaviour of Trace and Rare Earth Elements (REE) during Hydrothermal Alteration in the Rangan Area (Central Iran). J. Asian Earth Sci. 34, 123–134. doi:10.1016/j.jseaes.2008.04.005

74. Pirajno, F. (2009). Hydrothermal Processes and Mineral System. Berlin, Germany: Springer Netherlands

75. Pirajno, F. (2013). The Geology and Tectonic Settings of China's Mineral Deposits. Berlin, Germany: Springer Netherlands. doi:10.1007/978-94-007-4444-8

76. Qin, K.-z., Su, B.-x., Sakyi, P. A., Tang, D.-m., Li, X.-h., Sun, H., et al. (2011). SIMS Zircon U-Pb Geochronology and Sr-Nd Isotopes of Ni-Cu-Bearing Mafic-Ultramafic Intrusions in Eastern Tianshan and Beishan in Correlation with Flood Basalts in Tarim Basin (NW China): Constraints on a Ca. 280 Ma Mantle Plume. Am. J. Sci. 311, 237–260. doi:10.2475/03.2011.03

77. Qin, Y. S., Zhang, J. C., Ding, K. S., Xu, Y. X., Tang, D. M., Xu, X. W., et al. (2009). Semi-empirical Likelihood Confidence Intervals for the Differences of Quantiles with Missing Data. Acta Math. Sin.-English Ser. 25, 845–854. [in Chinese with English abstract]. doi:10.1007/s10114-009-6476-5

78. Reed, M. H. (1997). "Hydrothermal Alteration and its Relationship to Ore Fluid Composition," in Geochemistry of Hydrothermal Ore Deposits. Editor H. L. Barnes. third ed. (New York: Wiley), 1–29.

79. Reich, M., Deditius, A., Chryssoulis, S., Li, J.-W., Ma, C.-Q., Parada, M. A., et al. (2013). Pyrite as a Record of Hydrothermal Fluid Evolution in a Porphyry Copper System: A SIMS/EMPA Trace Element Study. Geochimica Et Cosmochimica Acta. 104, 42–62. doi:10.1016/j.gca.2012.11.006

80. Rui, Z. Y., Wang, F. T., Li, H. H., Dong, L. H., Wang, L., Jiang, L. F., et al. (2001). New Progress in the Eastern Tianshan Porphyry Copper belt. Chin. Geology. 28, 11–16. doi:10.3969/j.issn.1000-3657.2001.02.003 [in Chinese with English abstract]

81. Rui, Z. Y., Wang, L. S., and Wang, Y. T. (2002). Discussion on Metallogenic Epoch of Tuwu and Yandong Porphyry Copper Deposits in Eastern Tianshan Mountains, Xinjiang. Mineral. Deposits. 21, 16–22. doi:10.3969/j.issn.0258-7106.2002.01.003 [in Chinese with English abstract]

82. Şengör, A. M. C., Natal'in, B. A., and Burtman, U. S. (1993). Evolution of the Altaid tectonic collage and Paleozoic crustal growth in Eurasia. Nature 364, 209–304. doi:10.1038/364299a0

83. Shen, P., Pan, H. D., Dong, L. H., Yang, J. T., Shen, Y. C., Dai, H. W., et al. (2012). Caldera Complex, Hosted Rocks and Alteration of the Yandong Porphyry Copper deposit in Eastern Tianshan, Xinjiang. Acta Petrologica Sinica. 28, 1966–1980. [in Chinese with English abstract]

84. Shen, P., Pan, H., and Dong, L. (2014a). Yandong Porphyry Cu deposit, Xinjiang, China-geology, Geochemistry and Sims U-Pb Zircon Geochronology of Host Porphyries and Associated Alteration and Mineralization. J. Asian Earth Sci. 80, 197–217. doi:10.1016/j.jseaes.2013.11.006

85. Shen, P., Pan, H., Zhou, T., and Wang, J. (2014b). Petrography, Geochemistry and Geochronology of the Host Porphyries and Associated Alteration at the Tuwu Cu deposit, NW China: a Case for Increased Depositional Efficiency by Reaction with Mafic Hostrock? Miner Deposita. 49, 709–731. doi:10.1007/s00126-014-0517-4

86. Shu, L., Wang, B., Zhu, W., Guo, Z., Charvet, J., and Zhang, Y. (2011). Timing of Initiation of Extension in the Tianshan, Based on Structural, Geochemical and Geochronological Analyses of Bimodal Volcanism and Olistostrome in the Bogda

Shan (NW China). Int. J. Earth Sci. (Geol Rundsch). 100, 1647–1663. doi:10.1007/s00531-010-0575-5

87. Song, B., Li, J. Y., Li, W. P., Wang, K. Z., and Wang, Y. B. (2002). SHRIMP Dating of the Dananhu and Kezirkalasayi Granitoids Batholiths in Southern Margin of Tuha Basin and Their Geological Implication. Xingjiang Geology. 20, 332–345. [in Chinese with English abstract]

88. Song, X. X., and Zhang, J. K. (1986). Minor Elements in Pyrite of Various Genetic Types Form China. Bulletin of the Institute of mineral Deposites, Chinese Academy of Geological Sciences. Beijing: Geological Publishing House

89. Sun, G. H., Li, J. Y., Gao, L. M., and Yang, T. N. (2005). Zircon SHRIMP U-Pb Age of a Dioritic Pluton in the Harlik Mountain, Eastern Xinjiang, and its Tectonic Implication. Geol. Rev. 51, 463–469. doi:10.3321/j.issn:0371-5736.2005.04.015 [in Chinese with English abstract]

90. Sun, Y., Wang, J. B., Li, Y. C., Wang, Y. W., Yu, M. J., Long, L. L., et al. (2017). Recognition of Late Ordovician Yudai Porphyry Cu (Au, Mo) Mineralization in the Kalatag District, Eastern Tianshan Terrane, NW China: Constraints from Geology, Geochronology, and Petrology. Ore Geology. Rev. 100, 220–236. doi:10.1016/j.oregeorev.2017.07.011

91. Sverjensky, D. A. (1984). Europium Redox Equilibria in Aqueous Solution. Earth Planet. Sci. Lett. 67, 70–78. doi:10.1016/0012-821x(84)90039-6

92. Tan, W., Han, R. S., Wang, L., Liu, F., Guo, Y. X. Y., Wang, M. Z., et al. (2016). REE Geochemical of Gold-Polymetallic deposit in Beiya, Western Yunnan Province. J. Chin. Soc. rare earths. 34, 113–128. [in Chinese with English abstract]

93. Tan, W., Han, R. S., Wang, L., Liu, F., Wang, M. Z., and Guo, Y. X. Y. (2018). Iron Source of the Beiya Porphyry-type Au-Polymetallic deposit in Northwestern Yunnan, China. Geochimica. 47, 541–553. doi:10.19700/j.0379-1726.2018.05.007 [in Chinese with English abstract].

94. Tang, J. H., Gu, L. X., Zheng, Y. C., Fang, T. H., Zhang, Z. Z., Gao, J. H., et al. (2006). Petrology, Geochemistry and Genesis of the Na-Rich Volcanic Rocks of the Kalatag Area, Eastern Tianshan. Acta Petrologica Sinica. 22, 1150–1166. doi:10.3321/j.issn:1000-0569.2006.05.008 [in Chinese with English abstract]

95. Taylor, S. R., and Mclennan, S. M. (1985). The Continental Crust: Its Composition and Evolution. J. Geology. 94, 57–72. doi:10.1086/629067

96. Terakado, Y., and Fujitani, T. (1998). Behavior of the Rare Earth Elements and Other Trace Elements during Interactions between Acidic Hydrothermal Solutions and Silicic Volcanic Rocks, Southwestern Japan. Geochimica Et Cosmochimica Acta. 62, 1903–1917. doi:10.1016/s0016-7037(98)00109-4

97. Wang, F. T., Feng, J., Hu, J. W., Wang, L., Jiang, L. F., and Zhang, Z. (2001). The geological characteristics and discovery significance of Tuwu porphyry copper deposit, Xinjiang. Geology. China 28, 36–39. doi:10.1007/s00126-018-0859-4 [in Chinese with English abstract]

98. Wang, J. B., Wang, Y. W., and He, Z. J. (2006). Ore Deposits as a Guide to the Tectonic Evolution in the East Tianshan Mountains,NW China. Geology. China, 461–469. [in Chinese with English abstract]

99. Wang, J. R., Li, T. D., Tian, L. P., Yu, M., Wang, H. T., Zhao, Z. X., et al. (2010). Late Paleozoic Tectono-Magmatic Evolution in Bogda Orogenic Belt, Xinjiang: Evidence from Geochemistry of Volcanic Rocks. Acta Petrologica Sinica 26, 1103–1115. [in Chinese with English abstract]

100. Wang, K. R. (1989). Genesis Mineralogy of the Earth and Universe. Hefei, Anhui Education Press

101. Wang, Y.-H., Xue, C.-J., Liu, J.-J., Wang, J.-P., Yang, J.-T., Zhang, F.-F., et al. (2015). Early Carboniferous Adakitic Rocks in the Area of the Tuwu deposit, Eastern Tianshan, NW China: Slab Melting and Implications for Porphyry Copper Mineralization. J. Asian Earth Sci. 103, 332–349. doi:10.1016/j.jseaes.2014.09.032

102. Wang, Y., Chen, H., Baker, M. J., Han, J., Xiao, B., Yang, J., et al. (2018). Multiple Mineralization Events of the Paleozoic Tuwu Porphyry Copper deposit, Eastern Tianshan: Evidence from Geology, Fluid Inclusions, Sulfur Isotopes, and Geochronology. Miner Deposita. 54, 1053–1076. doi:10.1007/s00126-018-0859-4

103. Wang, Y. F., Chen, H. Y., Xiao, B., Han, J. S., and Yang, J. T. (2016). Porphyritic-overlapped Mineralization of Tuwu and Yandong Copper Deposits in Eastern Tianshan Mountains, Xinjiang. Mineral. Deposits. 35, 51–68. doi:10.16111/j.0258-7106.2016.01.004 [in Chinese with English abstract]

104. Wang, Y. L., Zhang, Z. W., Zhang, J. W., You, M. X., Li, K., and Wang, B. L. (2016). Mineralogy and Sr-Nd Isotope Characteristics of Baixintan Ni-Cu deposit in Eastern Tianshan Mountains, Xinjiang, and Mineralization Process. Acta Geologica Sinica. 90, 2747–2758. [in Chinese with English abstract]

105. Wang, Y. H., Xue, C. J., Liu, J. J., Wang, J. P., Yang, J. T., Zhang, F. F., et al. (2014). Geochemistry, Geochronology, Hf Isotope, and Geological Significance of the Tuwu Porphyry Copper deposit in Eastern Tianshan, Xinjiang. Acta Petrologica Sinica. 30, 3383–3399. [in Chinese with English abstract]

106. Wilkinson, J. J., Chang, Z., Cooke, D. R., Baker, M. J., Wilkinson, C. C., Inglis, S., et al. (2015). The Chlorite Proximitor: A New Tool for Detecting Porphyry Ore Deposits. J. Geochemical Exploration. 152, 10–26. doi:10.1016/j.gexplo.2015.01.005

107. Wu, H., Li, H. Q., Chen, F. W., Lu, Y. F., Deng, G., Mei, Y. P., et al. (2006). Zircon SHRIMP U-Pb Dating of Plagiogranite Porphyry in the Chihu Molybdenum-Copper District, Hami, East Tianshan. Geol. Bull. China 25, 549–552. doi:10.3969/j.issn.1671-2552.2006.05.003 [in Chinese with English abstract]

108. Xiao, B., Chen, H., Hollings, P., Wang, Y., Yang, J., and Wang, F. (2018). Element Transport and Enrichment during Propylitic Alteration in Paleozoic Porphyry Cu Mineralization Systems: Insights from Chlorite Chemistry. Ore Geology. Rev. 102, 437–448. doi:10.1016/j.oregeorev.2018.09.020

109. Xiao, B., Chen, H. Y., Wang, Y. F., and Yang, J. T. (2017). Zircon U-Pb and Molybdenite Re-os Dating of the Tuwu–Yandong Cu Deposit Belt of the Eastern Tianshan Mountains and its Geological Significance. Geotectonica et Metallogenia 41, 145–156. doi:10.16539/j.ddgzyckx.2016.06.020

110. Xiao, W.-J., Zhang, L. C., Qin, K. Z., Sun, S., and Li, J. L. (2004). Paleozoic Accretionary and Collisional Tectonics of the Eastern Tianshan (China): Implications for the continental Growth of central Asia. Am. J. Sci. 304, 370–395. doi:10.2475/ajs.304.4.370

111. Xiao, W., Windley, B. F., Sun, S., Li, J., Huang, B., Han, C., et al. (2015). A Tale of Amalgamation of Three Permo-Triassic Collage Systems in Central Asia: Oroclines, Sutures, and Terminal Accretion. Annu. Rev. Earth Planet. Sci. 43, 477–507. doi:10.1146/annurev-earth-060614-105254

112. Yang, K., and Scott, S. D. (1996). Possible Contribution of a Metal-Rich Magmatic Fluid to a Sea-Floor Hydrothermal System. Nature. 383, 420–423. doi:10.1038/383420a0

113. Yao, F. J. (2006). The Study of Spectral Feature and Extraction of Remote Sensing Alteration Anomalies for Metallic Ore Deposits in Gobi Arid Zone and Application in

mineral Resources. Master Thesis. Beijing:Chinese Academy of Geological Sciences.

114. Ye, L., Cook, N. J., Ciobanu, C. L., Yuping, L., Qian, Z., Tiegeng, L., et al. (2011). Trace and Minor Elements in Sphalerite from Base Metal Deposits in South China: A LA-ICPMS Study. Ore Geology. Rev. 39, 188–217. doi:10.1016/j.oregeorev.2011.03.001

115. Yu, M., Wang, Y., Wang, J., Mao, Q., Deng, X., Sun, Y., et al. (2019). The Mineralization of the Kalatage Arc, Eastern Tianshan, NW China: Insights from the Geochronology of the Meiling Cu-Zn(-Au) deposit. Ore Geology. Rev. 107, 72–86. doi:10.1016/j.oregeorev.2018.12.009

116. Zang, W., and Fyfe, W. S. (1995). Chloritization of the Hydrothermally Altered Bedrock at the Igarapé Bahia Gold deposit, Carajás, Brazil. Mineralium Deposita. 30, 30–38. doi:10.1007/bf00208874

117. Zhai, Y. S., Yao, S. Z., and Cai, K. Q. (2011). Ore deposit Science. Third edition. Beijing: Geological Publishing House. doi:10.1145/2063348.2063363

118. Zhang, L., Xiao, W., Qin, K., and Zhang, Q. (2006). The Adakite Connection of the Tuwu-Yandong Copper Porphyry belt, Eastern Tianshan, Nw china: Trace Element and Sr-Nd-Pb Isotope Geochemistry. Miner Deposita. 41, 188–200. doi:10.1007/s00126-006-0058-6

119. Zhang, R., Mao, Q. G., Yu, M. J., Fang, T. H., and Cheng, F. W. (2017). Geochemical Characteristics and Zircon U-Pb Ages of the Magmatite in the Qiongdukuke Ag-Pb-Zn Polymetallic deposit of Xinjiang. Geology. Exploration. 53, 270–282. [in Chinese with English abstract]

120. Zhang, W. D., Wu, X. B., Deng, X. H., Mao, Q. G., Zhang, H. Q., Yang, L. Y., et al. (2018). Fluid Inclusions Constraints on the Origin of the Xiaorequanzi Deposit in Eastern Tianshan. Diqiu Kexue Zhongguo Dizhi Daxue Xuebao/earth Sci. J. China Univ. Geosciences. 43, 3036–3048. doi:10.3799/dqkx.2018.150

121. Zhang, W., Lentz, D. R., Thorne, K. G., and Massawe, R. J. R. (2020). Late Silurian-Early Devonian Slab Break-Off beneath the Canadian Appalachians: Insights from the Nashwaak Granite, West-central New Brunswick, Canada. Lithos. 358-359, 105393. doi:10.1016/j.lithos.2020.105393

122. Zhao, Z. N., Wang, Y. H., Wang, J. P., Dong, L. S., Wang, H., Zhang, F. F., et al. (2014). Geochemistry and Geochronology of Diorite Porphyrite of the Tuwu Copper Deposit and its Geological Significance, Eastern Tianshan. Acta Mineralogica Sinica. 34, 332–349. doi:10.16461/j.cnki.1000-4734.2014.04.012 [in Chinese with English abstract]

123. Zhou, J. Y., Cui, B. F., Xiao, H. L., Chen, S. Z., and Zhu, D. M. (2001). The Kangguertag-Huangshan Collision Zone of Bilateral Subduction and its Metallogenic Model and Programs in Xinjiang, China. Volcanology Mineral. Resour. 22, 252–263.

124. Zhou, M.-F., Michael Lesher, C., Yang, Z., Li, J., and Sun, M. (2004). Geochemistry and Petrogenesis of 270 Ma Ni-Cu-(PGE) Sulfide-Bearing Mafic Intrusions in the Huangshan District, Eastern Xinjiang, Northwest China: Implications for the Tectonic Evolution of the Central Asian Orogenic belt. Chem. Geology. 209, 233–257. doi:10.1016/j.chemgeo.2004.05.005

125. Zhou, T. F., Zhang, L. J., Yuan, F., Fan, Y., and David, R. C. (2010). LA-ICP-MS In Situ Trace Element of Pyrite from the Xinqiao Cu-Au-S deposit in Tongling, Anhui, and its Constraints on the Ore Genesis. Earth Sci. Front. 17, 306–319. [in Chinese with English abstract]

6

MINERALOGY, GEOCHEMISTRY, AND GENESIS OF AGATES FROM CHIHUAHUA, NORTHERN MEXICO

Maximilian Mrozik[1,2], Jens Götze[1], Yuanming Pan[3] and Robert Möckel[4]

[1]Institute of Mineralogy, TU Bergakademie Freiberg, Brennhausgasse 14, 09599 Freiberg, Germany

[2]Geowissenschaftliche Sammlungen, TU Bergakademie Freiberg, Brennhausgasse 14, 09599 Freiberg, Germany

[3]Department of Geological Sciences, University of Saskatchewan, Saskatoon, SK S7N 5E2, Canada

[4]Helmholtz Institute Freiberg for Resource Technology, Chemnitzer Str. 40, 09599 Freiberg, Germany

ABSTRACT

The present study aimed to investigate the genesis and characteristics of some of the world-famous agate deposits in the state of Chihuahua, Mexico (Rancho Coyamito, Ojo Laguna, Moctezuma, Huevos del Diablo, Agua Nueva). Geochemical and textural studies of host rocks showed that all the studied deposits are related to the same rock type within the geological unit of Rancho el Agate andesite, a quartz-free latite that shows clear indications of magma mixing. As a result of their large-scale distribution and various differentiation processes, as well as transport separation, different textures and local chemical differences between rocks of different localities can be observed. These differences have also influenced the properties of SiO_2 mineralization in the rocks. The mixing of near-surface fluids from rock alterations with magmatic hydrothermal solutions led to the accumulation of various elements in the SiO_2 matrix of the agates, which were, on the one hand, mobilized during secondary rock alteration (Fe, U, Ca, K, Al, Si) and, on the other hand, transported with magmatic fluids (Zn, Sb, Si, Zr, Cr). Different generations of chalcedony indicate a multi-stage formation as well as multiple cycles of filling the cavities with fluids. The hydrothermal fluids are presumably related to the residual solutions of a *rhyolitic volcanism, which followed the latitic extrusions in the area and probably caused the formation of polymetallic ore deposits in the Chihuahua area. The enrichment of highly immobile elements indicates the involvement of volatile fluids in the agate formation. The vivid colors of the agates are almost exclusively due to various mineral inclusions, which consist mainly of iron compounds.*

Keywords: *Agate, Chalcedony, Trace elements, EPR spectroscopy, Silica minerals, Agate colors, Cathodoluminescence, Geology, Rancho Coyamito, Ojo Laguna, Moctezuma, Agua Nueva*

INTRODUCTION

Agates from the state of Chihuahua in northern Mexico are some of the most beautiful and most expensive agates worldwide and are, therefore, of great importance, especially for the collector market and the jewelry industry. They occur in the Chihuahuan Desert in spatially and often widely separated deposits, mainly within the voluminous, intermediate volcanic units of Sierra del Gallego [1]. In these locations, they are mostly exploited commercially by private, small-scale miners or small companies and are of considerable economic importance because of their high prices in the world market.

Despite the high prominence of the deposits, only a few studies have been carried out on the exact characteristics and genesis of their silica mineralization. Therefore, the present study aimed to carry out a comprehensive geological and mineralogical investigation into some of the most important agate deposits in the Chihuahua region for the first time. The chemical and mineralogical compositions of the host rocks, as well as their textures and alterations, were used to assign the individual deposits to geological units and to check whether all of the agates in the region could be related to the same volcanic event. Through various mineralogical and geochemical analyses of the agate samples, conclusions can be drawn about the formation mechanism and the relative formation time of the mineralization in the area in general and in the individual deposits. The aim was to reconstruct the geological and mineralogical–geochemical processes from the actual data and to develop a model that clarifies the mechanisms of agate formation in northern Chihuahua in more detail.

GEOLOGICAL BACKGROUND

From a physiographic point of view, the agate-rich deposits that were studied are located in the area of the flat ranges and plains of northern Mexico, between the Sierra Madre Occidental to the west and the Sierra Madre Oriental to the east [2]. This area is primarily characterized by NNW-trending mountain ranges of Cretaceous sediments and Tertiary volcanic rocks, separated by broad, endorheic, intramontane basins, so-called 'bolsons' [3]. This region is located at an elevation of 1200 m to 2400 m above sea level, with the terrain rising towards the south. Because of the predominantly arid desert climate, a typical cuesta landscape with extensive pediments and steep fault terraces formed, resulting in the formation of extensive basins between the escarpments during the predominant weathering of soft rock units [3]. Drainage in the near-perennially dry area is directed primarily into the intramontane basins during the short monsoon period in summer, sometimes resulting in nondurable lakes in late summer [4].

The geology of the studied area is dominated by late Mesozoic to Cenozoic volcanic units of the El Sueco-Sierra del Gallego complex (Figure 1), which is located between the calc-alkaline volcanic provinces of the Sierra Madre Occidental and the alkaline volcanic province of Trans-Pecos [3,5]. Because of its geographic location and specific geochemical composition, it represents the intermediate igneous province between these two major complexes [5]. Within this area, the volcanic complex is about 1200 m thick and can be subdivided into four major units according to Bockoven [3]:

(1). Tuffs and volcanoclastic sediments of the Libres Formation;

(2). Voluminous intermediate-to-mafic lava flows of the Rancho el Agate andesite;

(3). Rhyolitic units with lava flows, domes, and veins (Mesteño, Gallego, Agua Nueva, Carneros, and El Dos);

(4). Basalt flows and flow tuffs of the Milagro Basalt (compare Figure 1).

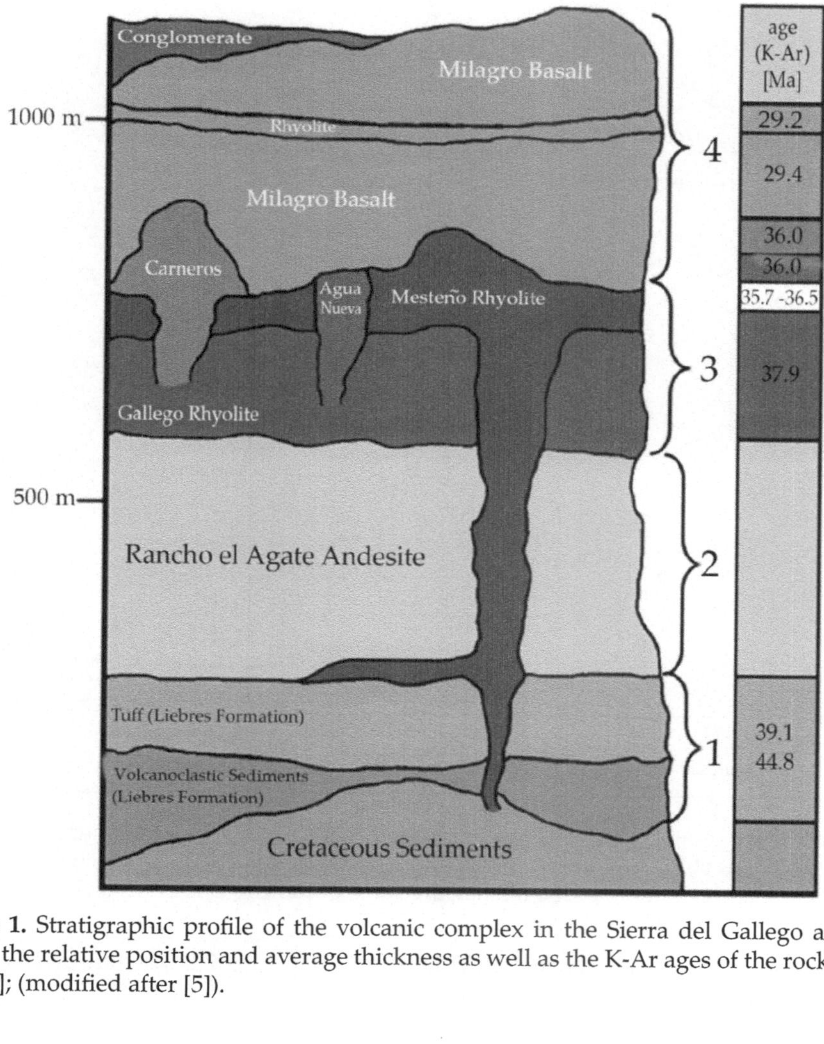

Figure 1. Stratigraphic profile of the volcanic complex in the Sierra del Gallego area. It shows the relative position and average thickness as well as the K-Ar ages of the rock units [in Ma]; (modified after [5]).

The subalkaline volcanism was dated to between 45 and 29 million years by Keller [5], with no volcanic activity for an extended period between 36 and 29.5 million years ago. No directly determined age data are available for the Rancho el Agate andesite, which is particularly relevant for this work. However, based on the potassium argon ages (K-Ar) of the overlying and underlying units, an age between 39 and 38 million years can be assumed [5].

Field observations indicate that all deposits are related to the same volcanic unit, probably even to the same lava flow of the so-called Rancho el Agate andesite. The agates occur mainly in former vesicular cavities, where they are paragenetically associated with iron oxides/hydroxides and carbonates. Pseudomorphs are also not uncommon. The current investigations indicated direct, hydrothermal silicification processes as well as secondary silicification by SiO_2, which was released during the host rock alteration.

MATERIALS AND METHODS

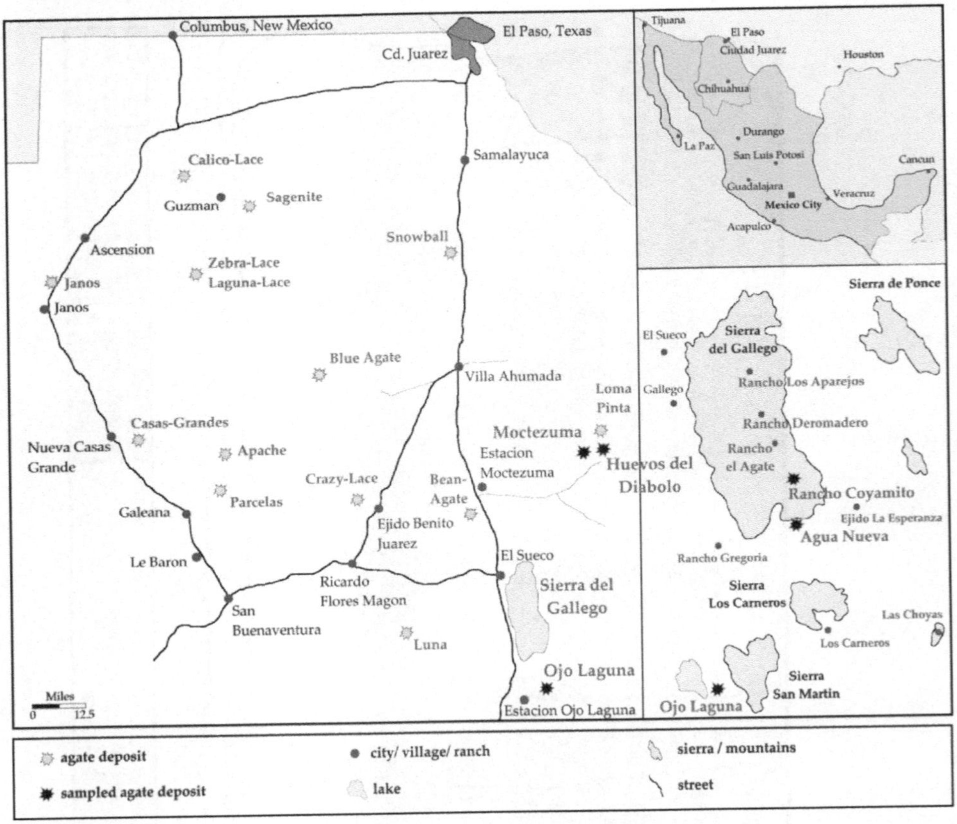

Figure 2. Map of agate deposits in the northern part of the state of Chihuahua. Marked are the deposits studied in this work and the Rancho el Agate Ranch, which, according to Bockoven [3], is to be seen as the main source area of Rancho el Agate andesite; (modified after [6]).

The samples were collected during a field trip in 2020. The studied deposits (Figure 2) are specifically located in Agua Nueva (AN), Rancho Coyamito (Coy),

Moctezuma (Mo), Huevos del Diabolo (HdD), and Arcoiris claim near Ojo Laguna (OL), with Rancho Coyamito being sampled at three different claims (the Los Alamos claim (CLA), the Japanese deposit (CLJ), and the La Fortuna claim (CLF)) and the Agua Nueva deposit at two locations (AN, ANZ). For each deposit, the studied material consists of both the representative host rock and associated agate and chalcedony samples. The mineralization and sample points in the agates were always considered to be representative for the locality. If possible, all visibly distinguishable chalcedony or quartz generations of different samples were separated in order to be able to compare them within the individual agate samples and between the individual deposits. The color, shape, and type of banding as well as the preparability of the individual chalcedony bands were considered during the selection.

Eight different rock samples (Table 1) were crushed manually to the size of <400 μm using a steel mortar. In addition, individual macroscopically recognizable bands of the agates were selected primarily according to color and texture criteria and then separated (Table 2). After the rough crushing, a McCrone percussion mortar with ZrO_2 grinding elements was used. The crushed agate samples were selected by hand under binoculars and the single-colored bands, i.e., sample points, were separated. Subsequently, the separated subsamples were milled by hand in an agate mortar.

Table 1. Description of the host rock samples

Sample	Color	Deposit	Occurence	Properties
AN	red-purple	Agua Nueva	1	porphyritic, fresh
ANZ	gray-purple	Agua Nueva	2	microcrystalline, aphyric
CLJ	gray	Coyamito	Japanese deposit	porphyritic, fresh
CLA	gray	Coyamito	Los Alamos claim	porphyritic, fresh
CLF	gray-red	Coyamito	La Fortuna claim	porphyritic, fresh, with big agate nodules
HdD	greenish-gray	Huevos del Diabolo	Huevos del Diabolo	aphyric, strongly altered
Mo	gray	Moctezuma	Moctezuma	porphyritic, fresh
OL	red	Ojo Laguna	Arcoiris claim	porphyritic, strongly altered, impregnated with chalcedony

The mineral composition of the agate host rocks was analyzed using the X-ray diffractometer Orion Comet P2 (XRD-Eigenmann GmbH, SEIFERT Analytical X-ray) with Co-Kα radiation and the semiconductor detector Meteor OD in the measuring range from 5 to 80° 2-Theta. A total of 20 wt.% corundum was added to each of the powdered rock samples as an internal standard for the quantification of X-ray amorphous phases. For single mineral measurements, separate mineral grains were crushed in an agate mortar and prepared on a polished silicon carrier using technical ethanol (96%). Clay mineral investigations were carried out on special texture preparations using the sedimentation method on a glass carrier. Measurements were taken on a X-ray diffractometer HZG4 (Seifert/Freiberger Präzisionsmechanik, Freiberg, Germany) using Cu-Kα radiation and a scintillation counter as the detector in the measuring range from 2 to 13° 2-theta. The evaluation of all measurements was performed using the Profex software [7].

Table 2. Description of the separated agate, chalcedony, and quartz samples

Sample	Color	Deposit	Occurence	Host Rock	Type
AAN_c	colorless	Agua Nueva	1	AN	chalcedony
AANZ	yellow	Agua Nueva	2	ANZ	agate
AOL_b	red-brown	Ojo Laguna	Arcoiris	OL	agate
AOL_w	white	Ojo Laguna	Arcoiris	OL	agate
ACLF_r	red	Coyamito	La Fortuna	CLF	agate
ACLF_w	white	Coyamito	La Fortuna	CLF	agate
AMo_g	yellow	Moctezuma	Moctezuma	Mo	agate
AMo_r	red	Moctezuma	Moctezuma	Mo	agate
AMo_q	colorless	Moctezuma	Moctezuma	Mo	quartz
ACLA_r	red	Coyamito	Los Alamos	CLA	agate
ACLA_w	colorless-white	Coyamito	Los Alamos	CLA	chalcedony
AHdD_b	dark-brown	Huevos del Diabolo	Huevos del Diabolo	HdD	agate
AHdD_bi	light-brown	Huevos del Diabolo	Huevos del Diabolo	HdD	agate
AHdD_q	colorless	Huevos del Diabolo	Huevos del Diabolo	HdD	quartz
ACJ_r	red	Coyamito	Japanese	CLJ	agate
ACJ_v	violet	Coyamito	Japenese	CLJ	agate

Polished thin sections (30 μm) of the rocks and of one agate per occurrence were prepared. These thin sections were examined using transmitted light microscopy on an Axio Imager.A1m polarization microscope, equipped with an AxioCam color 105 and the AxioVision software (ZEISS, Thornwood, NY, USA). Studies using scanning electron microscopes (SE, BSE, EDS) were also performed on the polished and carbon-coated thin sections. These investigations were performed using a JEOL 6400 SEM with EDX detector (JEOL Ltd., Akishima, Japan).

Cathodoluminescence investigations (CL) using a "hot cathode" cathodoluminescence microscope HC1-LM (LUMIC, Bochum, Germany) were performed on the polished and carbon-coated thin sections of agates and rock samples [8]. This instrument was operated at a 14 kV accelerating voltage and a current of 0.2 mA (10 μA/mm^2). Images were acquired using a Peltier-cooled digital video camera (OLYMPUS DP72, OLYMPUS Deutschland GmbH, Hamburg, Germany). CL spectra in the range of 370 to 1000 nm were measured under standardized conditions using an Acton Research SP-2356 digital spectrograph equipped with a Princeton Spec-10 CCD detector and WinSpec software (OLYMPUS Deutschland GmbH, Hamburg, Germany). Wavelength calibration was performed using an Hg-halogen lamp, a spot width of 30 μm, and a 5 s measuring time.

The precrushed and divided rock samples were milled (<63 μm) in an agate grinding pot using a Retsch vibratory disk mill. The powders were dried overnight at 105 °C, then annealed for 2 h at 950 °C to determine the loss on ignition. Subsequently, homogeneous fused beads were prepared from 1 g of the annealed sample powder and 8 g of lithium tetraborate. Measurements were carried out using a wavelength dispersive PANalyticalAxios Minerals spectrometer and WROXI package (PANalytical, Almelo, The Netherlands).

Trace-element analyses were carried out using 100 mg of powdered rock and agate samples dissolved with a hydrofluoric acid digestion containing nitric acid and mixed with 1 mL rhenium solution (100 μg/L) as an internal standard. Measurements were taken with a Perkin Elmer Sciex Elan 5000 quadrupole mass spectrometer (ICP-MS) with a crossflow nebulizer and a rhyton spray chamber (Perkin Elmer Inc., Baesweiler, Germany) [9].

The paramagnetic centers of the powder agate samples were analyzed using electron paramagnetic resonance (EPR) spectroscopy with a Bruker EMX spectrometer (Saskatchewan Structural Science Centre, Saskatoon, Canada). Measurements were performed on ~100 mg powdered sample material at both RT (295 K) and 30 K. RT-EPR measurements were operated at a microwave frequency of ~9.86 GHz, using three sets of experimental conditions: (1) a wide scan in the magnetic field of 0–700 mT at a microwave power of ~2.0 mW, a modulation frequency of 100 KHz, a modulation amplitude of 0.5 mT, and a spectral resolution of 0.05 mT; (2) a narrow scan in the magnetic field of 335–360 mT at a microwave power of ~0.21 mW, a modulation frequency of 100 KHz, a modulation amplitude of 0.1 mT, and a spectral resolution of 0.01 mT; and (3) a narrow scan in the magnetic field of 320–360 mT using a microwave power of ~21.0 mW, a modulation frequency of 100 KHz, a modulation amplitude of 0.1 mT, and a spectral resolution of 0.01 mT.

Powder EPR measurements at 30 K were taken on the same spectrometer equipped with an Oxford cryostat using liquid helium at a microwave frequency of ~9.38 GHz with two sets of experimental conditions: (1) a scan range in the magnetic field of 320–370 mT at a microwave power of ~0.21 mW, a modulation frequency of 100 KHz, a modulation amplitude of 0.1 mT, and a spectral resolution of 0.01 mT; and (2) a scan range in the magnetic field of 320–370 mT at a microwave power of ~21.0 mW, a modulation frequency of 100 KHz, a modulation amplitude of 0.1 mT, and a spectral resolution of 0.01 mT.

Raman spectroscopy was performed using an XplorRa-Plus Raman spectrometer (Horiba France SAS, Palaiseau, France) operated with a 532 nm laser. The measurements were taken on polished thin sections at 100× objective magnification in the ranges of 0 cm^{-1} to 1500 cm^{-1} and 3020 cm^{-1} to 4100 cm^{-1} Raman shift. In both intervals, 200 spectra with a 1 s acquisition time each were averaged per measurement point.

RESULTS

Host Rocks

Chemistry

The results of the X-ray fluorescence analyses are shown in Table 3. The loss on ignition of each sample ranged from 2.5 wt.% to 4.3 wt.%. All of the host rocks showed very similar chemical compositions, which were characterized by an intermediate SiO_2 content, a high iron content, and a constant alkali metal content. Only the sample from Ojo Laguna (OL) showed a significantly higher SiO_2 content. A clear correlation among the main elements was revealed exclusively in the negative correlation of the calcium–potassium ratio. The commonly used TAS classifications of Cox [10], LeBas [11], and Middlemost [12] classify almost all of the investigated samples as intermediate rocks in the trachyandesite field (Figure 3). Thus, the rocks plot near the boundary between alkaline and subalkaline volcanism, with the stronger subalkaline influence becoming particularly evident in the classification of LeBas [11]. A similar classification of the rocks as trachyandesites is shown in the Zr/TiO_2-SiO_2 plot [13] and in the Nb/Y-Zr/Ti plot [14], with the rocks in the latter being classified at the andesite/trachyandesite boundary. A more exact subdivision is possible based on Na–K ratios, in which, according to Shelley [15], all examined trachyandesites are to be classified as latites. This also indicates the R1-R2 classifications [16], in which the samples are classified among the trachyandesites predominantly as latites and quartz-latites. Ultimately, this classification can also be confirmed by the purely trace-element-based classifications, as in the Co-Th plot [17], in which the rocks plot in the field of strongly potassic, calc-alkaline latites. One exception in the common chemical classifications is the host rock of Ojo Laguna (OL). It can be classified as either dacite [11,12], rhyodacite [16], or very acidic andesite [10] according to the TAS and R1-R2 classifications, primarily because of its higher SiO_2 content and low alkali metal content.

Table 3. Chemical composition of the investigated host rocks. XRF analysis. [wt.%]

	AN	CLF	HdD	Mo	CLJ	CLA	ANZ	OL
LOI	2.93	2.51	3.02	4.24	3.05	3.52	3.75	4.27
Total	99.95	99.78	99.66	98.97	99.35	99.57	99.36	99.03
SiO_2	58.98	57.65	55.33	53.79	58.22	57.50	55.57	62.19
TiO_2	1.40	1.63	1.57	1.51	1.45	1.64	1.39	1.27
Al_2O_3	15.55	16.30	15.45	15.28	15.31	16.48	16.69	12.72
Fe_2O_3	8.04	7.50	11.31	8.86	6.83	7.51	7.93	9.12
Mn_3O_4	0.19	0.06	0.09	0.15	0.08	0.11	0.13	0.03

MgO	1.45	2.15	0.89	2.16	1.36	1.68	0.75	0.72
CaO	3.72	4.31	3.15	5.89	4.89	3.40	5.94	2.86
Na_2O	3.69	3.73	3.79	3.57	3.87	3.58	3.81	2.71
K_2O	3.35	3.29	4.10	2.90	3.64	3.47	2.65	2.54
P_2O_5	0.39	0.41	0.68	0.38	0.40	0.43	0.45	0.37
V_2O_5	0.02	0.02	0.02	0.03	0.02	0.02	0.03	0.02
SrO	0.03	0.04	0.04	0.05	0.03	0.03	0.07	0.03
ZrO_2	0.07	0.06	0.08	0.05	0.07	0.06	0.05	0.05
BaO	0.13	0.11	0.13	0.10	0.12	0.12	0.13	0.11
ZnO	0.01	0.01	0.01	0.01	0.01	0.01	0.01	0.01

a

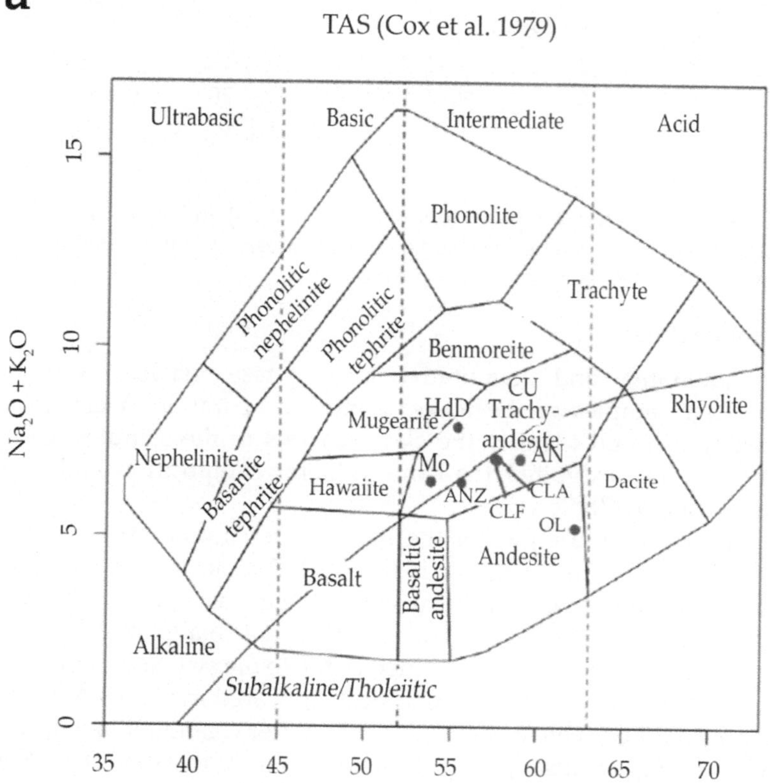

b

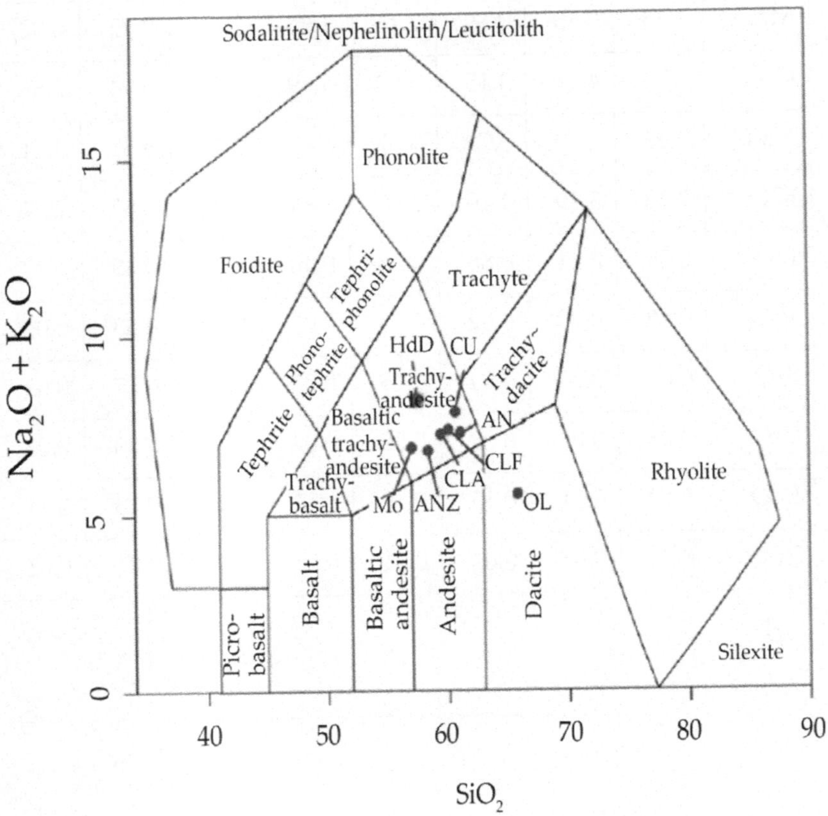

Figure 3. TAS-classification diagrams following Cox [10] (a) and Middlemost [12] (b) show that nearly all host rocks can be classified as trachyandesite. Only the sample from Ojo Laguna shows a different, more acidic composition.

Phase Composition

Both qualitative and quantitative X-ray phase analyses were performed on all powder samples of the rocks. Texture preparations were also made to better distinguish and identify the clay minerals in the samples. Qualitatively, the different rocks exhibited the same mineral composition and a significant amorphous content (Table 4).

The clay mineral analyses of the texture preparations showed a clear and symmetrical reflex in the range of 15.0 Å to 15.5 Å in the untreated (air-dried) samples. In addition to this dominant signal, there was a peak at 10.0 Å in almost all samples. After treatment with ethylene glycol, a change in the position of the peaks at about 1.5 Å was observed, which is due to the swelling of the sheet silicate layers. In addition, an indistinct, broad peak appeared in the 8.3 Å to 8.5 Å range. No changes could be seen at the 10 Å peaks. After treatment at 400 °C, peaks in the small angle region disappeared completely and intensities at 10 Å increased. In the samples in which no peak had previously been present in this area, a new, broad peak appeared. Judging from the X-ray diffractograms of the HdD sample, which show the typical 006 peak at 1.509 Å [18], and the microscopic observations, the 10 Å mica is celadonite. This is also supported by the missing 002 peak, which is suppressed by the strong scattering of the octahedral iron in celadonite [18]. The typical behaviors of swellable clay minerals indicate that they are smectites, which does not contain significant mixed-layer components [19]. The 006 peak, which was dominant in the 1.49 Å to 1.50 Å range, shows that the samples contained predominantly dioctahedral smectite [18]. Only in sample OL can a trioctahedral smectite be assumed from the distinct 006 peak in the range of 1.545 Å [18].

Table 4. Quantitative composition of the samples. Determined with XRD and the program Profex using 20 wt.% corundum as internal standard. [data in wt.%]

	AN	ANZ	CLJ	CLA	CLF	Mo	HdD	OL
Plagioclase	25.1	35.0	26.1	27.6	30.9	28.1	23.8	20.9
Albite	10.6	9.9	11.1	13.8	9.6	11.7	17.3	7.6
Sanidin-Na_{16}	16.1	13.0	16.0	18.5	20.3	12.1	13.4	12.6
Sanidin-Na_{67}	4.1	3.8	5.0	1.8	1.2	5.5	7.6	1.1
Augite	6.3	2.2	5.5	2.9	5.9	5.8	2.6	2.2
Fluorapatite	3.4	1.1	4.2	2.0	1.4	1.9	1.5	5.5
Hematite	3.0	2.1	2.1	1.8	3.0	0.8	0.9	4.8
Goethite	0.2	0.8	0.6	0.0	0.0	1.8	2.5	0.0
Rutile	0.3	0.3	0.1	0.0	0.1	0.0	0.4	0.2
Magnetite	0.0	0.0	0.0	0.0	0.0	0.8	2.0	0.0
Opal-CT	3.1	3.6	1.0	0.0	0.9	0.6	2.6	7.4
Cristobalite	3.9	4.9	5.1	2.9	5.1	4.4	4.3	2.1
Quartz	0.2	0.0	3.3	1.2	0.9	2.5	0.0	9.6
Calcite	0.2	2.3	2.2	0.0	0.3	1.9	0.0	1.0
Smectite	13.8	7.5	8.0	17.8	14.6	9.7	2.6	12.9
Celadonite	0.4	0.0	0.0	0.0	0.0	0.0	4.2	0.0
Amorphous content	9.3	13.5	9.6	9.6	5.9	12.4	14.4	12.3
Total	100.0	100.0	100.0	100.0	99.9	100.0	100.0	100.0

Microscopic Studies

Most of the rocks were rich in phenocrysts, while only two of the samples were completely or nearly aphyric. The majority displayed a rather uniform appearance with a porphyric-dominated texture. They comprised partly large phenocrysts of fresh plagioclase as well as single feldspar crystals, which showed a distinct resorption structure. The rock samples from the central sites of Rancho

Coyamito exhibited a fine-grained matrix with squat plagioclase crystals, which were on average 60–70 μm in size. The matrix structure was predominantly intergranular with a strong tendency to be intersertal and transitioned in some areas to felsitic, especially in the CLA sample (Figure 4c). Swallowtail plagioclase was concentrated in some zones inside the matrix of sample CLJ, where it was also combined with skeletal crystals. In the sample CLA, neither the swallowtail plagioclase nor the skeletal crystals occur. The samples from Moctezuma (Mo) and Agua Nueva 2 (ANZ) had intergranular- to intersertal-dominated matrix that included feldspar crystals at an average size of 100 μm (Figure 4b). Swallowtail plagioclase was found only very subordinately in the matrix of these samples.

The matrix of sample OL was also predominantly intersertal, only subordinately intergranular, and of a mainly red appearance (Figure 4d). Many of the matrix feldspars were almost exclusively swallowtail plagioclase or skeletal crystals and were on average 100 μm in length. The matrix of Agua Nueva, as well as that of Ojo Laguna, was intergranular to strongly intersertal and red in appearance.

In general, plagioclase phenocrysts within the rocks of Sierra del Gallego were found mainly as fresh, single-crystal individuals of up to 10 mm in size (Ø 1.5 mm) (feldspar generation 1) (Figure 5a). In the central areas of Rancho Coyamito and Agua Nueva, glomeroporphyritic aggregates as well as synneutic intergrowth occurred predominantly. In marginal deposits, these aggregates either did not appear (Mo) or occurred only subordinately in smaller agglomerates (OL), and phenocrysts were usually smaller and isolated. Synneutic intergrowth occurred less frequently.

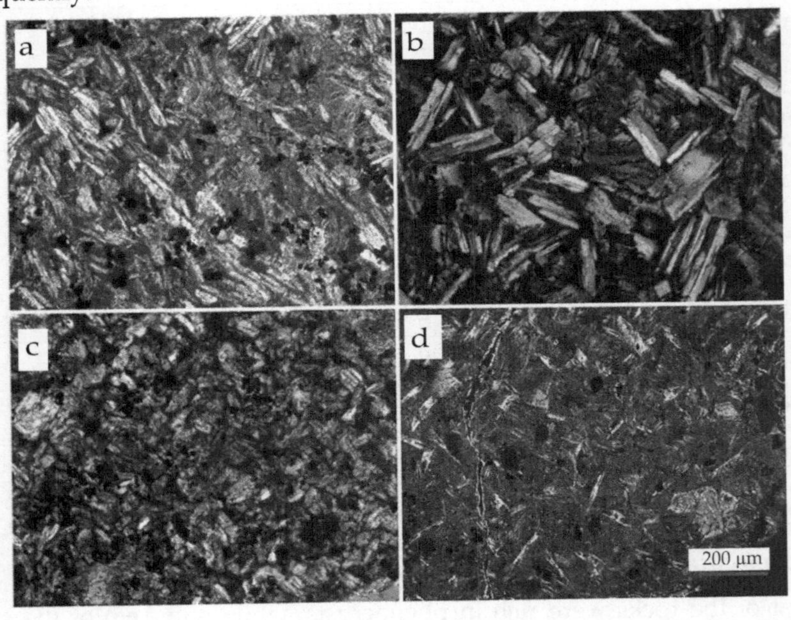

Figure 4. Micrographs in transmitted light (crossed polars) showing typical matrix textures in the studied samples. (a) Poorly trachytic matrix due to slight alignment of plagioclase in sample HdD; (b) intergranular, slightly glassy matrix of sample ANZ; (c) strong interlocking of plagioclase crystals in felsitic matrix of sample CLA; (d) strongly glassy, intersertal matrix with swallowtail plagioclase and skeletal crystals in sample OL.

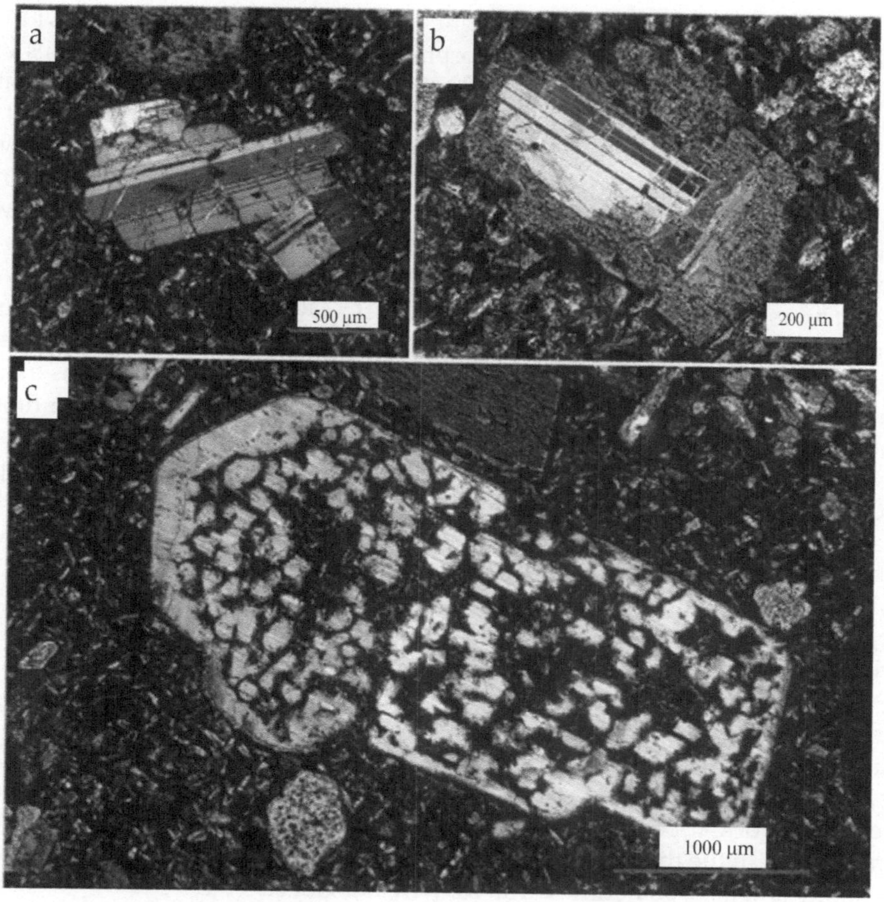

Figure 5. Micrographs in transmitted light (crossed polars) showing the different generations of feldspar in the host rock CLJ. (a) Fresh polysynthetically twinned plagioclase crystal of generation 1; (b) sieve textured plagioclase crystal with fresh core of generation 2; (c) idiomorphic feldspar crystal with large embayment-like resorption structures of generation 3; all images are from sample CLJ.

Glomeroporphyritic aggregates, in particular, usually showed no or only very weak resorption phenomena on feldspar crystals and often contained a lot of pyroxene. In contrast, some plagioclase phenocrysts occurred subordinately with only a small fresh area in the core of the crystal. The outer areas were completely altered and showed a typical fine-grained sieve structure (feldspar generation 2) (Figure 5b). A few of these crystals also had large embayments. A third feldspar generation was characterized by similar intense resorption phenomena and was often simple twinned. The crystals formed mostly idiomorphic individuals, with their edges apparently formed by epitaxially regrown feldspar. Resorption phenomena subordinately caused the typical fine-grained sieve structures but appeared much more frequently as large embayments (Figure 5c). Within the embayments, mostly glassy-dominated areas or crystallized matrix could be found. Large phenocrysts, in particular, sometimes showed several generations of resorption and epitaxial growth. Many feldspar crystals also possessed pronounced oscillatory zoning, which was still evident in even the highly resorbed individuals. In addition to the feldspars, hypidiomorphic-to-idiomorphic clinopyroxenes with sizes of up to 1000 µm (Ø 250 µm–400 µm) commonly occurred in the central deposits. The pyroxenes were mostly untwinned and commonly found as inclusions in the resorbed feldspar phenocrysts. While only a few crystals were fresh, most showed clear signs of alteration at the rim or along the cleavage planes. Some larger individuals were already completely altered, while others showed an incipient uralitization. Apatite occurred predominantly in long prismatic-to-acicular idiomorphic crystals in all of the porphyritic samples. It was found in the matrix as well as poikilitic in the phenocrysts. Overall, apatite occurred much more frequently in the central deposits, where the crystals were also significantly larger in size than in the marginal localities. In the latter ones, apatite was mainly concentrated in small crystals within the matrix. Opaque minerals were mostly xenomorphic, rarely hypidiomorphic, and mainly confined to the matrix structure. However, they often also formed large individuals of up to 600 µm. The samples of the central areas did not contain microscopic amygdales or vesicles, while these were common in the marginal deposits. In most cases, they were circular (sometimes oval), of very variable size, and filled with either calcite, celadonite, zeolite, or chalcedony. In the southern deposit of OL, vesicles occurred mainly as irregular-flat or angular cavities of up to 7 mm in size. Most of them were filled with calcite, and some with chalcedony.

The matrix of aphyric rocks from Huevos del Diabolo (HdD) can be described as predominantly microcrystalline intergranular-to-vitreous intersertal. It consists, to a large extent, of elongated feldspars with an average size of 100–200 µm. To a large extent, these reveal a weak alignment (trachytic microstructure) (Figure 4a). Overall, the rock appeared predominantly greenish, both macroscopically and in thin sections. In the thin sections, amygdales of up to 4 mm were found filled with celadonite and also, sometimes, iron (oxy-hydr)oxides.

Agate

Macroscopic and Microscopic Appearance

The color spectrum of the individual bands in the examined agates ranged from colorless and white over yellow, brown, pink, or red in different nuances to a strong violet, whereby the color could change laterally within chalcedony bands. The different colors were not characteristic of single occurrences, but certain color preferences could be associated with the different localities [1,6]. Thus, strong violet colors as well as dark red or yellow colors appeared predominantly in rocks from Rancho Coyamito, while Moctezuma, which is located to the north, was characterized primarily by rocks with pastel pink, yellow, or red tones (Figure 6) [1]. Similar differences could be seen in the color distribution of individual bands. While agates from Ojo Laguna showed almost exclusively sharply delineated and finely banded colors, the Coyamito agates displayed a less sharp, or smooth, color delineation [1,6]. These significant similarities and the fact that almost all colors and shapes occurred at all localities suggest that the deposits had similar formation processes. However, the varying predominance as well as subtle differences in the characteristics of color and banding indicate specific and slightly differing geochemical situations and conditions during agate formation among the individual deposits in Sierra del Gallego.

The agate samples appeared both macroscopically and in the polarization microscope as classical 'wall-lining' agates, in which chalcedony banding is parallel to the outer wall of the former rock cavity. Only this wall-parallel banding was found in the examined samples. The agates consisted exclusively of length-fast chalcedony. The quartzine (length-slow) variety was not observed in the investigated Chihuahua agates of the Rancho el Agate andesite. In addition, a typical crystallization series (from microcrystalline to spherulitic chalcedony to fibrous chalcedony and, finally, to macrocrystalline quartz, Figure 7) occurred several times in some cases in the same agate.

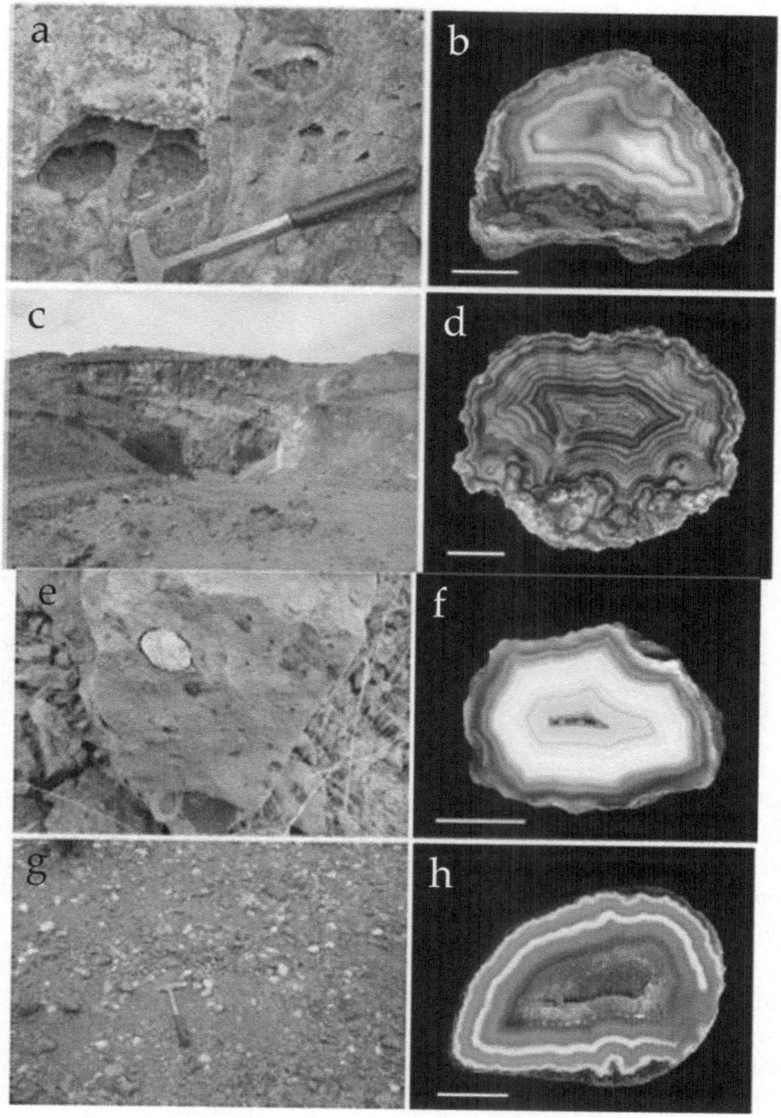

Figure 6. Investigated agate deposits and representative samples (scale bar for agates is 2 cm); (a) Upper part of the lava flows of the Rancho el Agate andesite at the Rancho Coyamito with empty vesicles and those filled with agate; (b) characteristic Coyamito agate; (c) strongly altered volcanic rocks and tuff layers at the Arcoiris claim, Ojo Laguna; (d) Laguna agate with typical fine banding; (e) vesicular andesite at the Moctezuma deposit; (f) Moctezuma agate; (g) weathering debris with loose agate geodes at Huevos del Diablo; (h) agate from Huevos del Diablo with characteristic brownish-yellow color banding; (i) reddish andesite from Agua Nueva containing agate geodes; (k) agate from Agua Nueva.

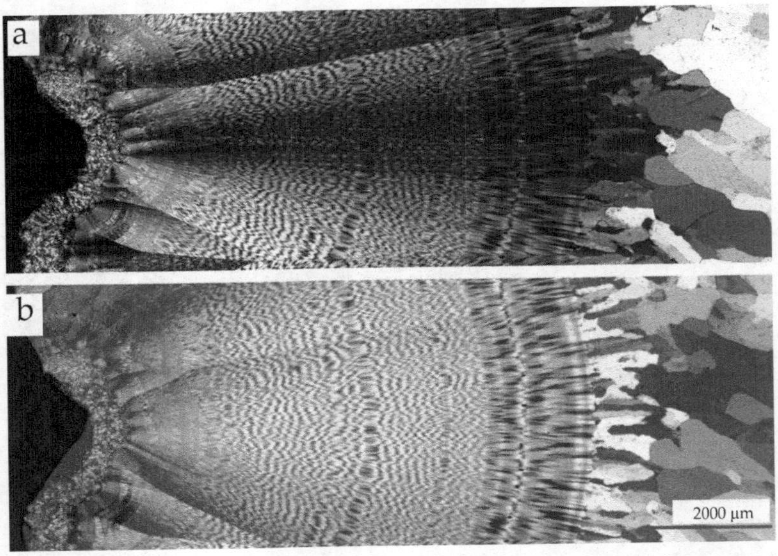

Figure 7. A complete agate thin section from Moctezuma depicted in transmitted light (crossed polars (a) with additional λ-compensator (b)). The typical sequence for wall-lining agates is clearly visible; from left to right: microcrystalline/spherulitic chalcedony to fibrous chalcedony to macrocrystalline quartz; in the lower image with λ-compensator, the continuous length-fast chalcedony is visible.

The initial microcrystalline or spherulitic chalcedony layer was usually only very thin and did not occur in all agates. Only the agate samples that had a strongly irregular boundary with the host rock or contained host rock fragments and initial mineral precipitates (mainly ferruginous) had a very dominant generation of microcrystalline chalcedony. The majority of the chalcedony in the agates consisted of long, fibrous chalcedony. Some of the individual chalcedony layers showed a "Runzelbänderung," which is developed differently in the individual bands of the agates, while it is usually more distinct in the central areas of the agates. Some single chalcedony layers that were contrastive in color in the thin section showed a very strong, pronounced "Runzelbänderung." In most specimen, the chalcedony layers were followed by macrocrystalline quartz. The quartz varied from being thin crusts to dominant layers with idiomorphic crystals. In some cases, these crystals were overgrown by further generations of chalcedony. In one agate, a second generation of chalcedony could be detected that did not overgrow macrocrystalline quartz but directly followed the long, fibrous chalcedony. This was interrupted by microcrystalline chalcedony, which showed a completely new spherulitic growth at the beginning of the layer and ended in long, fibrous chalcedony.

The agates showed a strongly varying spectrum of macroscopic and microscopic inclusions. These differed in their type and color as well as in their size and distribution.

White mineral inclusions were predominantly found in the outer areas of the agates, their abundance decreasing significantly towards the center of the mineralization.

Overall, iron-rich, yellow-red mineral inclusions were dominant. The major macroscopically recognizable minerals were arranged wall-parallel along the crystallization fronts of the chalcedony bands. Smaller inclusions were distributed irregularly within the individual chalcedony bands in addition to having a wall-parallel arrangement. Additionally, a Zr-bearing phase as well as several spheroidal and semi-transparent "bubbles" of up to 100 μm in size and of an orange-brown color occurred in the bands of some agates of Moctezuma and Huevos del Diabolo (Figure 8).

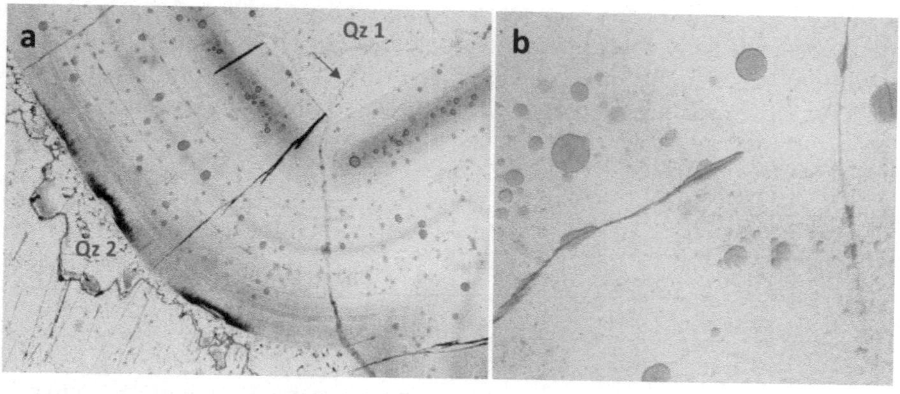

Figure 8. Micrographs in transmitted light showing orange-brown inclusions in an agate sample of Huevos del Diablo (HdD). In the overview image (a), the irregular distribution of the bubblelike inclusions is clearly visible, as well as the secondary fractures, which extend into quartz generation 1 (Qz 1) (red arrow) and end at generation 2 (Qz 2); In the detail image (b), a similarity between the bubblelike inclusions and those along the secondary fractures can be seen.

The "bubbles" were aligned with banding but occurred irregularly and seemed randomly distributed within these individual bands. They occurred only in individual chalcedony bands in the mentioned agates and were always anisotropic in their overall appearance under crossed polars. Some of these spheres additionally contained a darker colored core. A few of them were elliptically elongated along chalcedony fibers.

These inclusions occurred in one sample within a second chalcedony generation, which had grown on macrocrystalline quartz. There, they were associated with secondary fractures that also extended into the quartz. The fissures within the chalcedony were filled with identical reddish-brown inclusions. A similar optical effect and zoning of darker central areas and lighter transparent rims, as in the "bubbles," could be observed along the fissures. In the macrocrystalline quartz, the filling of these fissures with secondary minerals was not observed.

Trace Elements

The trace-element composition of the investigated agate samples is compiled in Table 5. Overall, the depletion of certain trace elements was detected in the agates compared to the related host rocks, in which the trace-element contents were mostly higher in chalcedony compared to the associated macrocrystalline quartz, which always had the lowest concentrations. In particular, the immobile trace elements Th and Ta could not be detected in any of the agate samples because the contents were below the detection limit. Furthermore, the elements Nb, Rb, Co, Sc, V, and Cs were detected in contents below 1 ppm in the agates. Other trace elements, such as Li, Cu, Sr, Y, and Ba, which were mostly present in concentrations below 5 ppm in the agates, have occurred in higher concentrations in some discrete bands of several agates.

Only a few trace elements were enriched in the silica matrix of the agates. Several hundreds of ppm were analyzed in the chalcedony matrix of the agates for Fe, Na, K and Ca (Table 5). Elevated concentrations were also detected for Al and Mg, which additionally showed a strong statistic scatter. A high aluminum content was especially found in white agate bands, while, in reddish agates, the iron content was usually more elevated as a result of their iron (oxy-hydr)oxide inclusions, which mainly occurred within these red areas of the agates.

Table 5. Trace-element contents (in ppm) of the separated chalcedony and macrocrystalline quartz bands analyzed using ICP-MS

	AAN_C	ACJ_r	ACJ_v	AOL_w	AOL_r	AHdD_b	AHdD_bi	AHdD_q
Li	0.933	0.364	0.322	1.34	1.34	1.43	3.21	10.0
Na	241	131	119	197	213	324	464	304
Mg	22.9	16	15.7	51	22.6	19.7	23.7	15
Al	53.2	34.3	25.1	78.2	57.5	207	96.7	19.5
K	114	66.3	67.6	109	143	240	227	98.6
Ca	268	223	250	388	639	205	464	175
Sc	0.151	<0.1	<0.1	0.109	<0.1	<0.1	0.164	<0.1
Ti	3.02	3.93	1.31	3.51	3.52	21.3	24.3	1.65
V	0.218	1.83	0.313	0.415	0.36	0.521	0.207	<0.1

Cr	17.4	1.96	28.3	40.8	24.3	50.3	21.0	11.3
Mn	3.55	1.99	3.74	13.1	4.29	54.6	11.6	2.05
Fe	456	748	256	455	471	604	522	77.9
Co	0.133	0.029	0.247	0.319	0.209	0.376	0.179	0.086
Cu	1.16	1.57	1.61	2.94	10.2	1.96	1.31	0.846
Zn	11.9	4.63	14.4	7.26	14	9.43	27.8	11
Cs	<0.01	<0.01	0.015	1.98	1.19	0.065	0.067	<0.01
Rb	0.593	0.262	0.185	0.982	0.706	1.58	0.596	0.215
Sr	4.21	2.84	13.2	16.0	15.0	21.2	142	1.56
Y	3.17	0.65	2.39	2.75	2.29	1.88	7.45	1.32
Nb	0.152	0.070	0.065	0.086	0.071	0.095	0.318	0.044
Sb	1.91	2.60	1.01	305	343	0.407	0.202	0.099
Ba	2.4	2.27	1.73	0.982	1.03	7.32	133	0.574
Ta	<0.1	<0.1	<0.1	<0.1	<0.1	<0.1	<0.1	<0.1
Th	<0.5	<0.5	<0.5	<0.5	<0.5	<0.5	<0.5	<0.5
U	16.5	31.5	8.31	5.29	8.87	11.5	8.25	2.26
	AMo_g	**AMo_r**	**AMo_q**	**ACLF_w**	**ACLF_r**	**ACLA_w**	**ACLA_r**	**AANZ**
Li	1.88	0.423	1.35	0.769	0.772	0.904	0.821	0.971
Na	372	557	394	273	192	243	346	139
Mg	24.5	17.9	17.6	139	27.6	23.2	19.2	39
Al	76.2	23.5	25.9	78.9	42.3	49.4	33.7	28.6
K	179	148	61	140	123	97.7	161	88.4
Ca	285	255	197	311	225	156	191	236
Sc	0.142	0.103	<0.1	0.268	<0.1	<0.1	<0.1	<0.1
Ti	2.55	<0.1	1.19	4.39	3.54	5.22	20.7	6.06
V	0.223	<0.1	<0.1	0.241	0.264	0.216	0.305	0.173
Cr	32.8	4.27	4.90	19.2	14.5	12.9	14.2	18.3
Mn	7.45	6.17	2.01	4.14	8.24	5.06	5.85	3.96
Fe	633	538	141	361	654	262	863	969
Co	0.256	0.045	0.048	0.208	0.115	0.151	0.130	0.167
Cu	2.26	6.96	3.16	4.7	3.97	1.86	1.2	4.07
Zn	6.53	14.8	4.44	13.4	16	4.25	187	6.69
Cs	0.040	0.044	0.025	0.192	0.118	0.097	0.191	0.227
Rb	0.411	0.239	0.131	0.599	0.616	0.475	0.768	0.620
Sr	35.1	2.34	1.82	3.91	2.95	1.76	2.41	4.33

Y	3.33	0.371	1.52	1.76	1.69	3.1	2.55	1.11
Nb	0.055	0.035	0.031	0.053	0.031	0.028	0.153	0.047
Sb	1.26	3.61	11.3	1.13	1.29	4.88	2.47	0.314
Ba	3.22	3.19	0.856	4.46	5.38	1.78	2.53	1.49
Ta	<0.1	<0.1	<0.1	<0.1	<0.1	<0.1	<0.1	<0.1
Th	<0.5	<0.5	<0.5	<0.5	<0.5	<0.5	<0.5	<0.5
U	13.3	24.4	10.8	13.9	30.6	7.18	32.5	19

Elevated concentrations were also measured for the elements Sb and U. Whereas uranium was enriched in almost all agate samples (up to >30 ppm), Sb reached high concentrations only in the agate samples from Ojo Laguna (>300 ppm). Such anomalous concentrations of Sb and U (also Cs, Sr) were also detected in the volcanic host rocks from Ojo Laguna, where they were up to 200 times and 20 times higher, respectively, compared to the host rocks of the other localities.

Contents of manganese and zinc were generally rather low and homogeneously distributed in all samples. Remarkably, there were pronounced Zn, Ba, and Sr anomalies in individual agate bands. Chromium occurred in some of the agate samples in high concentrations (up to 50.3 ppm), sometimes even higher than in the associated host rocks. Moreover, Cr and Co contents exhibited a straight, statistical correlation (Figure 9a).

a

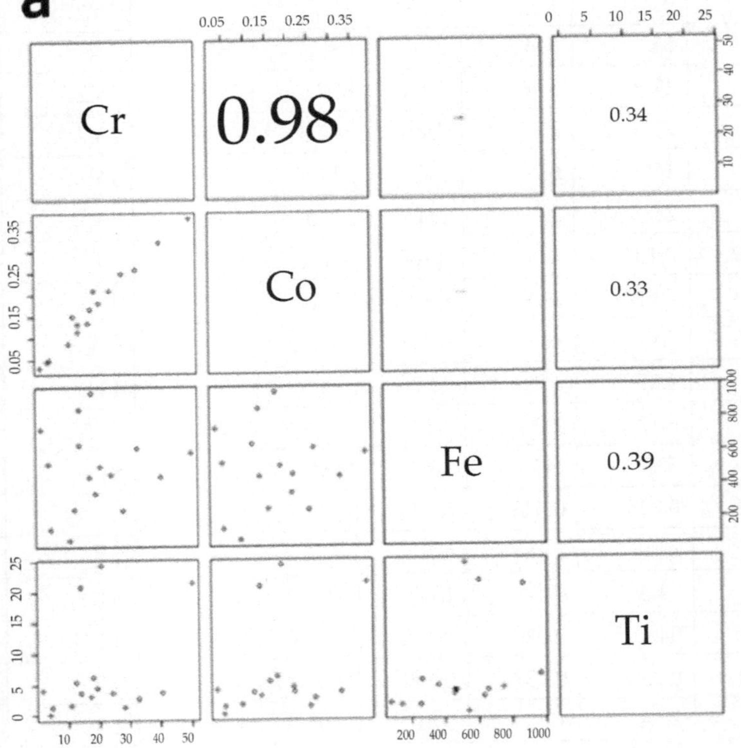

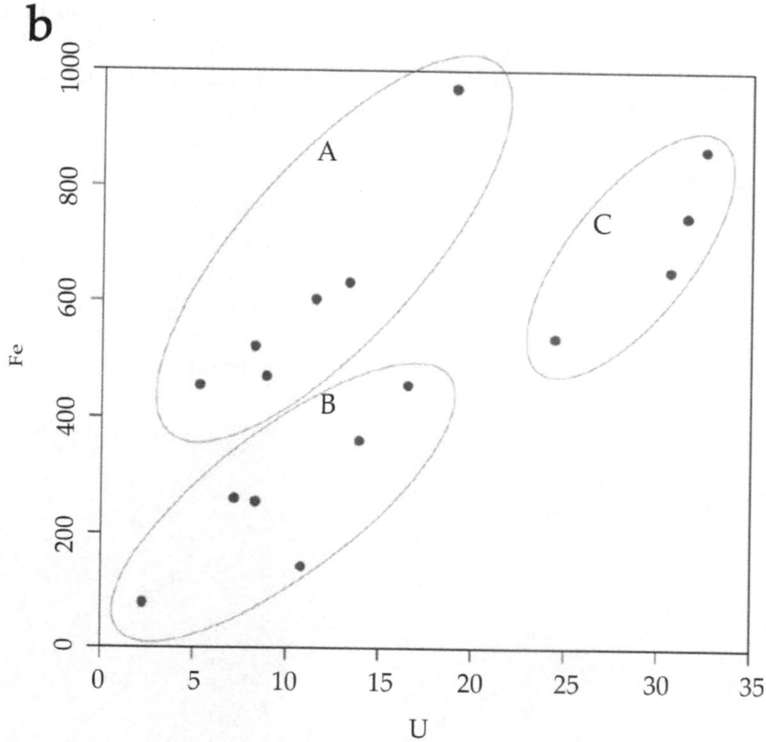

Figure 9. Correlation pattern of Cr, Co, Fe, and Ti of the agate samples (a). Note the perfect correlation of Cr and Co and Fe and U content in ppm (b). Marked are the differently colored agate samples: red agate (A), colorless/white agate and quartz (B), and brown/yellow agate (C).

Cathodoluminescence (CL)

The results of the CL investigations identified differing luminescence colors and associated chalcedony generations in the agates. The main CL emission bands were found at 450 nm (blue) and 650 nm (red) (Figure 10). The appearance of the ~450 nm (2.8 eV) emission band cou ld be related to two different defects in quartz. In Ti-poor quartz, the band is mainly associated with oxygen deficiency centers (ODC), but, in Ti-rich quartz, the intensity of the 450 nm band also shows a correlation with the concentration of trace Ti [20]. The red emission band at ~650 nm (1.95 eV) is attributed to the so-called non-bridging oxygen hole center (NBOHC) with a number of different precursors [21].

At least one generation of blue luminescent chalcedony occurred in most of the agates. If present, this chalcedony generation always marked the beginning of amygdale filling and was, therefore, generally present in the outermost regions. The emission spectra of this chalcedony type exhibited the pronounced band, which remained stable under prolonged electron irradiation, at 450 nm. In addition, in all agates, the red emission band appeared at 650 nm which had almost no influence on the visual luminescence color because of its low intensity. In a few samples, an indistinct peak at about 690–700 nm could be found below this broad band.

Furthermore, in some samples, green luminescence colors could be detected within some chalcedony bands. The spectra show several sharp lines between 500 nm and 600 nm (Figure 10), which can be associated with the luminescence of the uranyl ion (UO_2^{2+}) [22]. These lines mostly followed the blue chalcedony bands with dominant 450 nm luminescence emissions or formed alternating layers with them (Figure 10). Sometimes, the colors overlapped, in which cases both bands could be seen in different intensity ratios within the spectra. Green uranyl luminescence never appeared in the marginal areas of a mineralization.

Yellow luminescence with a pronounced broad band at about 570 nm was detected exclusively in the agate sample from Ojo Laguna (AOL), where it prevailed throughout chalcedony (Figure 10). The intensity of the band decreased continuously alongside prolonged electron bombardment. This CL emission was not detected in any of the other agate samples.

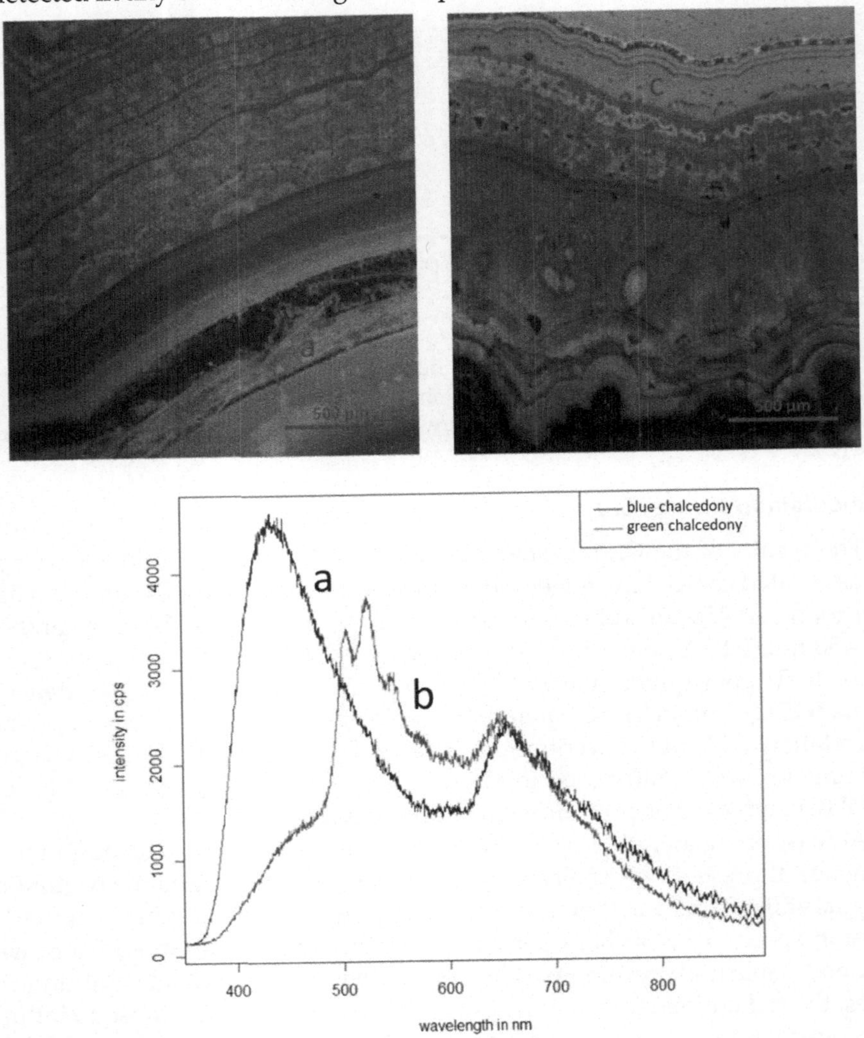

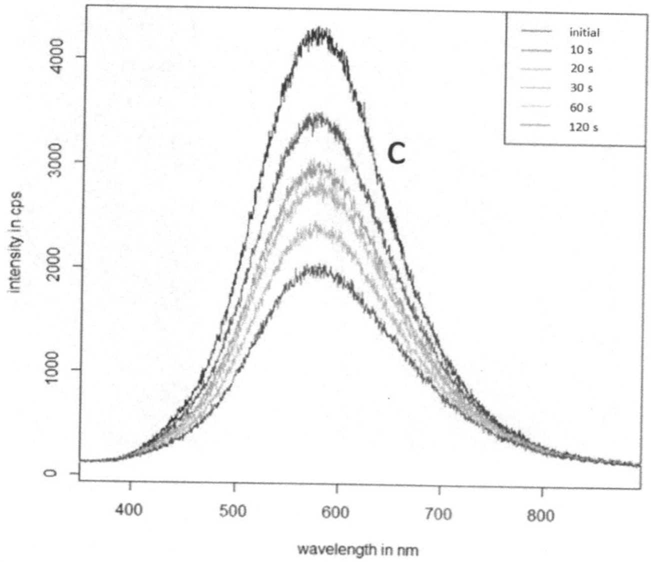

Figure 10. CL images of the blue (a), green (b), and yellow (c) luminescent chalcedony with associated spectra. The letters a-c represent measuring points and their associated spectra. Sample ACJ (left); sample AOL (right). Note the spherulitic growth in chalcedony at the transition from blue luminescent chalcedony to green luminescent chalcedony.

In the macrocrystalline quartz parts of the agates, blue, yellowish-green, and red luminescence prevailed. A stable blue emission band at about 450 nm and broad bands in the region of 650 nm to around 720 nm were mainly dominant in the spectra. In the spectra of the greenish luminescent quartz, the typical sharp emission peaks of uranyl ion [22] occurred between 500 nm and 600 nm. Quartz crystals with yellow CL exhibited a transient broad band between 500 nm and 600 nm, which disappeared during electron bombardment, whereas the partially overlapping red and blue bands remained stable. The strongly pronounced sector zoning under CL within the quartz crystals of all samples, which otherwise appeared homogeneous in polarized light, as well as oscillatory growth zoning, is also remarkable.

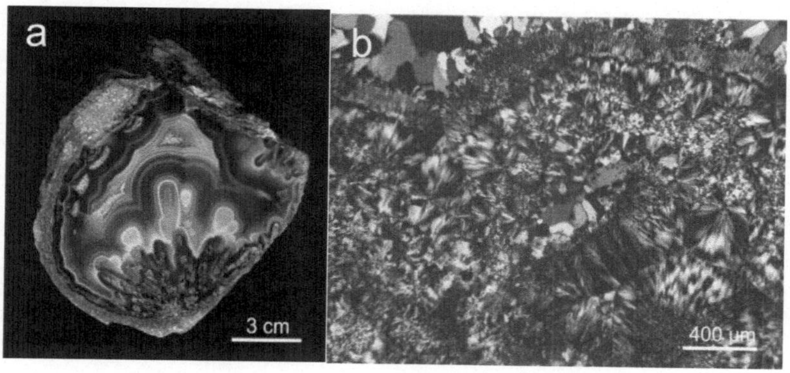

Figure 11. (a) Typical pseudomorphs in an agate from Rancho Coyamito; (b,c) micrographs of the pseudomorphic micro-structure in transmitted light (crossed polars (b)) and CL (c); the pre-existing minerals were almost completely replaced by chalcedony and macrocrystalline quartz; only small relics of former calcite were detectable by CL (orange dots, see arrows).

A common feature in agates, especially those found near Rancho Coyamito, was the presence of pseudomorphs of chalcedony/agate after carbonates. The initial minerals were, in most cases, completely replaced by chalcedony or macrocrystalline quartz. Only small relics of the initial carbonates (mainly calcite) were detected during CL measurements (Figure 11).

Electron Paramagnetic Resonance (EPR)

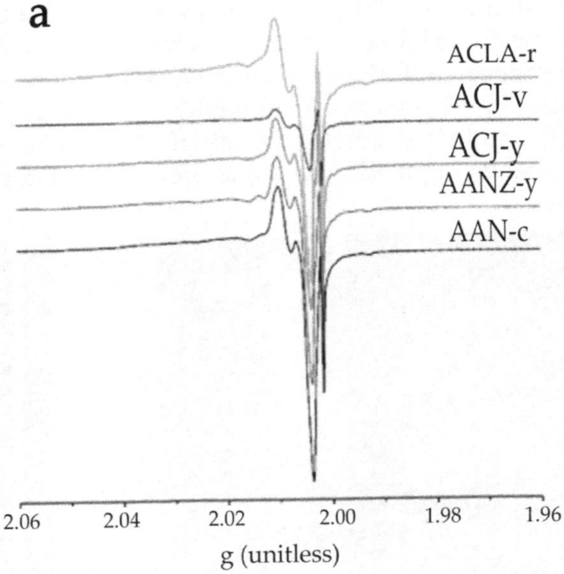

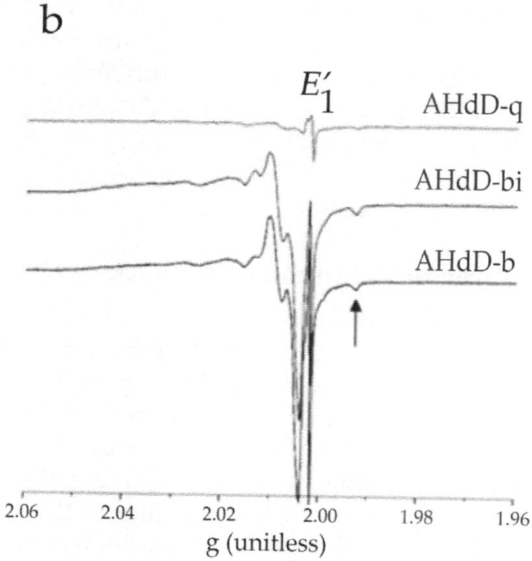

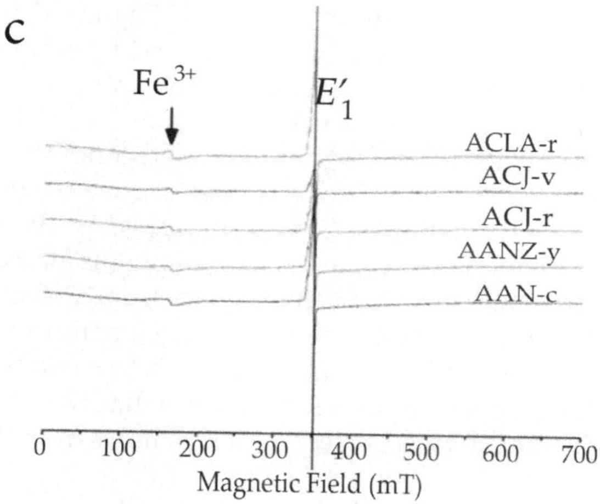

Figure 12. EPR spectra of some chalcedony samples (a) as well as a comparison of chalcedony and macrocrystalline quartz from Huevos del Diabolo (b). The E'_1 defect center, another signal at g = 1.993 (arrow), presumably related to various Ge-related electron centers, as well as the clearly lower intensity of these signals in the macrocrystalline quartz samples are notable; measuring condition: (2) at RT. Powder EPR-spectra wide scans (c) showing the rhombic Fe^{3+} center at g = 4.28 and radiation-induced defects, including the E'_1 center, at g = ~2.00; measuring condition: (1) at RT.

The EPR spectra of all examined chalcedony samples showed a weak signal of the rhombic Fe^{3+} center at g = 4.28 as well as various radiation-induced lattice defects, including the pronounced E'_1 center at g = ca. 2.00, in the wide scan measured at room temperature (Figure 12). From the narrow scans, it is obvious that the studied macrocrystalline quartz samples displayed lower intensities of these defects (Figure 12), showing that the defect density in chalcedony was much higher than in the associated macrocrystalline quartz. Furthermore, the "hole centers" in chalcedony also differed from those in quartz and include an unknown defect at $g_1 = 2.010$, $g_2 = 2.006$, and $g_3 = 2.003$. Additional signals were present at the effective g values of ~2.001, 1.998, and 1.993, and could possibly be assigned to various Ge-related electron centers, such as $[GeO_4/M^+]^0$, where M = Li or Na [23,24] (Figure 12).

DISCUSSION

Characteristics and Formation of Volcanic Host-Rocks

The mineralogical and chemical compositions of the investigated volcanic rocks show that the studied agates are related to the same host rock within the unit of the Rancho el Agate andesite. In addition, similarities between the major- and trace-element data indicate that all deposits belong to the same lava flow. This assumption is strengthened by the common geochemical classifications. Based on the mineral compositions determined using XRD, all samples can be classified as quartz-poor to quartz-free latite according to the Streckeisen diagram. Small differences in their quantitative-phase compositions can be explained by the differing textures of the rocks and their varying amorphous contents.

The distint chemical composition of the Ojo Laguna deposit, which led to a different classification, can probably be explained by primary and secondary effects. Both an initial deviant magma composition and a secondary overprint of the volcanic host rocks may be responsible for its higher SiO_2 content as well as for differences in its alkali metal ratios and elevated concentrations of some trace elements (e.g., Sb, U, Cs, Sr). In general, the Ojo Laguna magma appears to have been slightly more acidic than the others, as indicated by the yellow CL with an emission band at 570 nm. Such characteristic yellow luminescence of chalcedony has previously been detected exclusively in agates from acidic volcanic host rocks and is therefore a strong indication concerning the geological and physico-chemical environment of the mineralization [25]. Nevertheless, the strong equivalence between the major element compositions indicates that all of the studied volcanic host rocks belong to the intermediate series of lava flows from the Rancho el Agate andesite.

Bockoven [3] described a vesicular texture in the whole rock unit (at least twelve lava flows), which was most pronounced in the two youngest and thickest flows. Therefore, the occurrence of agate is theoretically possible in different lava units with slightly differing chemical compositions. This assumption was confirmed by Keller [5], who assumed several source channels, which are supposed to be mainly located on the eastern flank of Sierra del Gallego, for the Rancho el Agate andesite in their study area. The isolated location of the Ojo Laguna deposit in Sierra San Martin, which is both geographically and morphologically separated from Sierra del Gallego, reinforces the assumption of an independent source in this area. The chalcophile character of antimony additionally suggests the involvement of hydrothermal fluids in the agate formation [26], which may have also affected the composition of the host rock.

The various matrix textures in the different samples, from nearly completely intergranular to almost completely intersertal as well as single felsitic areas, indicate very differentiated cooling mechanisms. The whole unit can be classified as a near-surface formation, in which an extensive lava flow with high fluid lava cannot be excluded [3] but is rather implausible as a result of the lack of flow textures in the rocks of most localities.

The different contents and sizes of the phenocrysts additionally indicate a transport separation of phenocryst-rich and -poor magma fractions during lava flow in addition to differentiated cooling at the surface. The samples that were situated furthest from the main source area of the rock unit near Rancho el Agate, as suggested by Keller [5] and Bockoven [3], had the highest amorphous contents. Likewise, they had the highest tendency to contain skeletal crystals, swallowtail plagioclase, and a glassy matrix (Ojo Laguna, Huevos del Diabolo). In contrast, the central localities, especially on the eastern flank of Sierra del Gallego, were characterized by intergranular-to-felsitic textures with many phenocrysts and large apatite crystals. This is in agreement with the work of Bockoven [3], who described the occurrence of long prismatic apatite crystals within the Rancho el Agate andesite only in phenocryst-rich samples, which mainly occur in the vicinity of the possible source center of the magma near the Rancho el Agate. Therefore, the textures are primarily geographically specific properties of the rocks that were formed by differentiation and transport processes during the emplacement of the lava and depend significantly on their distance from the magma source.

As already assumed by Bockoven [3], the presence of resorbed feldspar phenocrysts indicates a magma mixing. Some feldspars were already crystallized during magma mixing and were partially resorbed by this process as a result of their original composition [15]. These were predominantly simple or untwinned, idomorphic-to-hypidiomorphic individuals that exhibited strong resorption phenomena in the form of large embayment structures (feldspar generation 3). The former phenocrysts were no longer stable in the hybrid magma as a result of their chemical compositions and the temperature differences, which led to a strong resorption of the crystals [15]. The subsequent epitaxial crystallization of intermediate feldspar formed the idiomorphic outer rims that are present today [15]. Generation two feldspar, which contained unresorbed, calcium- and sodium-rich, polysynthetically twinned cores, exhibited a strong sieve texture in the external regions of the crystals. These phenocrysts either formed after magma mixing or were present before mixing and, as a result of their composition, were more stable to the hybrid magma than generation three. The sieve texture seems to be a decompression effect from the magma ascent [15,27,28].

Another indication of magma mixing is the occurrence of apatite accumulations along the dissolution structures of feldspar generation three (Figure 13), which were absent in the sieve-textured crystals of generation two. The mass crystallization of apatite was, in this case, due to the rapidly decreasing solubility of the mineral alongside its decreasing temperature and increasing SiO_2 content [29,30]. Apatite mineralization accumulates mainly on already crystallized phenocrysts, which themselves are unable to incorporate phosphorus [31]. In this particular case, this happened mainly along the edges of resorbed generation three feldspar crystals. This association also indicates that the present embayment structures are dissolution phenomena and not dendritic growth, as demonstrated in other deposits [32].

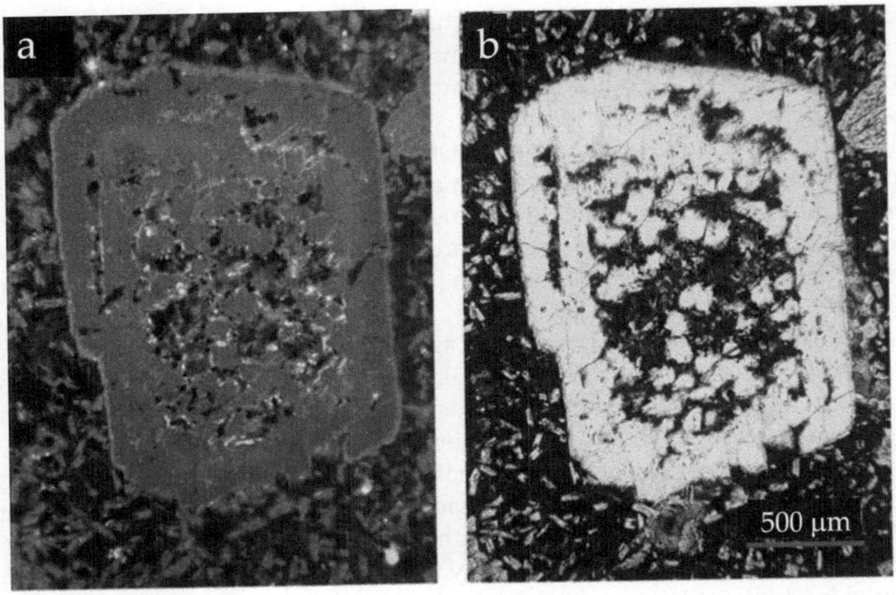

Figure 13. Micrographs in CL (a) and transmitted light (b) of generation three feldspar phenocryst with prominent precipitates of microcrystalline apatite along the resorption rims (yellow spots in CL) formed by magma mixing inside the idiomorphic crystal.

The fresh, polysynthetically twinned, calcium-rich phenocrysts of feldspar generation one are relicts from one of the source magmas that, because of their chemical composition and the high melting point of anorthite, could not be dissolved by either magma mixing or rapid decompression [15]. A later formation of these crystals during the outgoing crystallization of the magma can be ruled out, since no evidence of poikilitic texture was found in the individual crystals. Furthermore, the frequently observed synneusis of phenocrysts is exclusively related to the higher fluid stages of magma under turbulent conditions [31]. Glomeroporphyritic aggregates, which consist mainly of fresh plagioclase crystals, pyroxene phenocrysts, and opaque minerals, occur exclusively in the porphyritic rocks with predominantly intergranular or felsitic matrix as well as large apatite crystals, as is the case in Rancho Coyamito and Agua Nueva. The lower mobility of these aggregates and their predominant origination in the deeper parts of the magma chamber indicate that these rocks solidified not far from the source vent [3].

Origin of Silica for Agate Formation

Agates form when silica fills formerly gas or liquid bubbles in lava. In this study, these vesicular cavities formed before the actual agate formation during the ascent and emplacement of the magma and were fixed in the rock during the cooling process. Most of these amygdales were irregularly distributed and concentrated in the upper part of the lava flow (Figure 6a). Only a few are clearly flattened and follow a weak flow texture. This is consistent with the matrix samples examined, which also contain only minor or no evidence of a flow texture.

The yellow CL signal of the agate from Ojo Laguna provides a clear indication of the formation conditions of this agate deposit. Agates with the dominant 570 nm luminescence emission band form exclusively during the alteration of acidic volcanic host rocks [25]. These formation conditions are associated with a rapid crystallization that is mostly below 250 °C and under an oxygen deficit [25]. A relation between E'_1 defect centers and the intensity of the 570 nm luminescence band, discussed in previous publications [33,34,35], could not be observed in the Mexican samples. On the other hand, the Laguna agates showed rather low concentrations of E'_1 defect centers compared to the other studied samples.

The presence of extremely high antimony contents (>300 ppm) in the agates from Ojo Laguna is, among Mexican agates, a unique feature. High antimony contents were detected not only in the chalcedony itself, but also in the host rock in significant amounts (~90 ppm). The clearly lower contents in the host rock show that the element is enriched during the formation process of these agates. This is also confirmed by the other agate samples, in which Sb always shows elevated values compared to the host rock. However, only in the Ojo Laguna samples could antimony be detected locally (REM-EDS) within reddish iron oxide inclusions. Antimony (as a chalcophile element) is almost exclusively accumulated hydrothermally [26]. Therefore, it can be assumed that a later hydrothermal influence occurred in the Ojo Laguna area. In general, high antimony contents in quartz have been detected in association with hydrothermal Sb-quartz mineralization [36,37]. The Sb content of quartz, as a result of its lack of correlation with other trace elements, is apparently determined solely by the content of antimony in the fluid [36,37,38,39]. No economically relevant antimony deposits are yet known within the Sierra del Gallego area [40]. In the frequently occurring Pb-Zn-Ag-Cu deposits in the state of Chihuahua, however, high amounts of Sb, which occurs primarily in galena and replaces lead with coupled substitution together with silver [41], were detected. Thus, a connection between a hydrothermal formation of agates and possible polymetallic deposits seems plausible.

Also remarkable is the high chromium content of all analyzed agate samples. As an immobile element, chromium is only rarely accumulated in agate or in chalcedony [39]. High contents (up to 9500 ppm [42]) have been thus far detected mainly in SiO_2 mineralization, which is spatially as well as genetically bound to chromium-bearing, ultramafic rocks or chromium deposits [43]. No association between the high chromium content and chromiferous rocks or deposits was found in the Chihuahua area [5,40]. However, this lack of association occurs in other localities as well. For example, Powolny [44] also described chromium-bearing agates (up to 140 ppm) without the typical green color of mtorolite from an alkaline and low-chromium series of the Borowno quarry in Poland. However, chromium can be mobilized via the hydrothermal alteration of volcanic rocks, amongst others, by the dissolution of chromite [45], resulting in a depletion of the element in these rocks [46]. Chromium in the agates can therefore be interpreted as another indication of the involvement of hydrothermal solutions during their formation. It remains unclear from which rocks the chromium was mobilized, since the surrounding rocks are poor in chromium. While there was no evidence of chromium within the mineral inclusions of the agates, some of the CL spectra additionally showed an emission line at about 690 nm. At this position (678 nm–694 nm), the sharp R lines of Cr^{3+} can be found [47,48]. Therefore, the CL spectra indicate an incorporation of Cr^{3+} within the chalcedony structure.

Microstructure and Color of Agates

Most of the analyzed trace elements are found within the numerous mineral inclusions occurring in the SiO_2 matrix of agate, while only a few of them can substitute small amounts of silicon within the structures of quartz or chalcedony (e.g., Ga, Ge, Ti, Al, Fe, B) [23,49]. The substitution of Si ions with U in a quartz lattice is unlikely due to the different crystal-chemical properties, instead, the data from Pan et al. [50] favor their incorporation as a uranyl–silicate complex in microcrystalline silicas.

White colors are common in some Agates from Chihuahua, while the frequency of white bands decreased clearly from the rim to the core. In the current samples, exclusively, a small part of the white bands was clearly colored, while the remaining part was almost completely transparent and did not contain any inclusions (Figure 14). White chalcedony bands had significantly lower iron contents compared to colored bands, as well as slightly elevated concentrations of Mg, Al, Sc, and Ca in comparison to colorless, transparent chalcedony. In general, it is assumed that the white color is related to chalcedony with large particles and low porosity [49].

However, an plate edge-like structure, which occurs in many white, especially, old agates worldwide [49,51,52], was not detected in any of the Mexican agates and can be excluded for the current samples based on their observed micro-structures and young ages. Since only minor differences in the trace-element contents, but no significant structural defects, could be detected in the CL or EPR spectra, it can be assumed that the coloration of white agate bands is mainly caused by different, not more closely specified, Al-Ca-Na-bearing mineral inclusions.

Within the red-colored agates, mineral inclusions were usually smaller and homogenously distributed inside chalcedony bands, which generally resulted in a more homogeneous color distribution within the bands. This distribution in the chalcedony bands, additionally occurred in the pink-, brown-, and yellow-colored agates, being most pronounced in the latter two.

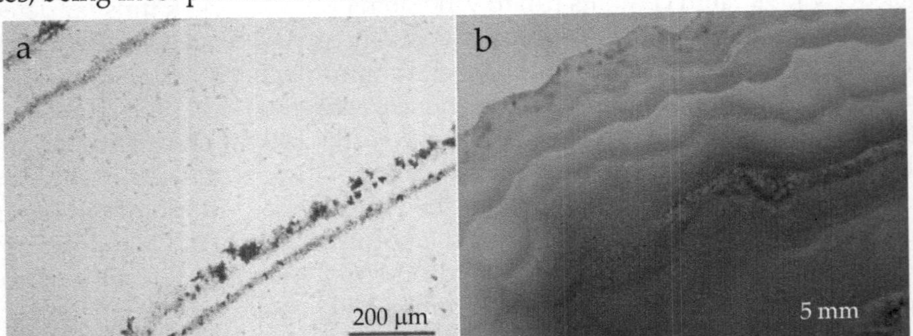

Figure 14. Micrograph in transmitted light shows different inclusions that are accumulated only within a small part of each chalcedony band in an agate sample of the La Fortuna Claim, Rancho Coyamito (a). The same distribution is also macroscopically visible in the polished section (b).

This is probably due to the significantly smaller particle size of the coloring components (Figure 15a,d,f) [53]. In the trace element analyses, high iron contents, which correlate weakly with the intensity of the respective color shade, and somewhat lower Al contents were particularly noticeable. Agates in intense red or yellow shades had the highest iron contents.

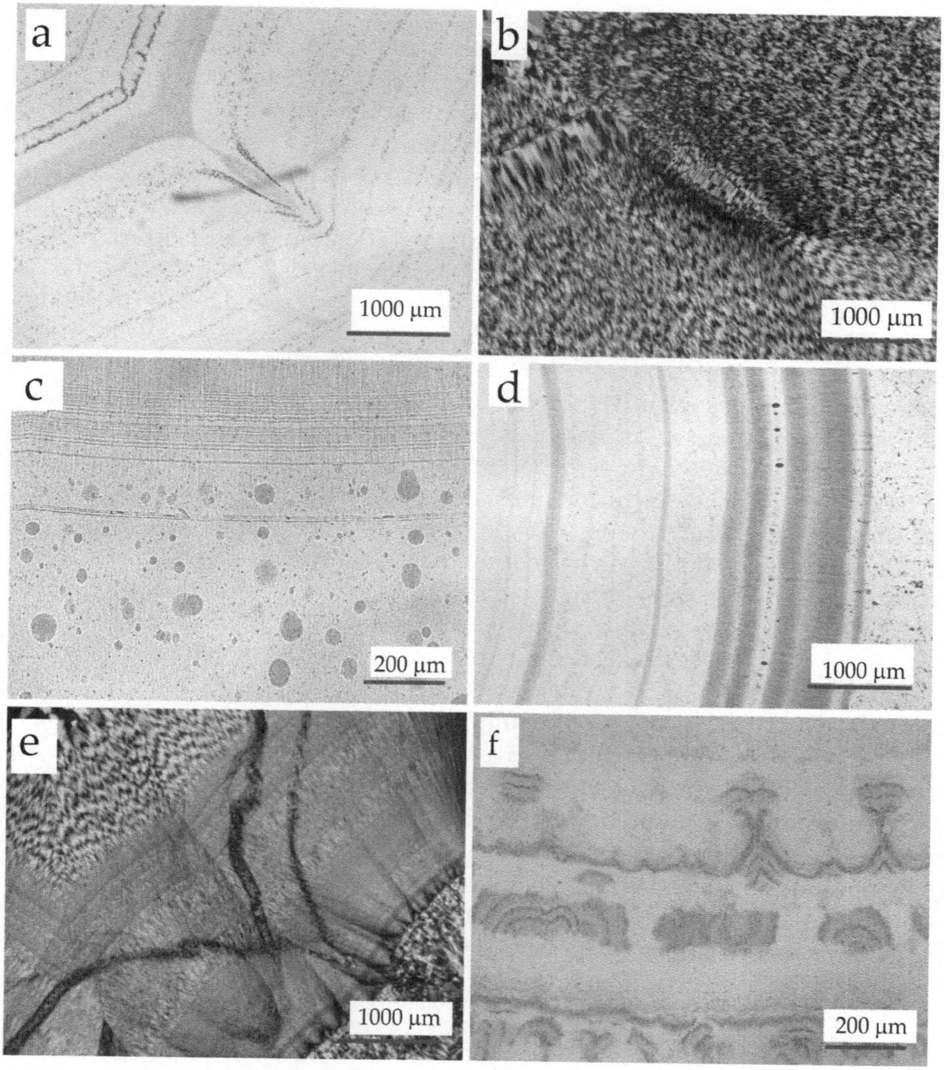

Figure 15. Typical micro-structures of the investigated agate samples ((a,c,d,f); transmitted light; (b,e) crossed polars). (a,b) Escape tube in a sample from Coyamito with obvious material displacement caused by the rapid escape of gas/liquid; in polarized light it is visible that the surrounding chalcedony fibers were only partially deformed by the process, and material of the inner bands was crystallized in the channel; (c) micrograph of a sample from Huevos del Diablo showing the wall-parallel arrangement of micro-inclusions, while the large roundish and semi-transparent inclusions occurred mainly irregularly arranged within the individual bands; (d) distinctly colored wall-parallel banding in an agate from Moctezuma; in addition, similar semi-transparent inclusions, as in agates from HdD, were present within the bands; (e) secondary chalcedony mineralization, crosscutting fibers arranged perpendicular to the crack surfaces in an agate from Ojo Laguna; (f) very tiny

micro-inclusions concentrically arranged within thin bands of an agate from the Japanese deposit of the Rancho Coyamito; the inclusions were seemingly cut off by the banding and formed shapes reminiscent of "micro-escape tubes" in some places (uppermost band).

These high iron contents were also associated with high uranium contents, which, however, did not correlate exclusively with the iron contents. Uranium is mostly concentrated in the dark red-colored agates, while lower uranium contents were determined in the yellow and brown areas. The correlation between uranium and iron seems to be an effect of similar transportation mechanisms within the aqueous fluids either as a silicon complex [22,54] or as a result of the adsorption of uranium on colloidal iron particles [55,56,57]. During the crystallization of agates, uranium is incorporated as a uranyl–silica complex within the structure of SiO_2 [50], while iron forms different iron oxide/hydroxide inclusions within the SiO_2 matrix.

The mineral inclusions, themselves, were characterized using methods of REM-EDS and Raman spectroscopy. In different agate samples, mainly iron oxide/hydroxide, mineral inclusions were found. Most of them were purely iron bearing, while some samples from Ojo Laguna contained a significantly high amount of antimony. Iron mineral inclusions consisted of two different mineral phases, which existed in different transitional stages from each other. Raman spectroscopy of the more yellowlike inclusions revealed mainly the modes of goethite, while the spectra of the transparent, mostly spherical and bubblelike inclusions (Figure 15c) additionally exhibit the chalcedony matrix. This was shown in the form of the typical Raman modes of the α-quartz as well as the additional mode at 502 cm^{-1} (Figure 16), which could be assigned to moganite [58]. The more reddish-appearing inclusions in the agates, as well as some central regions of the transparent "bubbles," contain only hematite. Especially in the latter one, but also in other inclusions, transitional states of goethite and hematite were observed. However, intense red colors were predominantly caused by the incorporation of hematite. Iron hydroxides, such as goethite, predominated in yellow and brown bands. Different compositions and mixtures of iron oxide, hydroxide, and oxyhydroxide phases cause the differing color shades. Since manganese was present in relevantly high amounts only in brown agate areas, it can be assumed that manganese-rich compounds and minerals contributed to the color effect in these agate bands. However, neither manganese nor any manganese-bearing phase could be detected locally with REM-EDS or Raman spectroscopy.

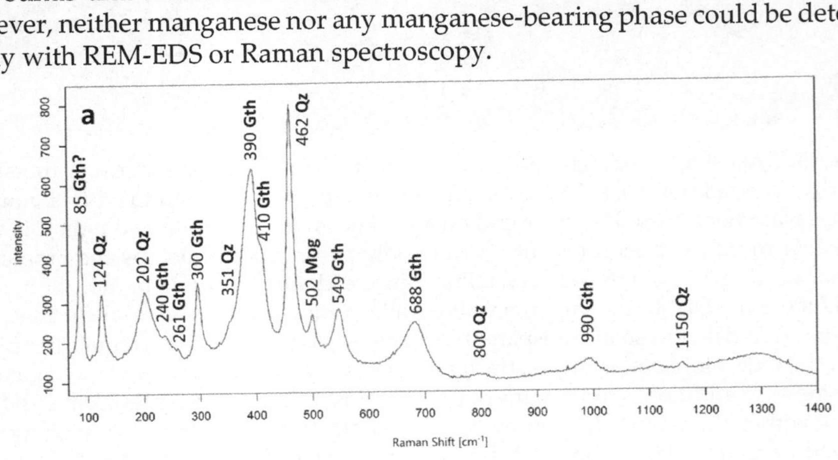

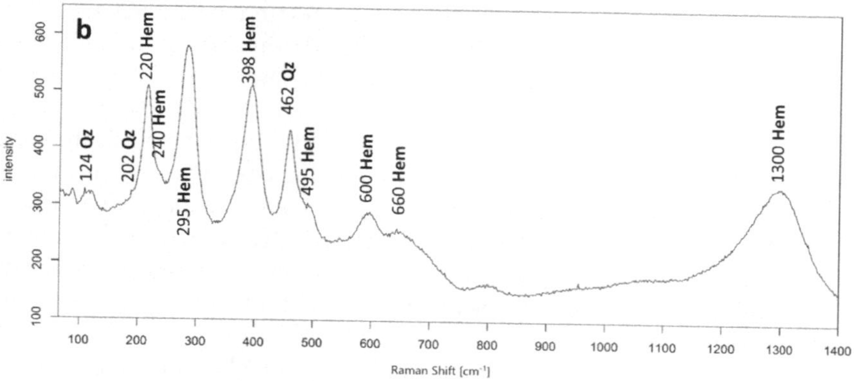

Figure 16. Raman spectra of analyzed micro-inclusions in sample AHdD. The spectra of light brown inclusions contain mainly goethite while the matrix of the chalcedony with the modes of quartz and moganite is also visible (a). The spectra of dark red inclusions contain modes of hematite and quartz (b).

The particle size of the inclusions also had a decisive influence on the color impression, indicating that different color shades can also be caused by mineral inclusions of different sizes (especially iron minerals), which are chemically and structurally identical [49,53]. This cannot be excluded as an explanation for the yellowish agates, which showed a very finely dispersed distribution of inclusions. Considering the increased Na content and only slightly lower Fe content in the pink sample compared to the red samples, its color effect seems to be caused by a mixture of white and red mineral inclusions [1,59].

Agate Formation Process

The initial cavities of the andesite were subsequently filled with Si-rich fluids. Likely during a hiatus from volcanic activity or the fading rhyolitic volcanism, the exposure of emplaced units, primarily the Rancho el Agate andesite, led to an increased near-surface alteration of the rocks. During the alteration and extensive weathering of primary rock-forming minerals, SiO_2 was mobilized in the voluminous rock units that were accumulated within aqueous solutions. Within trace-element data from the agate samples, there were indications (Al, Fe, K, Na, Ca, and U) of the formation of agate from such near-surface weathering solutions [49]. Since thorium is strongly immobile under weathering conditions, unlike uranium, the measured U–Th ratio in the agates argues for mobilization under weathering conditions [46,60]. The enrichments of chromium, antimony, zirconium, and zinc, among others, in the mineralization indicate the involvement of hydrothermal solutions during the agate formation process. In particular, the very high antimony contents in Ojo Laguna agate indicate intense hydrothermal activity resulting in an enrichment of the predominantly chalcophile element during agate formation.

The state of Chihuahua is one of the most deposit- and mineral-rich states in Mexico, and several (hydrothermal) ore deposits are known to exist in the area of Sierra del Gallego [61]. In the vicinity of the studied deposits of agate, polymetallic Pb-Zn-Ag-Cu deposits are mainly found, but isolated uranium and manganese deposits also occur [40]. Most of these deposits are related to Cretaceous sediments or directly to the volcanic rock units [40,61]. The fluid origin of polymetallic deposits has not been conclusively determined. Megaw [62] assumed that hydrothermal polymetallic ore deposits were formed in association with felsitic intrusions. These intrusions can be related both geochemically and temporally to different felsitic/rhyolitic intrusions near various polymetallic ore deposits throughout all of northern Mexico, implying a similar genesis and magma source for these rocks [62]. This also suggests the involvement of similar intrusions or extrusions in the formation of the agates and explains the high antimony contents and enrichments in the mineralization as a result of the participation of residual magmatic solutions during their formation.

The ages of the mentioned felsitic intrusions correlate approximately with the ages of rhyolites in the Sierra del Gallego area [61,62], suggesting that the rhyolites are a potential source of hydrothermal solutions. The involvement of Mesteño Rhyolite in the formation of Las Choyas Geodes, which occur regionally within the Libres Formation, has been assumed for a long time in this context [51]. For the studied agates, various trace-element concentrations suggest that both near-surface weathering by meteoric waters and interaction with hydrothermal solutions resulted in the agate formation.

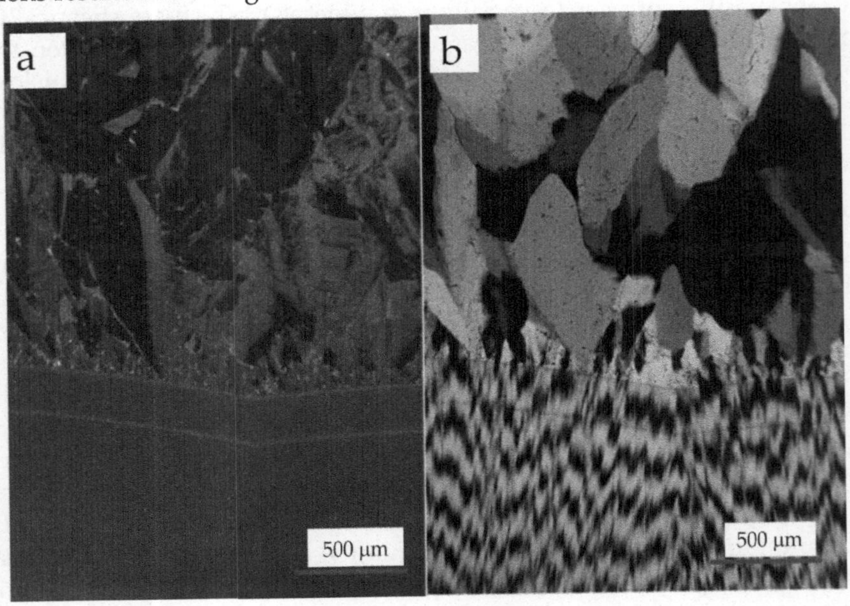

Figure 17. Micrographs of chalcedony and macrocrystalline quartz in an agate from Huevos del Diablo in CL (a) and transmitted light (crossed polars, (b)) Strong sector zoning between yellow-greenish and blue luminescent areas within the macrocrystalline quartz grains, which appear homogeneous in polarized light, is visible; a decrease in the greenish-yellow luminescence color as a result of longer electron bombardment can be seen in the lower right part of the CL image.

For the formation of the agates themselves, the migration of silica-enriched waters into vesicular cavities by diffusion is particularly of decisive importance [49], but suitable secondary pathways to the individual vesicles, such as fine cracks and veins, have also been found during field observations. The accumulation and condensation of monomeric silicic acid results first in the formation of dimers and oligomers, while changing physicochemical conditions (pH, Eh, temperature, changes in chemical composition, etc.) cause the precipitation of amorphous SiO_2. The subsequent crystallization starts from the margins of cavities [49,63]. As the SiO_2 concentration in the silica-bearing fluids decreases, macrocrystalline quartz finally crystallizes from the undersaturated residual solutions [49]. Sector zoning, which was detected in the CL images of macrocrystalline quartz (Figure 17), suggests rapid crystallization under disequilibrium conditions [64]. The chalcedony, itself, is normally strongly disordered, especially in young agates such as in the studied samples from Mexico [49,52]. However, actual Raman spectroscopy investigations have shown a significant amount of moganite within agate samples, which additional to the young age of the mineralization suggests a lower crystallinity [65].

Most of the micro-inclusions of coloring impurities were initially precipitated together with SiO_2 and arranged during crystallization using a kind of "self-cleaning process" along the crystallization fronts in the later chalcedony [49]. This process results in the typical wall-parallel arrangement of micro-inclusions in many agates. The deformation of this banding along so-called "escape tubes" shows that the mineral inclusions were already fixed before the final crystallization process (Figure 15a,b). Accordingly, a degassing through these escape tubes seems to have happened in the transition from amorphous SiO_2-gel to crystalline chalcedony, since the chalcedony fibers along this structure show little to no deformation. However, in certain agate samples, there are also indications that some of the impurities entered the agate later by infiltration after the crystallization of chalcedony.

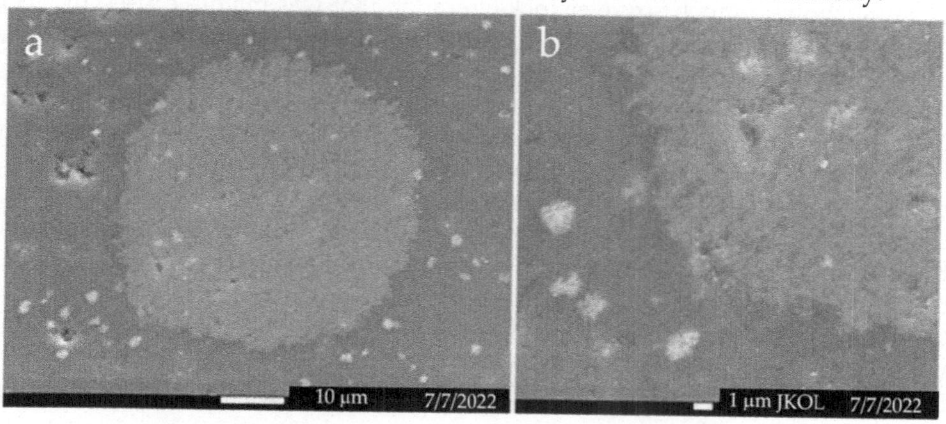

Figure 18. REM–Micrographs of the "bubble" mineral inclusions in an agate sample from Huevos del Diabolo. Goethite and hematite form big accumulations within the matrix of the chalcedony while small Zr-bearing inclusions are randomly arranged within the matrix (a). The detailed image shows the distribution within the pore spaces of the chalcedony for the Fe-minerals as well as the bright Zr-bearing inclusions (b).

This is shown by micro-inclusions of goethite and hematite in an agate sample from Huevos del Diablo, where an identical mineralization was detected both in the pore spaces (Figure 18) of selected chalcedony bands and in secondary cracks within the same agate (Figure 8). Because another generation of macrocrystalline quartz was precipitated after this infiltrative formation of inclusions, it can be assumed that not all agates were formed in a continuous process from a single silica supply. In the agates of Huevos del Diabolo, an additional Zr-bearing phase could be identified which seems to be zircon, according to the first analyses (Raman spectroscopy, EDS). The occurrence of this phase within secondary iron mineral accumulations indicates an earlier formation of the Zr-mineral (Figure 18). This multistage formation, which indicates multiple infillings of the cavities with SiO_2-containing fluids or a crystallization interruption [66], is also emphasized by CL studies. Clearly differing chalcedony generations, each initiated by a separate spherulitic growth at its boundary with the next generation, could be detected (Figure 10). Another indication of multiple silica infill in some agates are secondary cracks, which cross-cut former chalcedony layers that have been subsequently filled with a new generation of chalcedony (Figure 15e). Different generations may have also be formed as a result of a rapid change in physicochemical conditions, e.g., a secondary supply of hydrothermal fluids.

CONCLUSIONS

The present study aimed to carry out a comprehensive mineralogical and geochemical investigation into some of the world-famous agate deposits in the Chihuahua region of Mexico. The volcanic host rocks of the agates are intermediate trachyandesites according to the classical TAS classifications and can be classified geochemically as quartz-poor latite, which belong to the intermediate unit of Rancho el Agate andesite. The agates from Ojo Laguna also belongs to the unit of the Rancho el Agate andesite but have slightly higher SiO_2 content. In addition, this unit shows a much stronger rock alteration that was mainly caused by deep hydrothermal waters in addition to the general near-surface weathering that caused slight alterations in all host rocks. Geochemical data indicate that these hydrothermal fluids are probably closely related to the fluids that were involved in the formation of widespread polymetallic ore deposits in the Chihuahua region.

The micro-textures of the investigated rocks range from intersertal and intergranular over trachytic to felsitic and have differing phenocryst contents and sizes. They represent local-specific textures of an intermediate unit formed close to the surface. The different textures have been formed by various differentiation and transport processes during lava extrusion and the final emplacement of this unit. On the other hand, they are an expression of the different cooling mechanisms within sections of lava flow. Different characteristics of the rock-forming minerals indicate that the intermediate trachyandesitic magma was formed by the mixing of acidic magma with basic (basaltic) magma.

Overall, the volcanic rocks are strongly vesicular, especially in the upper parts of the unit, with partly large amygdaloid cavities. The later infill of silica-rich solutions resulted in the formation of agates. Mineralogical and geochemical data indicate that, in the investigated agate occurrences within the area of Sierra de Gallego, both secondary rock alteration and hydrothermal solutions have led to agate formation. These conditions resulted in the extensive mobilization of Si and other trace elements (Fe, U, Al, Ca, K, etc.). The hydrothermal activity in the area, which was mainly caused by outgoing rhyolitic volcanism, provided additional element-enriched fluids (Si, Sb, Zn, etc.) and the necessary heat for the improved transportability of SiO_2 in aqueous solutions. Likewise, the migration of hydrothermal fluids may have led to mobilization of additional elements, such as Cr and Zr, from host rocks. Volatile fluids acted as transport media, enabling the transport of immobile trace elements and silicon. The migration of fluids occurred both on fractures and by diffusion on intergranular pore spaces, which caused an enrichment inside the vesicular cavities and later crystallization. Thus, the different chalcedony generations, and also the appearance of micro-inclusions, show that several cycles of crystallization and fluid replenishment took place in the cavity.

The predominant similarities between agates from spatially separated deposits prove a fundamental connection in the formation processes of all agates within the area. Depending on the local influences of host rocks and hydrothermal systems and the influence of different fluids, agates with slightly different characteristics were formed. For example, Ojo Laguna agates seem to have formed with significantly higher influences from hydrothermal fluids. This can be seen in the texture of the secondary rock as well as in the trace elements of the agates. Significant amounts of antimony can be associated with a metal-bearing hydrothermal fluid, seemingly related to the fluids that caused the polymetallic ore deposit formations in the area, while the yellow CL of quartz in this deposit normally indicates rapid crystallization in an oxygen-deficient environment under low temperatures.

The colors of the agates are mainly due to mineral micro-inclusions. These were previously formed during crystallization in the chalcedony via self-cleaning processes. In addition, there are also agates in which inclusions were formed by subsequent infiltration. These inclusions are different iron (oxy-hydr)oxides which cause differing yellow, red, and brown colors as a result of their different crystallinity and crystallite sizes. White colors are also exclusively caused by foreign mineral inclusions and are not related to the formation and texture of the chalcedony. The common violet color of agates from Rancho Coyamito is not caused as a result of mineral inclusions but seems to be a special defect in the agate's structure. This unique color, which can predominantly be found in Mexico, is the subject of further research.

AUTHOR CONTRIBUTIONS

J.G. collected the studied samples and provided the geological data. M.M., J.G., Y.P. and R.M. conducted different analytical measurements and evaluated the mineralogical and geochemical data. M.M. and J.G. compiled the data, and M.M. wrote the final version of the manuscript. All authors have read and agreed to the published version of the manuscript.

ACKNOWLEDGMENTS

We are grateful to Andres Carrillo (Chihuahua, Mexico) and Joshua Ritter (Dresden, Germany) for their support during field work and for providing sample material. Ronny Ziesemann, Michael Magnus, Reinhard Kleeberg, Ulf Kempe, Ulrike Fischer, and Alexander Plessow (Freiberg, Germany) are thanked for their help during the analytical work and with sample preparation. Reviews of two anonymous reviewers improved the quality of the manuscript significantly.

REFERENCES

1. Cross, B.L. The Agates of Northern Mexico; Burgess International: Edina, MN, USA, 1996; 201p.
2. Morán-Zenteno, D.J. The Geology of the Mexican Republic, 1st English ed.; AAPG Studies in Geology; American Association of Petroleum Geologists: Tulsa, OK, USA, 1994; Volume 39, 160p.
3. Bockoven, N.T. Petrology and Volcanic Stratigraphy of the El Sueco Area, Chihuahua, Mexico. Doctoral Dissertaition, The University of Texas at Austin, Austin, TX, USA, 1976.
4. Mandujano, M.C.; Pisanty, I.; Eguiarte, L.E. Plant Diversity and Ecology in the Chihuahuan Desert; Springer International Publishing: Berlin/Heidelberg, Germany, 2020; 327p.
5. Keller, P.C.; Bockoven, N.T.; McDowell, F.W. Tertiary volcanic history of the Sierra del Gallego area, Chihuahua, Mexico. Geol. Soc. Am. Bull. 1982, 93, 303–314.
6. Zenz, J. Achate; Bode-Verlag: Haltern, Germany, 2005; 656p.
7. Döbelin, N.; Kleeberg, R. Profex: A graphical user interface for the Rietveld refinement program BGMN. J. Appl. Crystallogr. 2005, 48, 1573–1580.
8. Neuser, R.D.; Bruhn, F.; Götze, J.; Habermann, D.; Richter, D.K. Kathodolumineszenz: Methodik und Anwendung. Zent. Für Geol. Und Paläontologie Teil 1 1995, 1, 287–306.
9. Monecke, T.; Bombach, G.; Klemm, W.; Kempe, U.; Götze, J.; Wolf, D. Determination of trace elements in quartz standard UNS-SpS and in natural quartz by ICP-MS. Geostand. Newsl. 2000, 24, 73–81.
10. Cox, K.G.; Bell, J.D.; Pankhurst, R.J. The Interpretation of Igneous Rocks; Springer eBook Collection; Springer: Dordrecht, The Netherlands, 1979; 450p.
11. LeBas, M.J.; Maitre, R.W.L.; Streckeisen, A.; Zanettin, B. A Chemical Classification of Volcanic Rocks Based on the Total Alkali-Silica Diagram. J. Petrol. 1986, 27, 745–750.
12. Middlemost, E.A. Naming materials in the magma/igneous rock system. Earth-Sci. Rev. 1994, 37, 215–224.
13. Winchester, J.A.; Floyd, P.A. Geochemical discrimination of different magma series and their differentiation products using immobile elements. Chem. Geol. 1977, 20, 325–343.
14. Pearce, J.A. A User's Guide to Basalt Discrimination Diagrams. In Trace Element Geo-Chemistry of Volcanic Rocks—Applications for Massive Sulphide Exploration, ShortCourse Notes; Geological Association of Canada: St. Johns, NL, Canada, 1996; Volume 12, 113p.
15. Shelley, D. Igneous and Metamorphic Rocks under the Microscope: Classification, Textures, Microstructures and Mineral Preferred Orientations; Chapman and Hall: London, UK, 1997; 445p.
16. de La Roche, H.; Leterrier, J.; Grandclaude, P.; Marchal, M. A classification of volcanic and plutonic rocks using R 1 R 2-diagram and major-element analyses—Its relationships with current nomenclature. Chem. Geol. 1980, 29, 183–210.
17. Hastie, A.R.; Kerr, A.C.; Pearce, J.A.; Mitchell, S.F. Classification of Altered Volcanic Island Arc Rocks using Immobile Trace Elements: Development of the Th–Co Discrimination Diagram. J. Petrol. 2007, 48, 2341–2357.
18. Brindley, G.W.; Brown, G. Crystal Structures of Clay Minerals and Their X-ray Identification; The Mineralogical Society of Great Britain and Ireland: London, UK, 1984; Volume 5, 495p.

19. Moore, D.M.; Reynolds, R.C. X-Ray Diffraction and the Identification and Analysis of Clay Minerals, 2nd ed.; Oxford University Press: Oxford, UK, 1997; 378p.
20. Götze, J.; MacRae, C.; Pan, Y.; Wilson, N.C.; Torpy, A.; Audédat, A. The 450 nm (2.8 eV) cathodoluminescence emission in quartz and its relation to structural defects and Ti contents. Am. Mineral. 2023, in press.
21. Stevens-Kalceff, M.A. Cathodoluminescence microcharacterization of point defects in α-quartz. Mineral. Mag. 2009, 73, 585–606.
22. Götze, J.; Gaft, M.; Möckel, R. Uranium and uranyl luminescence in agate/chalcedony. Mineral. Mag. 2015, 79, 985–995.
23. Götze, J.; Pan, Y.; Müller, A. Mineralogy and mineral chemistry of quartz—A review. Mineral. Mag. 2021, 85, 639–664.
24. Mashkovtsev, R.I.; Balitsky, V.S.; Pan, Y. EPR characteristics of radiation-induced defects in Ge-rich α-quartz. Solid State Sci. 2022, 125, 106833.
25. Götze, J.; Pan, Y.; Stevens-Kalceff, M.; Kempe, U.; Müller, A. Origin and significance of the yellow cathodoluminescence (CL) of quartz. Am. Mineral. 2015, 100, 1469–1482.
26. Goldschmidt, V.M. Geochemistry; Clarendon Press: Oxford, UK, 1954; 730p.
27. Nixon, G.T.; Pearce, T.H. Laser-interferometry study of oscillatory zoning in plagioclase; the record of magma mixing and phenocryst recycling in calc-alkaline magma chambers, Iztaccihuatl Volcano, Mexico. Am. Mineral. 1987, 72, 1144–1162.
28. Pearce, T.H.; Russell, J.K.; Wolfson, I. Laser-interference and Nomarski interference imaging of zoning profiles in plagioclase phenocrysts from the May 18, 1980, eruption of Mount St. Helens, Washington. Am. Mineral. 1987, 72, 1131–1143.
29. Watson, E.B.; Green, T.H. Apatite/liquid partition coefficients for the rare earth elements and strontium. Earth Planet. Sci. Lett. 1981, 56, 405–421.
30. Watson, E.; Capobianco, C.J. Phosphorus and the rare earth elements in felsic magmas: An assessment of the role of apatite. Geochim. Et Cosmochim. Acta 1981, 45, 2349–2358.
31. Vernon, R.H. A Practical Guide to Rock Microstructure, 2nd ed.; Cambridge University Press: Cambridge, UK, 2018; 431p.
32. Hibbard, M.J. The magma mixing origin of mantled feldspars. Contrib. Mineral. Petrol. 1981, 76, 158–170.
33. Rink, W.J.; Rendell, H.; Marseglia, E.A.; Luff, B.J.; Townsend, P.D. Thermoluminescence spectra of igneous quartz and hydrothermal vein quartz. Phys. Chem. Miner. 1993, 20, 353–361.
34. Götze, J.; Plötze, M.; Fuchs, H.; Habermann, D. Defect structure and luminescence behaviour of agate; results of electron paramagnetic resonance (EPR) and cathodolumines-cence (CL) studies. Mineral. Mag. 1999, 63, 149–163.
35. Götze, J. Chemistry, textures and physical properties of quartz—Geological interpretation and technical application. Mineral. Mag. 2009, 73, 645–671.
36. Rusk, B.; Möckel, R.; Götze, J. Cathodoluminescent Textures and Trace Elements in Hydrothermal Quartz. In Quartz: Deposits, Mineralogy and Analytics; Götze, J., Möckel, R., Eds.; Springer: Berlin/Heidelberg, Germany, 2012; pp. 307–329.
37. Monnier, L.; Salvi, S.; Pochon, A.; Melleton, J.; Béziat, D.; Lach, P.; Bailly, L. Antimony in quartz as a vector to mineralization: A statistical approach from five Variscan Sb occurrences (France). J. Geochem. Explor. 2021, 221, 106705.

38. Pacák, K.; Zachariáš, J.; Strnad, L. Trace-element chemistry of barren and ore-bearing quartz of selected Au, Au-Ag and Sb-Au deposits from the Bohemian Massif. J. Geosci. 2019, 64, 19–35.

39. Götze, J. Mineralogy, geochemistry and genesis of agate–revealing the mystery of agate formation. In Quartz2021; Natural History Museum, University of Oslo: Oslo, Norway, 2021; pp. 1–26.

40. Staude, J.-M. Map of metallic mining districts and mineral deposit types in Chihuahua, Mexico. US Geol. Surv. 1993, 93–718.

41. Lueth, V.W.; Megaw, P.K.M.; Pingitore, N.E.; Goodell, P.C. Systematic Variation in Galena Solid-Solution Compositions at Santa Eulalia, Chihuahua. Mex. Econ. Geol. 2000, 95, 1673–1687.

42. Hyrsl, J. Chrome chalcedony. J. Gemmol. 1999, 26, 364–370.

43. Hyrsl, J. Chrome Chalcedony from Tanzania. J. Gemmol. 2016, 35, 189–191.

44. Powolny, T.; Dumańska-Słowik, M.; Sikorska-Jaworowska, M.; Wójcik-Bania, M. Agate mineralization in spilitized Permian volcanics from "Borówno" quarry (Lower Silesia, Po-land)—Microtextural, mineralogical, and geochemical constraints. Ore Geol. Rev. 2019, 114, 103130.

45. Arai, S.; Akizawa, N. Precipitation and dissolution of chromite by hydrothermal solutions in the Oman ophiolite: New behavior of Cr and chromite. Am. Mineral. 2014, 99, 28–34.

46. Salaün, A.; Villemant, B.; Gérard, M.; Komorowski, J.-C.; Michel, A. Hydrothermal alteration in andesitic volcanoes: Trace element redistribution in active and ancient hydrothermal systems of Guadeloupe (Lesser Antilles). J. Geochem. Explor. 2011, 111, 59–83.

47. Götze, J.; Hanchar, M. Atlas of Cathodoluminescence Textures; Geological Association of Canada: St. John's, NL, Canada, 2018; 248p.

48. Pagel, M.; Barbin, V.; Blanc, P.; Ohnenstetter, D. Cathodoluminescence in Geosciences: An Introduction; Springer: Berlin/Heidelberg, Germany, 2000; 514p.

49. Götze, J.; Möckel, R.; Pan, Y. Mineralogy, Geochemistry and Genesis of Agate—A Review. Minerals 2020, 10, 1037.

50. Pan, Y.; Li, D.; Feng, R.; Wiens, E.; Chen, N.; Götze, J.; Lin, J. Uranyl binding mechanism in microcrystalline silicas: A potential missing link for uranium mineralization by direct uranyl co-precipitation and environmental implications. Geochim. Et Cosmochim. Acta 2021, 292, 518–531.

51. Moxon, T. A re-examination of water in agate and its bearing on the agate genesis enigma. Mineral. Mag. 2017, 81, 1223–1244.

52. Moxon, T. Agate: A study of ageing. Eur. J. Mineral. 2002, 14, 1109–1118.

53. Schwertmann, U. Relations between Iron Oxides, Soil Color, and Soil Formation. Soil Color 1993, 31, 51–69.

54. Porter, R.A.; Weber, W.J. The interaction of silicic acid with iron(III) and uranyl ions in dilute aqueous solution. J. Inorg. Nucl. Chem. 1971, 33, 2443–2449.

55. Krawczyk-Bärsch, E.; Arnold, T.; Reuther, H.; Brandt, F.; Bosbach, D.; Bernhard, G. Formation of secondary Fe-oxyhydroxide phases during the dissolution of chlorite—Effects on uranium sorption. Appl. Geochem. 2004, 19, 1403–1412.

56. Ames, L.L.; McGarrah, J.E.; Walker, B.A.; Salter, P.F. Uranium and radium sorption on amorphous ferric oxyhydroxide. Chem. Geol. 1983, 40, 135–148.

57. Bruno, J.; de Pablo, J.; Duro, L.; Figuerola, E. Experimental study and modeling of the U(VI)-Fe$(OH)_3$ surface precipitation/coprecipitation equilibria. Geochim. Et Cosmochim. Acta 1995, 59, 4113–4123.

58. Götze, J.; Nasdala, L.; Kleeberg, R.; Wenzel, M. Occurrence and distribution of "moganite" in agate/chalcedony: A combined micro-Raman, Rietveld, and cathodoluminescence study. Contrib. Mineral. Petrol. 1998, 133, 96.

59. Zenz, J. Achate III; Bode-Verlag: Lauenstein, Germany, 2011; 656p.

60. Mernagh, T.P.; Miezitis, Y. A Review of the Geochemical Processes Controlling the Distribution of Thorium in the Earth's Crust and Australia's Thorium Resources; Geoscience Australia: Canberra, Australia, 2008; 48p.

61. Keller, P.C. Geology of the Sierra Del Gallego Area, Chihuahua, Mexico; The University of Texas at Austin: Austin, TX, USA, 1977.

62. Megaw, P. Geology and Geochemistry of the Santa Eulalia Mining District, Chihuahua, Mexico; The University of Arizona: Austin, TX, USA, 1990.

63. Harder, H. Agates-formation as a multi component colloid chemical precipitation at low temperatures. Neues Jahrb. Für Mineral. Mon. 1993, 1, 31–48.

64. Götze, J. Application of cathodoluminescence microscopy and spectroscopy in geosciences. Microsc. Microanal. 2012, 18, 1270–1284.

65. Moxon, T.; Ríos, S. Moganite and water content as a function of age in agate: An XRD and thermogravimetric study. Eur. J. Mineral. 2004, 16, 269–278.

66. Moxon, T.; Palyanova, G. Agate Genesis: A Continuing Enigma. Minerals 2020, 10, 953.

7

MULTI-SCALE X-RAY COMPUTED TOMOGRAPHY ANALYSIS TO AID AUTOMATED MINERALOGY IN ORE GEOLOGY RESEARCH

Mathis Warlo[1], Glenn Bark[1], Christina Wanhainen[1], Alan R. Butcher[2], Fredrik Forsberg[3], Henrik Lycksam[3] and Jukka Kuva[2]

[1]Division of Geosciences and Environmental Engineering, Department of Civil, Environmental and Natural Resources Engineering, Luleå University of Technology, Luleå, Sweden

[2]Geological Survey of Finland (GTK), Espoo, Finland

[3]Division of Fluid and Experimental Mechanics, Department of Engineering, Sciences and Mathematics, Luleå University of Technology, Luleå, Sweden

ABSTRACT

Ore characterization is crucial for efficient and profitable production of mineral products from an ore deposit. Analysis is typically performed at various scales (meter to microns) in a sequential fashion, where sample volume is reduced with increasing spatial resolution due to the increasing costs and run times of analysis. Thus, at higher resolution, sampling and data quality become increasingly important to represent the entire ore deposit. In particular, trace metal mineral characterization requires high-resolution analysis, due to the typical very fine grain sizes (sub-millimeter) of trace metal minerals. Automated Mineralogy (AM) is a key technique in the mining industry to quantify process-relevant mineral parameters in ore samples. Yet the limitation to two-dimensional analysis of flat sample surfaces constrains the sampling volume, introduces an undesired stereological error, and makes spatial interpretation of textures and structures difficult. X-ray computed tomography (XCT) allows three-dimensional imaging of rock samples based on the x-ray linear attenuation of the constituting minerals. Minerals are visually differentiated though not chemically classified. In this study, decimeter to millimeter large ore samples were analyzed at resolutions from 45 to 1 μm by AM and XCT to investigate the potential of multi-scale correlative analysis between the two techniques. Mineralization styles of Au, Bi-minerals, scheelite, and molybdenite were studied. Results show that AM can aid segmentation (mineralogical classification) of the XCT data, and vice versa, that XCT can guide (sub-)sampling (e.g., for heavy trace minerals) for AM analysis and provide three-dimensional context to the two-dimensional quantitative AM data. XCT is particularly strong for multi-scale analysis, increasingly higher resolution scans of progressively smaller volumes (e.g., by mini-coring), while preserving spatial reference between (sub-) samples. However, results also reveal challenges and limitations with the segmentation of the XCT data and the data integration of AM and XCT, particularly for quantitative analysis, due to their different functionalities. In this study, no stereological error could be quantified as no proper grain separation of the segmented XCT data was performed. Yet, some well-separated grains exhibit a potential stereological effect. Overall, the integration of AM with XCT improves the output of both techniques and thereby ore characterization in general.

Keywords: *X-ray computed tomography, Automated mineralogy, Mineral segmentation, Ore, Trace metals*

INTRODUCTION

Mineralogical investigations are crucial in both the mining industry and ore geology research to understand the fundamental processes behind the formation of ore deposits. This type of investigation is necessary to optimize exploration for new deposits and production efficiency of existing mines. As knowledge of ore deposits increases, details become increasingly relevant. Over the last few decades, analytical techniques (including automated ones) have been continuously developed and improved such that they now enable study of ore deposits down to the nanoscale (Reich et al., 2017). Several automated scanning electron microscopy systems have been developed (QEMSCAN, MLA, Mineralogic, TIMA, AMICS, INCA Mineral), collectively referred to as Automated Mineralogy (AM), to rapidly scan and classify samples (e.g., polished thin sections, epoxy mounts) concerning their mineralogy based on backscattered electron (BSE) imaging and energy dispersive x-ray spectroscopy (EDS; Schulz et al., 2020). This allows quantification of geological and geometallurgical parameters such as mineral composition, mineral association, grain size distribution, degree of liberation, etc. Yet, AM has several drawbacks. 1) Cost and time of analysis increase with spatial resolution, thus generally only small sample areas (mm^2-cm^2) and a limited number of samples are analyzed. A sample selection suitable for its purpose is therefore critical. 2) Sample preparation is destructive and requires cutting and grinding of a larger sample (e.g., drill core). This causes loss of material and hence loss of potentially important information. 3) AM is limited to two-dimensional analysis, and thus requires that the geologist translates the two-dimensional observations into the third dimension (Baker et al., 2012). For rocks of complex mineralogy, textures, and structures, this can be challenging. In addition, quantitative data is subject to stereological error (Lätti & Adair 2001; Sutherland 2007), which is especially pronounced for preferably oriented mineral grains (low variability in the cutting plane relative to the crystal lattice) and trace minerals (nugget effect). Despite their low abundance, trace minerals can be significant to a mining venture if they contain metals of high economic value, either as main commodity or as by-products (e.g., Au, Ag), critical metals (e.g., Critical Raw Material; European Commission 2020), or penalty elements (e.g., As, Hg, U, Sb). In addition, detailed mineralogical knowledge of an ore deposit supports its ore genetic classification, which is an important input into any mineral exploration campaign. Hence, good control on trace metal mineral occurrence and distribution in an ore deposit is crucial.

An excellent tool to address the mentioned drawbacks is X-ray Computed Tomography (XCT), which allows the visualization of samples in 3D. A sample is placed between an x-ray source and a detector and rotated 360° around a vertical axis. A cone-shaped x-ray beam penetrates and is attenuated by the sample. The detector collects the transmitted x-rays, which provide information on the x-ray linear attenuation. Based on the variation of the x-ray attenuation as the sample rotates an x-ray linear attenuation coefficient is calculated for each voxel (3D pixel) in the sample. This coefficient is displayed as a distinct grey scale value in the XCT dataset, which is a cubical matrix of grey scale values. The x-ray linear attenuation coefficient depends on the density and element composition of the material analyzed and thus allows differentiation between minerals if their attenuation coefficients are sufficiently different. A detailed description of the XCT technique and its application in geoscience is provided in several reviews (e.g., Baker et al., 2012; Cnudde and Boone 2013; Godel 2013; Kyle and Ketcham 2015; Hanna and Ketcham 2017; Guntoro et al., 2019a; Wang and Miller 2020; Withers et al., 2021). XCT is non-destructive and applicable over a wide range of scales (meter to nanometer; Wang and Miller 2020), enabling scanning of entire drill core sections and to preserve their mineralogical, textural, and structural information prior to subsample production (thin section, rock chip, etc.) for 2D micro-analysis. The XCT data can aid in selecting regions of interest for micro-analysis and in correlating surface features of subsamples with their original position in the drill core. Drill cores are ideally suited for XCT analysis, due to their cylindrical shape. This allows consistent sample thickness between x-ray source and detector as the sample rotates 360° during analysis. Further, the cut-off-material of subsamples, the subsamples themselves, and smaller regions of interest therein can subsequently be re-scanned with higher resolution XCT for more detailed analysis, e.g., the study of individual mineral grains (Godel 2013). In the oil industry, XCT analysis of drill cores is performed routinely (Butcher 2020), e.g., to quantify the pore volume and its interconnectivity. Further, mini-cores are subsequently drilled from the larger cores for higher-resolution analysis (Butcher 2020). However, XCT analysis and mini-coring are yet uncommon in the mining industry. This may partly be due to the typically more complex mineralogy and texture of ore-forming rocks compared to oil-bearing sandstones and shales.

A major challenge with XCT is quantitative mineral analysis. This requires segmentation of the various minerals from the grey scale images. Difficulties arise since the grey scale value of a voxel, which represents an x-ray linear attenuation coefficient, is not only dependent on the x-ray attenuation of a mineral. Various artefacts influence x-ray attenuation across a sample, including but not limited to: 1) beam hardening, an increase in average beam energy along a ray path caused by the higher absorption of low-energetic (soft) x-rays compared to high-energetic (hard) x-rays within a sample. Beam hardening results in a gradient of the reconstructed x-ray linear attenuation coefficient towards the center of the sample (cupping effect, Cnudde and Boone 2013). 2) Streak artefacts, bright and dark streaks emanating from a highly attenuating phase into the surrounding material due to beam hardening (De Man et al., 1998; Kyle and Ketcham 2015). 3) X-ray attenuation reconstruction errors vertically away from the centre of a sample due to the conical shape of the x-ray beam (cone-beam effect; Cnudde and Boone 2013). 4) Ring artefacts caused by differences in sensitivity of the detector elements (Cnudde and Boone 2013; Kyle and Ketcham 2015). In addition, voxels may intersect multiple phases and represent mixed attenuation coefficients (Partial Volume Effect; Ketcham and Carlson 2001; Hanna and Ketcham 2017). A voxel may contain contributions from surrounding voxels due to machine- and acquisition-dependent factors, an effect known as blurring (Hanna and Ketcham 2017). Moreover, minerals of solid solution series and chemically zoned mineral grains may exhibit a range of x-ray linear attenuation coefficients, dependent on the exact chemical composition (Kyle and Ketcham 2015). All this causes the voxels of a mineral grain to range in grey scale values and to overlap with the grey scale values of other mineral grains, particularly if their theoretic x-ray linear attenuation coefficients are similar. Visual inseparability in the XCT dataset may be a result. The x-ray linear attenuation coefficient of a mineral varies with x-ray energy (Berger et al., 2010; Godel 2013; Reyes et al., 2017). The variation is not the same for all minerals, thus visual separability of some minerals may be increased by altering the x-ray energy used for analysis. One solution to improve separability of all minerals is to combine multiple scans at different energies. Nevertheless, segmentation of XCT data in general is much more limited compared with e.g. AM.

For segmentation of XCT data several methods exist, varying in complexity and computational intensity (Godel 2013; Wang and Miller 2020). The simplest is grey scale thresholding based on peaks in the histogram. This method is fast but due to the above-mentioned overlap between minerals prone to errors (Godel 2013; Wang and Miller 2020). It can be improved through combination with an edge intensity filter (e.g., Sobel) and the watershed algorithm (Godel 2013; Guntoro et al., 2019a; Wang and Miller 2020). This 'histographic segmentation' is also rapid and works well for samples with distinct grain boundaries but struggles with fine-grained and complex textures. Superior results are achieved by use of machine learning, yet these require sufficient training data, heavy computation power, and dedicated time spent on optimization (Godel 2013; Reyes et al., 2017; Guntoro et al., 2019b). In the end, the limitations of too similar x-ray linear attenuation coefficients apply to all segmentation methods. Nevertheless, some quantification of mineral volumes, association, grain sizes, orientation, etc. is possible and may supplement 2D quantitative data.

In this study, multi-scale (45, 30, 4, and 1 μm) XCT and AM analyses were performed on polished and epoxy mounted drill core rock chips of ore from the Liikavaara Östra Cu-(W-Au) deposit, northern Sweden. Correlation and integration of both techniques across the various scales were investigated. Focus was on the modal mineralogy, detection of heavy trace minerals, and textures. The benefits and challenges of multi-modal and multi-scale analysis for ore characterization are discussed.

GEOLOGICAL BACKGROUND

The samples investigated in this study are from the Liikavaara Östra Cu-(W-Au) deposit in northern Sweden. It is an intrusion-related vein-style deposit hosted by Paleoproterozoic *meta*-volcanoclastic rocks (Warlo et al., 2020). Boliden AB has scheduled the deposit for production in the coming years. Copper, hosted in chalcopyrite, will be the main commodity, and Au and Ag will be by-products. Molybdenum and the critical metals W and Bi are also enriched, though only Mo is currently considered a potential future by-product. The ore minerals are mainly hosted by quartz-(tourmaline-calcite) veins often within or in proximity to aplite dykes (Warlo et al., 2020). The precious metals are of diverse mineralogy. Gold occurs natively and as a Au-Ag-alloy; Ag also occurs as tellurides and sulfides (Warlo et al., 2020). The Au and Ag minerals are fine-grained (mainly <10 μm for Au, and <50 μm for Ag). They are found as inclusions within, and in contact with, chalcopyrite and gangue (mainly quartz and other sulfides). Gold is also associated with micron-sized grains of native Bi (Warlo et al., 2020). The only Mo-bearing mineral is molybdenite, grains of which are typically tens of microns to a few millimeter in size and associated with quartz, sulfides, scheelite, and calcite (Warlo et al., 2020). Previous analysis by scanning electron microscopy with energy dispersive spectrometry (SEM-EDS) and synchrotron radiation x-ray fluorescence mapping (SR-XFM), showed molybdenite to be rich in micron-to nano-inclusions (Warlo et al. submitted). The inclusions are trapped in cavities between striations in the molybdenite grains and are chemically diverse but mainly composed of Fe-oxides, Fe-silicates, and Bi-(Pb-Se), but also Au-Ag, and Ag-Te. In addition, SR-XFM revealed the existence of lattice-bound impurities of Re, Se, and W in molybdenite (Warlo et al. submitted). Scheelite is the only major W-bearing mineral and commonly confined to the quartz-dominated part of quartz-(tourmaline-calcite) veins. Scheelite grains are mostly larger (several millimeters in width) than associated mineral grains and typically fractured (Warlo et al., 2020).

SAMPLES AND METHODS

The samples were a ~20 cm long half drill core piece, four polished and epoxy-mounted cylindrical rock chips of ~30 mm diameter and 12 mm in height, and two mini cores of 4 mm diameter and 10 mm in height (Figure 1). The samples were cut from three drill cores that intersect the ore zone of the Liikavaara Östra Cu-(W-Au) deposit. Drill core logging and routine chemical assays from Boliden AB on metals of economic, environmental, and metallurgical interest, guided the sampling. The focus was on representative samples for the mineralization of Au, Mo, and W. Four styles of mineralization were sampled:

1) Scheelite-bearing aplite dyke.
2) Sulfide-rich quartz-(calcite-tourmaline) vein. The chemical assays for this structure were 6 ppm Au and 10.4 ppm Ag content, over a 1.3 m section.
3) A second aplite dyke that contains thin quartz veins rich in molybdenite and scheelite.
4) Contact between a quartz vein and host biotite schist. Molybdenite mineralization occurs along the contact within the quartz vein.

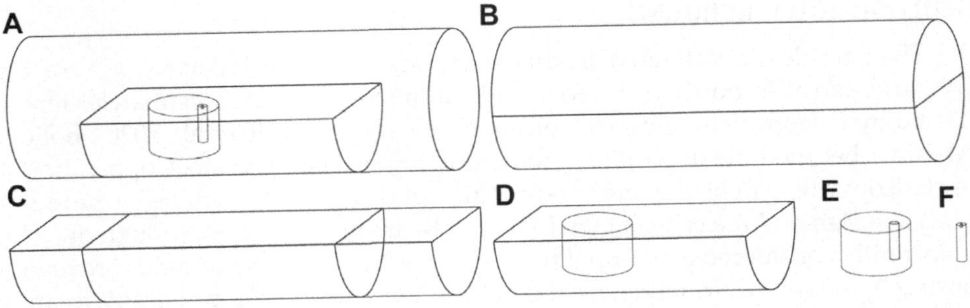

Figure 1. Schematic drawings of the sampling and analysis methodology of this study. (A) Overview of all a sample types in their relation to the original drill core. (B) A drill core is cut in half. One half core is analyzed by whole rock XRF and the other half is logged. (C) The logged half core is subsampled. In this study, an XCT scan at 45 μm voxel size was performed on one subsampled half core (2). (D) A rock chip is cut from the half core sample, polished, and epoxy-mounted. In this study, XCT scans at 30 μm voxel size were performed on four rock chips and 30 μm pixel size AM scans on their polished surface (1–4). (E) A mini core is drilled from the rock chip sample. In this study, XCT scans at 4 μm voxel size were performed on two mini cores and 1 μm pixel size AM scans on their polished surface prior to drilling (2, 3). (F) A cylindrical volume of interest within the mini core is selected for XCT analysis. In this study, XCT scans at 1 μm voxel size were performed in volumes of interest in two mini cores (2, 3).

For each mineralization style, one epoxy-mounted rock chip was prepared. For sample 2, the rock chip was cut from the half drill core piece also analyzed in this study. The two mini cores were drilled from the rock chips of samples 2 and 3, respectively (Figure 1E). The creation of mini cores was necessary to achieve a short distance between the x-ray source and the samples, and thereby allow high-resolution XCT analysis.

Automated Mineralogy

The polished surface of each of the four epoxy-mounted rock chips was scanned by AM using a Zeiss Sigma 300 VP with two EDS detectors and the Mineralogic software at Luleå University of Technology (Table 1). A pixel size (beam sampling distance) of 30 μm was chosen; a trade-off between resolution and time spent on analysis. Two regions of interest (4 mm diameter) in the rock chips of samples 2 and 3, subsequently drilled for extraction of the mini cores, were additionally scanned at 1 μm pixel size (Table 1). The same custom-made mineral library in the Mineralogic software was used for initial mineral classification in all scans. Element filters for Au, Ag, and Bi were placed at the top of the mineral library to identify sub-pixel sized grains of trace metal minerals recorded as mixed phase spectra (Warlo et al., 2019). Results were processed in the Mineralogic Explorer software. Here, mineral maps were stitched and montaged and quantitative data on bulk mineralogy, mineral association, metal deportment, were extracted. The accuracy and precision of the mineral library was assessed by manual spot-checking the BSE image and respective EDS spectra of the identified minerals. For Au and Ag minerals, every pixel found was reviewed. Based on this assessment, the mineral library was then fine-tuned for each scan and data reclassified.

Table 1. Experimental conditions of AM and XCT. For XCT analysis, the half drill core piece was analyzed on a GE phoenix v|tome|x s at GTK, Espoo, Finland. All other samples were analyzed on a Zeiss Xradia 510 Versa at LTU, Luleå, Sweden

Automated Mineralogy							
Sample	**S1**	**S2**	**S3**	**S4**	**S2 ROI**	**S3 ROI**	
Voltage	20 kV	20 kV	20 kV	20 kV	20 kV	20 kV	
Aperture	60 µm	60 µm	60 µm	60 µm	120 µm	120 µm	
BSE Pixel Size	1.87 µm	1.87 µm	1.87 µm	1.87 µm	250 nm	250 nm	
EDS Step Size	30 µm	30 µm	30 µm	30 µm	1 µm	1 µm	
No. of Scan Fields	294	306	305	301	305	299	
Scan Area	~570 mm^2	~570 mm^2	~570 mm^2	~570 mm^2	~13 mm^2	~13 mm^2	
Run time	06:44:16 h	07:13:22 h	7:02:48 h	06:48:23 h	62:27:54 h	63:25:44 h	
X-ray Computed Tomography							
Sample	**Half drill core (S2)**	**Rock chip (S1 & S2)**	**Rock chip (S3 & S4)**	**Mini core (S2)**	**Mini core (S3)**	**Mini core ROI (S2)**	**Mini core ROI (S3)**
Voltage	175 kV	160 kV	160 kV	80 kV	140 kV	160 kV	160 kV
Current	250 µA	62.5 µA	62.5 µA	87.5 µA	42.9 µA	62.5 µA	62.5 µA
Power	43.75 W	10 W	10 W	7 W	6 W	10 W	10 W
Exposure Time	3 × 333 ms	2 × 7 s	5 s	1.2 s	1 s	5 s	6 s
Objective		0.4x	0.4x	4x	4x	4x	4x
Bin	1	2	2	2	2	2	2
Filters	0.5 mm Cu	Zeiss HE5	Zeiss HE2	Zeiss HE1	Zeiss HE1	Zeiss LE5	Zeiss LE5
No. of projections	1,600	1,601	1,601	1,601	1,601	3,201	3,201
Voxel size	45 µm	30 µm	30 µm	4 µm	4 µm	1 µm	1 µm

Table 2. List of calculated x-ray linear attenuation coefficients of all minerals identified by AM in this study. The colors in the left column (Mineral) refer to the respective colors of the minerals in the AM mineral maps. The numbers in the columns are the x-ray linear attenuation coefficients of the minerals for the respective beam energy of each scan (175, 160, 140 and 80 kV). The coefficients were calculated using the MXLAC tool (Bam et al., 2020). For mineral groups (e.g., amphiboles, epidote) one mineral was picked to represent the whole group. For broad groups (e.g., silicates, Fe-oxides/carbonates) no values were calculated. The XCT data were segmented into classes (mineral groups) based on a combination of grey-scale thresholding and an edge intensity filter. To determine the minerals comprising each segmented class, the data were visually compared with the AM mineral maps. Each mineral from the original AM maps was then appointed to its corresponding class (mineral group) in the segmented XCT data and simplified AM mineral maps created. For minerals where visual comparison of the AM and XCT data was inconclusive to appoint it to a class, the calculated x-ray linear attenuation coefficient relative to that of the other minerals was used as reference. Color coding of the cells with numbers represents the simplification of the AM quantitative maps for correlation with the segmented XCT data. Uncolored cells indicate that the minerals were not present in the respective AM maps. Variable contrast between minerals in the XCT data due to variable beam energies, tomographic reconstruction, and beam hardening correction between XCT scans is the reason for the differences in mineralogy of the segmented classes between scans

	Sample 1	Sample 2				Sample 3			Sample 4
Mineral	Epoxy-mounted rock chip (160 kV)	[1]Half drill core (175 kV)	Epoxy-mounted rock chip (160 kV)	[2]Mini core (80 kV)	[1]Mini core (ROI) (160 kV)	Epoxy-mounted rock chip (160 kV)	Mini core (140 kV)	[1]Mini core (ROI) (160 kV)	Epoxy-mounted rock chip (160 kV)
Unclassified	-	-	-	-	-	-	-	-	-
Albite	0.57	0.55	0.57	0.83	0.57	0.57	0.61	0.57	0.57
Quartz	0.58	0.56	0.58	0.84	0.58	0.58	0.62	0.58	0.58
Tourmaline	0.63	0.61	0.63	0.87	0.63	0.63	0.67	0.63	0.63
Plagioclase	0.64	0.61	0.64	0.98	0.64	0.64	0.69	0.64	0.64
K-Feldspar	0.65	0.62	0.65	1.01	0.65	0.65	0.70	0.65	0.65
Muscovite	0.68	0.65	0.68	1.02	0.68	0.68	0.72	0.68	0.68
Anorthite	0.69	0.66	0.69	1.10	0.69	0.69	0.75	0.69	0.69
Dolomite	0.72	0.69	0.72	1.15	0.72	0.72	0.78	0.72	0.72
Chlorite	0.76	0.72	0.76	1.30	0.76	0.76	0.84	0.76	0.76
Calcite	0.83	0.78	0.83	1.46	0.83	0.83	0.92	0.83	0.83
Biotite	0.86	0.82	0.86	1.42	0.86	0.86	0.94	0.86	0.86
	0.92	0.87	0.92	1.56	0.92	0.92	1.01	0.92	0.92
Silicates	-	-	-	-	-	-	-	-	-
Apatite	1.05	0.99	1.05	1.92	1.05	1.05	1.17	1.05	1.05
Ankerite	1.13	1.05	1.13	2.15	1.13	1.13	1.28	1.13	1.13
Fe-Oxides/ Carbonates	-	-	-	-	-	-	-	-	-
Epidote	1.43	1.32	1.43	2.81	1.43	1.43	1.63	1.43	1.43
	1.76	1.63	1.76	3.50	1.76	1.76	2.01	1.76	1.76
Ilmenite	2.64	2.41	2.64	[2]5.61	2.64	2.64	3.06	2.64	2.64
Pyrite	2.97	2.72	2.97	6.28	2.97	2.97	3.42	2.97	2.97
Pyrrhotite	3.12	2.84	3.12	6.78	3.12	3.12	3.62	3.12	3.12
Chalcopyrite	3.20	2.87	3.20	7.59	3.20	3.20	3.77	3.20	3.20
Magnetite	3.52	3.20	3.52	[2]7.72	3.52	3.52	4.11	3.52	3.52
Hematite	3.53	3.21	3.53	[2]7.72	3.53	3.53	4.12	3.53	3.53
FeO	-	-	-	[2]-	-	-	-	-	-
Sphalerite	3.98	3.64	3.98	8.98	3.98	3.98	4.57	3.98	3.98
Arsenopyrite	7.05	6.43	7.05	15.75	7.05	7.05	8.15	7.05	7.05
Zircon	9.94	9.11	9.94	18.15	9.94	9.94	9.07	9.94	9.94
Tin-minerals	-	-	-	-	-	-	-	-	-
Scheelite	11.46	10.50	11.46	25.09	11.46	11.46	13.13	11.46	11.46
Molybdenite	11.70	10.69	11.70	26.38	11.70	11.70	13.44	11.70	11.70
Galena	25.13	23.07	25.13	54.43	25.13	25.13	28.68	25.13	25.13
Cassiterite	27.14	24.80	27.14	[2]53.66	27.14	27.14	31.20	27.14	27.14
Hessite	35.71	32.60	35.71	80.79	35.71	35.71	41.10	35.71	35.71
Bi-telluride	37.53	34.36	37.53	82.06	37.53	37.53	43.00	37.53	37.53
Bismuth	39.48	36.20	39.48	84.80	39.48	39.48	45.14	39.48	39.48
Silver	45.73	41.74	45.73	103.84	45.73	45.73	52.65	45.73	45.73
Electrum	-	-	-	-	-	-	-	-	-
Gold	65.18	60.32	65.18	134.72	65.18	65.18	72.67	65.18	65.18

[1]No AM was performed, hence no detailed information on mineralogy available.

[2]The AM data of the mini core of sample 2 became corrupt post analysis, which caused issues with reclassification. The marked minerals are wrongly classified. However, impact on the quantitative data was minimal.

For comparison with XCT, the detailed mineral maps were subsequently simplified based on the limiting number of segmented classes in the XCT scans. The Mineral X-Ray Linear Attenuation Coefficient Tool (MXLAC) by Bam et al. (2020) was used to calculate the mineral x-ray linear attenuation coefficient of each mineral for the respective beam energy used in the XCT analysis. All minerals identified by AM where ranked accordingly. For certain mineral groups (e.g., amphibole, epidote), the attenuation coefficient of only one mineral from the group was calculated to represent the whole group. For broader-defined groups (e.g., silicates, Fe-oxides/carbonates), no values were calculated but the approximate position in the ranking estimated. The coefficients were then used in conjunction with visual comparisons of AM and XCT images to assign the minerals to the classes segmented in the XCT scans (Table 2).

Since Mineralogic Explorer does not offer an option for analysis of subregions within a scan, the simplified quantitative maps were converted into grey scale images, imported into ORS Dragonfly, and re-segmented based on their histograms. This allowed overlay of the AM data with the XCT data and quantitative analysis of the same region. For the epoxy-mounted rock chip scans, a circular/cylindrical (AM/XCT) region was chosen, which maximized area/volume while excluding the border between rock and epoxy. For the high-resolution scans, an area/volume that maximized the overlap between the 1 μm pixel size mineral maps and the XCT scans of the mini cores was chosen.

The AM data of the 1 μm pixel size scan of sample 2 became corrupted after reclassification with the updated mineral library. Results were still accessible but further reclassification was no longer possible. This caused issues with simplification of the mineralogy for comparison with XCT. Some minerals initially colored the same could no longer be separated. This was the case for Fe-oxides (ilmenite, magnetite, hematite, FeO, Fe-oxides/carbonates) and tin minerals (cassiterite, tin minerals). They were therefore treated together according to the mineral of highest concentration (Fe-oxides/carbonates and tin minerals, respectively). Quantitatively, the miss-classified Fe-oxides covered ~0.2 area % and cassiterite <0.01 area % in the whole mineral map. Since no equivalent quantitative data could be inferred for the region compared with XCT, no correction was performed. Regardless, impact on the quantitative comparison was minimal.

X-Ray Computed Tomography

The half drill core piece was scanned by XCT at 45 μm voxel size using a GE phoenix v|tome|x s at the Geological Survey of Finland, Espoo, Finland (Table 1). The four epoxy-mounted rock chips were scanned at 30 μm voxel size using a Zeiss Xradia 510 Versa at Luleå University of Technology (Table 1). Two rock chips each (samples 1 and 2, and 3 and 4) were stacked with their polished surfaces face to face, fixed in place with tape, and analyzed together. This was to minimize the cone-beam effect on the polished surfaces of the samples. The two mini drill cores were also analyzed with the Zeiss Xradia 510 Versa at 4 μm voxel size. In addition, a cylindrical volume of interest of 1 mm in both diameter and height within each 4 mm drill core was scanned at 1 μm voxel size. The Zeiss Xradia system has multiple detector objectives enabling sample imaging at a number of resolution and field of view pairings. Here, the rock chip samples were scanned using a

×0.4-objective, while the mini drill core samples were scanned using a ×4-objective, specified more in detail in Table 1.

The tomographic reconstruction was carried out using filtered back projection, and beam hardening correction applied where necessary. However, due to the complex mineralogy within the samples (large range of x-ray linear attenuation coefficients) optimization for all minerals was not possible. Hence, corrections and filters commonly were applied in favor of the higher x-ray linear attenuation coefficients (denser minerals) as these were in focus in this study.

The XCT data were processed and analyzed using the software ORS Dragonfly (version 2021.1). The 'Unsharp' filter was applied to increase edge contrast of grains, followed by the 'Mean Shift' filter to smoothen the noise introduced by the 'Unsharp' filter. Segmentation was performed on a cylindrical subvolume in each data set, so that: 1) the epoxy mount was excluded 2) depth-related grey scale gradients (e.g., cone-beam effect), did not affect segmentation significantly, 3) each horizontal slice was of equal area, and 4) the top slice overlapped with the respective mineral map from AM. The only exception was the half drill core piece where segmentation was performed on a half cylinder partly including the contact with the surrounding air.

The 'Histographic Segmentation' plug-in in the ORS Dragonfly software was used for mineral segmentation. This plug-in includes a combination of grey scale thresholding and the sobel filter to define classes (groups of minerals of similar x-ray linear attenuation coefficient) and determine markers in the sobel image for the watershed algorithm (Guntoro et al., 2019a). Individual grains were determined based on a 6-point particle connection (shared voxel faces).

Heavy trace minerals in the samples of mineralization style 2 were of too low volume to be identified accurately by 'Histographic Segmentation'. Hence, the trace minerals were segmented by manual grey scale thresholding. The 'Histographic Segmentation' (and grey scale thresholding) were performed six times for each sample to get an estimate of the standard deviation for the various segmented phases.

The segmented XCT data were used for quantitative analysis. The mean bulk volumes of the segmented mineral classes and their respective area percentage in each horizontal slice were calculated for the six segmentations of each scan. The mean data of the slices overlapping with the mineral maps were used for quantitative comparison between XCT and AM. In addition, the grey scale and segmented XCT images were visually inspected concerning texture and structure of the ore samples.

RESULTS

The AM scans provide detailed mineral maps of the 2D surface of the rock chip samples, complemented with 3D-spatial data from XCT analysis. Below are the results of scans of the four samples: 1) 'scheelite- and Bi-telluride-bearing aplite', 2) 'gold-bearing sulfide-rich quartz vein', 3) 'molybdenite- and scheelite-bearing quartz veins in aplite', and 4) 'molybdenite-rich contact between a quartz vein and biotite schist', presented and illustrated in detail.

Scheelite- and Bi-Telluride-Bearing Aplite (Sample 1)

The AM scan shows the contact between biotite schist and an intruding aplite dyke (Figure 2). The aplite dyke contains several up to ~4.5 mm sized scheelite grains together with chalcopyrite and pyrrhotite. Several sub-millimeter inclusions of Bi-tellurides occur in the pyrrhotite. Some ambiguity on pixel classification exists in very fine-grained regions, e.g., the biotite schist, due to mixed EDS spectra. Overall, mineral classification and resolution of textures agree with previous observations from reflected light microscopy and SEM-EDS. In the XCT scan only four classes (mineral groups) can be visually separated, 1) silicates and carbonates, 2) sulfides and oxides, 3) scheelite, and 4) heavy minerals (mostly Bi-tellurides; Figure 2, Table 3). Comparison of the top slice of the segmented XCT data with the same area in the AM mineral map shows considerable quantitative differences (Table 3).

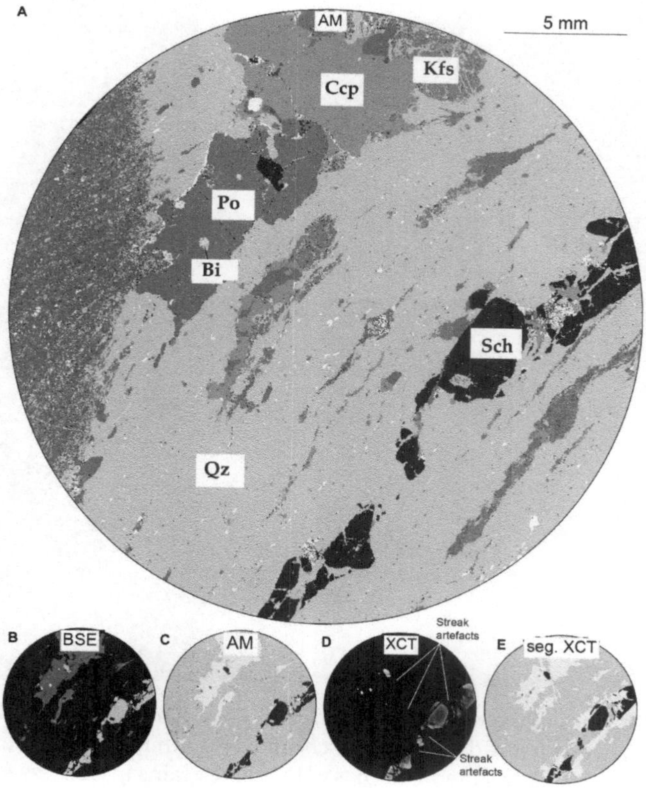

Figure 2. AM and XCT images of a polished rock chip of a scheelite- and Bi-telluride-bearing aplite (sample 1). For color coding of the minerals see Table 2. (A) 30 μm pixel size AM mineral map. (B) 1.87 μm pixel size AM BSE image. (C) 30 μm pixel size simplified AM mineral map. (D) 30 μm voxel size XCT scan. (E) 30 μm voxel size segmented XCT scan. Streak artefacts of scheelite and Bi-mineral grains in the XCT scan (D) cause segmentation of scheelite grain boundaries as heavy minerals, quartz as sulfides (bright streaks), and sulfides as quartz (dark streaks). Abbreviations: Ccp – chalcopyrite, Kfs – k-feldspar, Po – pyrrhotite, Sch – scheelite, Qz – quartz.

Although total difference in area between the techniques are largest for the silicates/carbonates (~11%), the relative difference is much larger for the heavy minerals (~70%) due to their overall low abundance (Table 3). The sulphides/oxides are high in both total and relative area difference. The differences can be explained mainly by the effect of streak artefacts in the XCT data, caused by the scheelite and heavy mineral grains, on the histographic segmentation. This results in spurious aureoles of sulfides/oxides around scheelite and heavy mineral grains (Figures 2, 3). Additionally, some scheelite grains have brighter rims than cores, which causes segmentation of the rims into the class of heavy minerals (Figures 2, 3). Segmentation of regions without scheelite grains appear satisfactory. Texturally, the unaided eye and AM indicate a preferred planar orientation of scheelite and sulfides within the quartz vein (Figure 2).

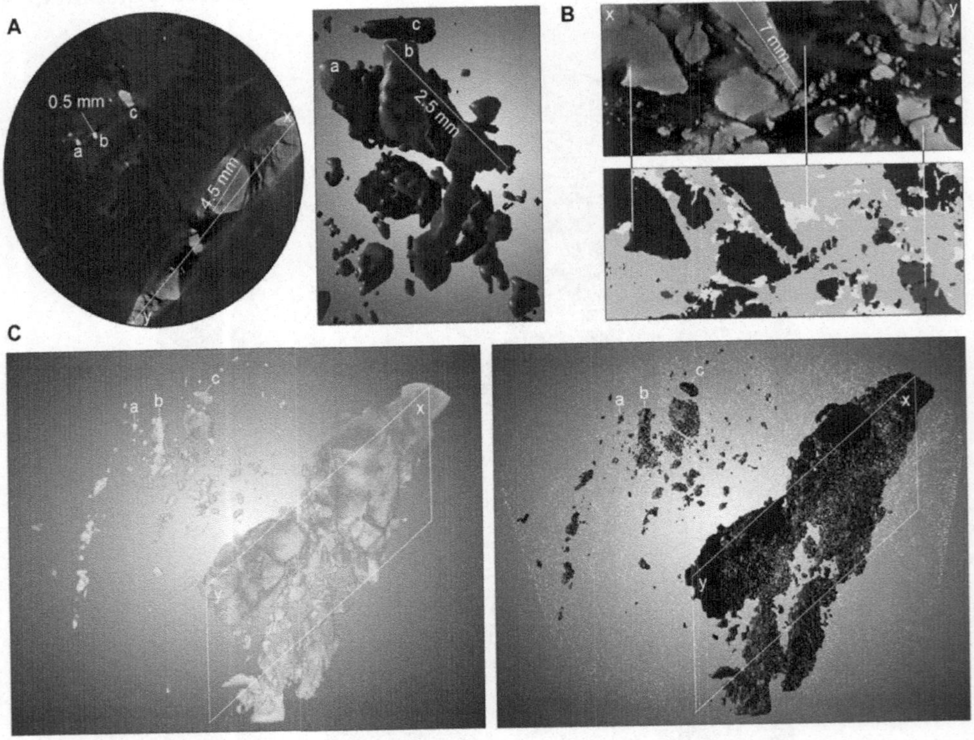

Figure 3. XCT images of the rock chip of sample 1. For color coding of the minerals, see Table 2. (A) Grains of Bi-telluride (a, b) and scheelite (c) within pyrrhotite in the top slice of the sample (LEFT). 3D segmented volumes of the same grains in 3D viewed at a close to 90° pitch (RIGHT). Especially grain 'b' has a much larger size than may be presumed from the 2D slice. (B) Profile along the scheelite vein. Scheelite grains are partly elongated towards depth and fractured. Variation in the grey scale due to streak artefacts (TOP) results in the margins or entire grains of scheelite to be segmented as heavy minerals and parts of quartz as sulfides (BOTTOM). (C) 3D-rendered volumes of scheelite and heavy minerals (Bi-tellurides) based on the x-ray attenuation data (LEFT) and segmentation (RIGHT). The yellow and pink dots in the segmented image (RIGHT) show the rough outline of the sulfides and quartz, respectively. The large heavy minerals in the scheelite vein are likely wrongly segmented scheelite grains (see (B)).

XCT confirms this and shows continuity of the scheelite vein deep into the sample. Scheelite grains are relatively large (~7 mm) and elongated towards depth but strongly fractured (Figure 3). Heavy minerals mainly form inclusions within the sulfides. The XCT images clearly display the dissemination of heavy mineral inclusions and the problem with grain size estimations based on 2D sections.

Table 3. Comparison of the area fraction of the classes segmented in the rock chip of samples 1 and 2 between the AM mineral map and the corresponding XCT slice. Additionally, the maximal difference in percent between the XCT slice used for comparison with AM and the following four slices is shown. Average and Max values refer to all slices within a specific segmented volume (rock chip, mini core) of an XCT data set. The XCT values are the mean values of six segmentations ($n = 6$) of the same volume, with the standard deviation in brackets. A minimal error (<0.2 area %) may occur for the 1 μm AM data of sample 2 due to file corruption (see section Automated Mineralogy)

Rock Chip (Sample 1)	**Heavy Minerals (vol%)**	**Scheelite (vol%)**	**Sulfides/Oxides (vol%)**	**Silicates/Carbonates (vol%)**
30 μm AM	0.09	5.14	14.6	79.9
30 μm XCT	0.16 (0.042)	5.84 (0.15)	25.1 (1.25)	68.9 (1.32)
Average	0.33 (0.071)	3.68 (0.059)	16.2 (0.62)	79.8 (0.66)
Max	0.69 (0.16)	7.01 (0.068)	25.3 (1.24)	91.0 (0.34)
Max. Difference in % (5 Slices)	+8.67 (7.28) −19.7 (12.0)	+2.65 (0.82) −0	+0.44 (0.21) −6.64 (2.55)	+2.31 (0.95) −0.27 (0.071)
Relative Difference in % (30 μm AM)	68.6 (45.0)	13.68 (2.87)	71.8 (8.55)	−13.8 (1.66)
Rock Chip (Sample 2)	**Heavy Minerals (vol%)**	**Sulfides/Oxides (vol%)**	**Silicates/Carbonates (vol%)**	**Quartz/Albite (vol%)**
30 μm AM	0.01	48.6	30.8	20.4
30 μm XCT	0.002 (0.0007)	43.8 (0.57)	37.3 (1.85)	18.9 (1.95)
Average	0.001 (0.0009)	46.0 (0.30)	33.8 (1.29)	20.2 (1.35)
Max	0.012 (0.002)	51.3 (0.29)	38.6 (1.43)	33.0 (2.25)
Max. Difference in % (5 Slices)	+16.1 (24.0) −72.6 (40.8)	+2.14 (0.60) −0.017 (0.041)	+2.53 (3.14) −1.19 (1.42)	+0.088 (0.20) -6.31 (5.63)
Relative Difference in % (30 μm AM)	−74.1 (8.63)	-9.79 (1.18)	21.0 (5.99)	-7.31 (9.55)
45 μm XCT	0	40.4 (0.81)	38.2 (1.93)	21.4 (2.36)
Average	0.0006 (0.0003)	46.5 (0.70)	36.0 (1.30)	17.4 (1.68)
Max	0.011 (0.002)	52.1 (0.62)	41.5 (1.33)	24.1 (2.68)
Max. Difference in % (5 Slices)	+0 −0	+2.20 (0.34) −0.21 (0.13)	+0 -9.18 (1.27)	+12.2 (1.32) −0
Relative Difference in % (30 μm AM)	−100	−16.8 (1.67)	23.8 (6.26)	−5.17 (11.60)
Mini Core (Sample 2)				
1 μm AM	0.06	69.9	[1]5.04	[1]25.0
30 μm AM	0.04	70.6	8.33	21.0
Relative Difference in % (1 μm AM)	−22.3	0.92	65.1	-15.8
4 μm XCT	0.0003 (0.0002)	65.0 (0.65)	[1]10.9 (3.65)	[1]24.1 (3.50)
Average	0.002 (0.00009)	14.7 (0.50)	40.6 (1.71)	44.8 (1.75)
Max	0.029 (0.001)	65.8 (0.29)	67.7 (1.77)	71.0 (1.79)
Max. Difference in % (5 Slices)	+200 (44.7) −0	+1.33 (0.71) −0	+0 −31.0 (3.33)	+11.0 (4.39) -0
Relative Difference in % (1 μm AM)	-99.4 (0.29)	-7.09 (0.92)	116 (72.3)	−3.17 (14.0)
30 μm XCT	0	54.8 (1.20)	29.3 (4.17)	15.9 (3.07)
Average	0.00005 (0.0001)	10.9 (0.14)	71.4 (6.13)	17.7 (6.01)
Max	0.006 (0.014)	55.3 (0.98)	95.2 (4.18)	38.6 (9.15)
Max. Difference in % (5 Slices)	+0 −0	+0.88 (0.52) −5.28 (1.09)	+24.7 (12.5) −0.62 (1.45)	+0.80 (2.16) −28.1 (21.9)
Relative Difference in % (1 μm AM)	−100	−21.6 (1.72)	481 (82.7)	−36.2 (12.3)
45 μm XCT	0	53.3 (8.67)	30.7 (8.41)	16.0 (2.21)
Average	0	9.38 (1.59)	60.1 (6.07)	30.5 (5.88)
Max	0	53.3 (8.67)	82.9 (4.56)	49.5 (7.26)
Max. Difference in % (5 Slices)	+0 −0	+0 −13.3 (1.72)	+5.89 (5.84) −1.58 (2.65)	+38.2 (5.97) -0
Relative Difference in % (1 μm AM)	−100	−23.8 (12.4)	509 (167)	−35.8 (8.87)

[1]*Tourmaline moved from silicates/carbonates to quartz/albite.*

The largest Bi-telluride grain observed on the polished surface by AM (~0.5 mm) is intersected at an edge and grows considerably larger with depth in the sample (~2.5 mm) (Figure 3). This also exceeds previously reported maximal grain sizes of Bi-minerals (0.5 mm) at the Liikavaara Östra Cu-(W-Au) deposit (Warlo et al., 2020).

Gold-Bearing Sulfide-Rich Quartz Vein (Sample 2)

The XCT scan of the half drill core shows a complex intergrowth of sulfide and gangue minerals (Figure 4). The undifferentiated sulfides mainly occur to one side of the core. The gangue can be differentiated into a dark grey fine-grained matrix replacing larger fractured grains of even darker grey. Very bright heavy mineral grains occur scattered within the gangue. These bright phases are likely galena and/or scheelite, based on the relatively large grain sizes compared with other heavy minerals in the Liikavaara Östra Cu-(W-Au) deposit (Warlo et al., 2020). To target the Au mineralization, the rock chip was taken from the sulfide-dominated part of the drill core. Although the XCT scan of the half core shows only few bright phases in this region (Figure 4), this is to be expected considering the generally small grain size of Au minerals (Warlo et al., 2020).

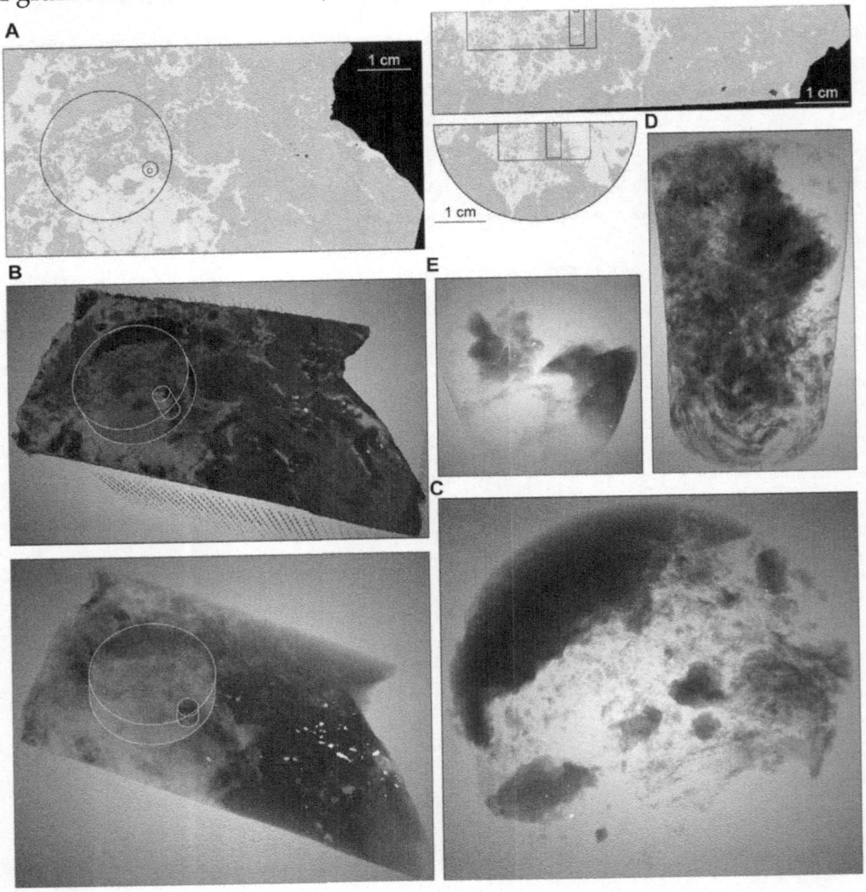

Figure 4. XCT images of the drill core piece, rock chip, and mini core of sample 2. For color coding of the minerals, see Table 2. (A) Segmented XCT images of the various scans (45, 30, 4, and 1 μm voxel size) along the three spatial axes. (B) Opaque (TOP) and transparent (BOTTOM) 3D-rendered XCT images of the drill core piece, rock chip, and mini core. The black and dark grey phases correspond mainly to silicates and carbonates, the medium grey phases to sulfides, and the bright phases to heavy minerals (e.g., scheelite, galena, Au- and Bi-minerals). (C) 3D-rendered XCT image of the rock chip (30 μm voxel size).

The sulfides are rendered transparent to better reveal the heavy mineral inclusions (bright grains). (D) 3D-rendered XCT image of the mini core (4 μm voxel size). The sulfides are rendered transparent to better reveal the heavy mineral inclusions (bright grains). (E) 3D-rendered XCT image of the volume of interest within the mini core (1 μm voxel size). The sulfides are rendered transparent to better reveal the heavy mineral inclusions (bright grains).

AM of the surface of the rock chip resolves the mineralogy in better detail compared with XCT (Figure 5). A fine-grained matrix of mainly biotite, chlorite, ankerite, quartz, and Fe-oxides/carbonates, with ubiquitous brecciated subhedral tourmaline, replaces large (>15 mm) quartz grains. Sub-to euhedral pyrite crystals grow within and replace massive chalcopyrite, which permeates the fine-grained silicate matrix. Thin calcite-filled cracks occur throughout the rock. Very fine mineral grains of <30 microns of precious metals Au and Ag, and Bi form scattered inclusions mainly in pyrite and quartz. In the XCT data of the rock chip, four mineral classes (similarly to the half drill core scan) are segmented (Figure 5, Table 3): 1) quartz and albite, 2) silicates and carbonates, 3) sulfides and oxides, and 4) heavy minerals. The high beam energy (160 kV) necessary to penetrate the sample sufficiently, does not allow visual separation of pyrite and chalcopyrite due to their overlapping x-ray attenuation coefficient at that energy (Figure 5; Godel 2013; Kyle and Ketcham 2015; Reyes et al., 2017). Similarly, no differentiation between the dense minerals, including Au-, Ag- and Bi-minerals, is possible. On the other hand, quartz and albite are somewhat distinguishable from other silicates and carbonates unlike in sample 1.

The fine-grained character of large parts of the sample cause a pronounced partial volume effect and significant grey scale overlap between the classes. This lowers image resolution compared to AM (both have a 30 micron pixel/voxel size), and causes over-segmentation of particularly silicates/carbonates at the expense of quartz/albite and sulfides/oxides (Figure 5, Table 3). Further, fine textures like carbonate-filled cracks are hardly resolved by the segmentation. Concerning detection of heavy trace metal minerals, both techniques are well suited. For AM, simple element filters enable effective detection and classification of trace metal mineral grains even below pixel size due to their unique element composition (Figure 5, Table 3; Warlo et al., 2019). For XCT, the much higher x-ray linear attenuation coefficient of the heavy trace metal minerals compared with the other minerals in the sample allow rapid segmentation via grey scale thresholding, although not between the different heavy trace metal minerals. In comparison, AM is more effective than XCT in the detection of heavy minerals. The area classified as heavy minerals is ~70% less in the XCT slice compared to the corresponding AM mineral map (Table 3). Yet, due to the benefit of XCT to analyze volumes, the overall number of heavy mineral grains detected by XCT is much larger than for AM (Figure 4). For example, the amount of heavy minerals in the top XCT slice is only 0.5% of the segmented volume of heavy minerals across all slices. To study the effect of voxel size on quantification, the AM mineral map was also compared with the corresponding XCT slice of the half drill core scan (Table 3). The discrepancy of sulfides/oxides is larger than between AM and the 30 micron XCT scan, whereas it is lower for quartz/albite. Notably, no heavy minerals are classified for the half drill core XCT slice (Table 3).

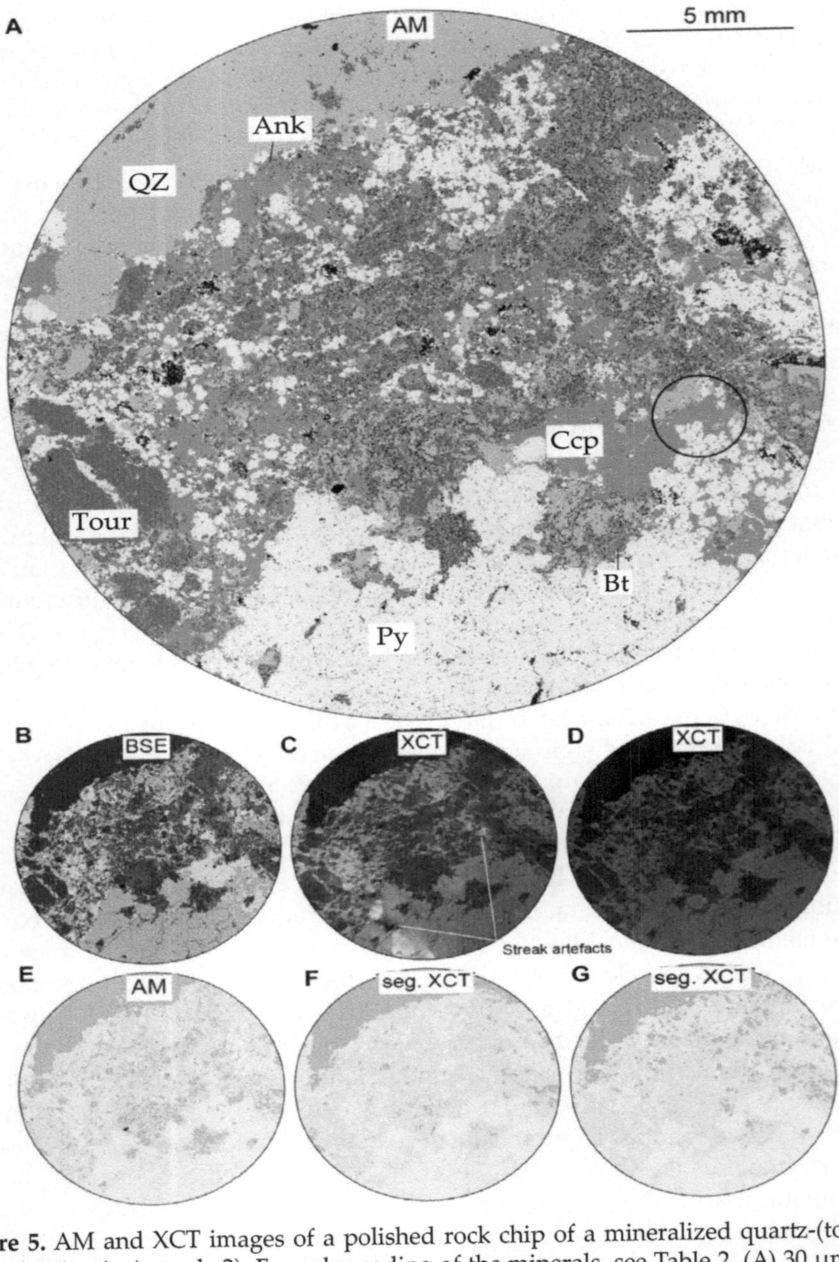

Figure 5. AM and XCT images of a polished rock chip of a mineralized quartz-(tourmaline-calcite) vein (sample 2). For color coding of the minerals, see Table 2. (A) 30 μm pixel size AM mineral map. The black circle indicates the overlapping region between the 1 μm pixel size AM mineral map and the 4 μm voxel size XCT scan (see Figure 6). (B) 1.87 μm AM BSE image. (C) 30 μm voxel size XCT scan. The image displays some bright and dark spots, streak artefacts from scheelite in the rock chip of sample 1 because of stacking of the samples for XCT analysis. (D) 45 μm voxel size XCT scan. (E) Simplified 30 μm pixel size AM mineral map. (F) 30 μm voxel size segmented XCT scan. (G) 45 μm voxel size segmented XCT scan. Abbreviations: Ank – ankerite, Bt – biotite, Ccp – chalcopyrite, Py – pyrite, Tour – tourmaline, Qz – quartz.

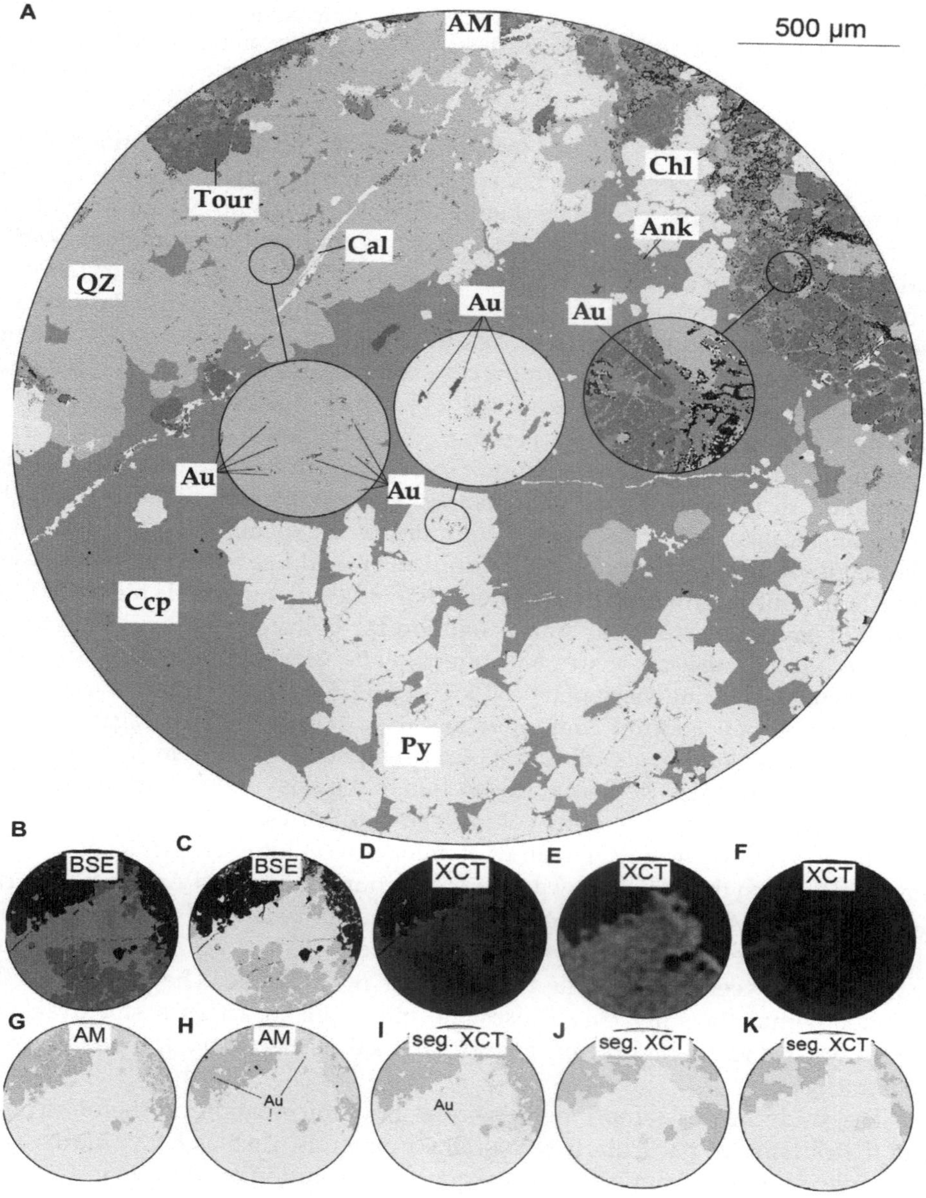

Figure 6. AM and XCT images of a region of interest in the polished rock chip of sample 2. For color coding of the minerals, see Table 2. (A) 1 μm pixel size AM mineral map. The circles show enlarged images of the small circles. Gold inclusions are found in a number of minerals but are very fine-grained. The marked grains represent only a portion of the Au grains detected in the map. (B) 1.87 μm pixel size AM BSE image. (C) 250 nm pixel size AM BSE image. (D) 4 μm voxel size XCT scan. (E) 30 μm voxel size XCT scan. (F) 45 μm voxel size XCT scan. (G) 1 μm pixel size simplified AM mineral map. The same gold grains as in (A) are detected but not visible in the map due to the small size of the image. (H) 30 μm pixel size simplified AM mineral map. Three Au grains are detected. (I) 4 μm voxel size segmented XCT scan. One Au grain is detected. (J) 30 μm voxel size segmented XCT

scan. (K) 45 μm voxel size segmented XCT scan. The loss in detail between the 4 μm XCT scan (D, I) and the 30 μm (E, J) and 45 μm (F, K) XCT scans is apparent. Abbreviations: Ank – ankerite, Cal – calcite, Ccp – chalcopyrite, Chl – chlorite, Py – pyrite, Tour – tourmaline, Qz – quartz.

In a small area within the AM mineral map, two pixels are identified as Au: one in pyrite, the other in adjacent quartz. This region of interest was subsequently scanned by AM at 1 μm pixel size (Figures 5, 6). It reveals numerous <30 μm Au grains (and Ag, Bi) within the pyrite and quartz, but also tourmaline, ankerite, and chlorite (Figure 6). Similarly, the XCT scan at 4 μm voxel size of the mini core, drilled through this region of interest, shows many more dense particles not seen in the coarser scans (Figure 4). Yet, the Au grains identified by AM are not resolved. The smaller size of the mini core compared with the rock chip allows XCT analysis at lower beam energy (80 kV), which makes visual separation of chalcopyrite and pyrite possible (Figure 6). Yet, grey levels are still too close for meaningful histographic segmentation. The change in beam energy also causes tourmaline to dominantly be classified together with quartz rather than the group of silicates/carbonates. Table 3 summarizes the quantitative comparison of this region of interest between the 1 μm pixel size AM scan and all other scans. Expectedly, but still noteworthy, the detection of heavy minerals is higher for AM than XCT and increases with scan resolution for both techniques. The detected amounts for silicates/carbonates and quartz/albite vary significantly, partly due to the change in beam energy (160–80 kV) for XCT analysis and in consequence ambiguous classification of some minerals (e.g., tourmaline). Yet again, overall detection of heavy minerals (all slices) is much higher for XCT than AM (Figure 4). For the 4 μm voxel size XCT scan, the portion of the class of heavy minerals in the top XCT slice is only 0.01% of the volume across all slices.

The 1 μm voxel size XCT scan of a cylindrical volume (1 mm height, 1 mm diameter) within the mini core was centered around the largest Au grain in pyrite (identified by AM). This scan resolves the Au grain and many more dense particles with depth, undetected by the 4 μm voxel size scan (Figure 4). Conversely, visual separation between chalcopyrite and pyrite is not possible. The position of the cylinder within the mini core causes the mini core to act as a shield against the x-rays. Hence, a high beam energy (160 kV) is necessary for proper x-ray penetration.

The study of this sample demonstrates how information on the occurrence and distribution of particularly fine-grained minerals, such as heavy trace metal minerals, is highly reliant on the technique and resolution of analysis.

Molybdenite- and Scheelite-Bearing Quartz Veins in Aplite (Sample 3)

The AM mineral map shows the aplite dyke to consist mainly of K-feldspar, plagioclase, and quartz with minor biotite and muscovite (Figure 7). Sulfides (pyrite, pyrrhotite, and chalcopyrite), tourmaline, and calcite occur disseminated. In the sample, three quartz vein connect at an angle (Figure 7). Molybdenite mineralization occurs in one vein but not in the others. Conversely, scheelite occurs in one of the other veins but not in the molybdenite-filled vein. Chalcopyrite, pyrite, and pyrrhotite are found in all three veins. In the XCT data, differentiation of the aplite dyke is hardly possible due to beam hardening correction optimized for the highly attenuating scheelite and molybdenite rather than the silicates and carbonates (similar to sample 1).

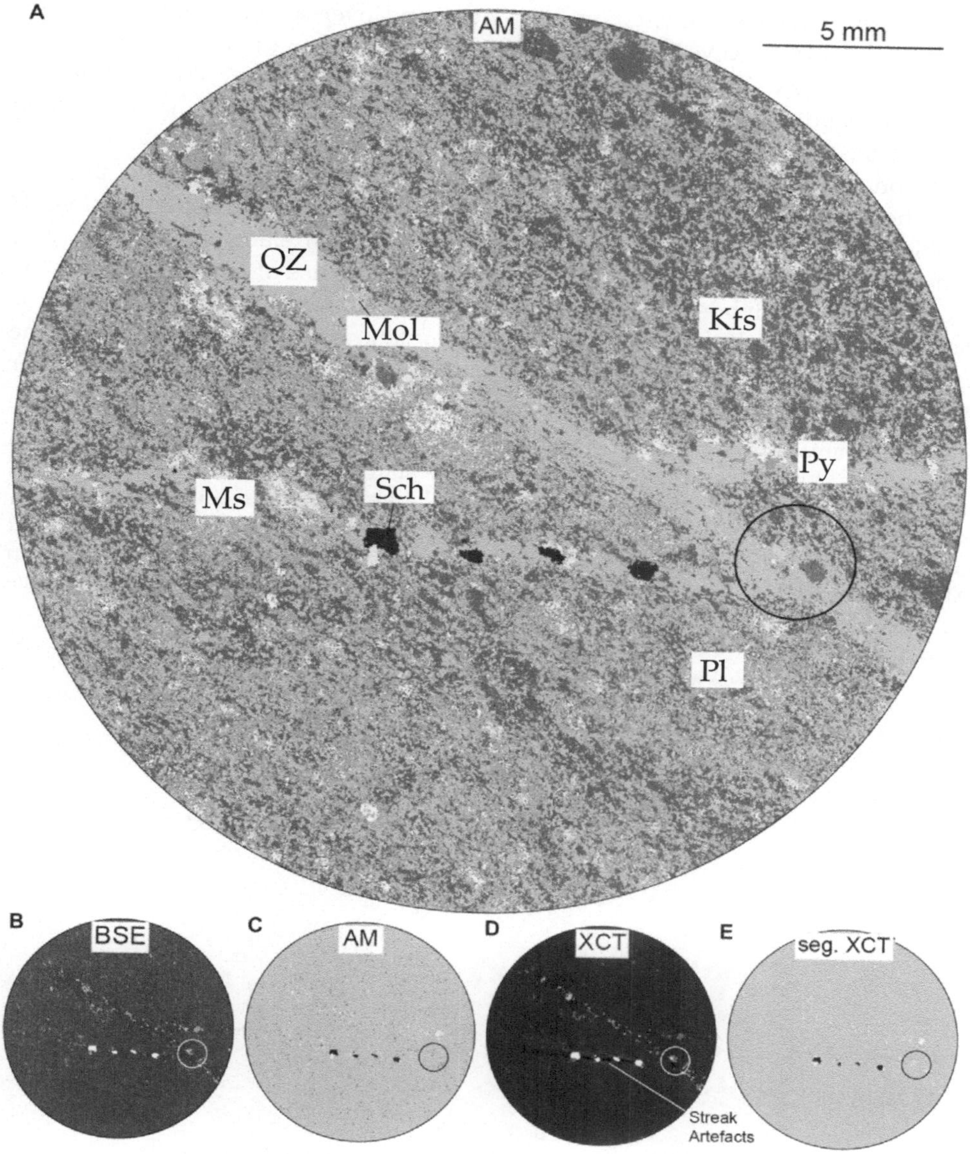

Figure 7. AM and XCT images of a polished rock chip of scheelite- and molybdenite-mineralized quartz veins in an aplite dyke (sample 3). For color coding of the minerals, see Table 2. (A) 30 μm pixel size AM mineral map. The black circle indicates the overlapping region between the 1 μm pixel size AM mineral map and the 4 μm voxel size XCT scan (see Figure 9). (B) 1.87 μm pixel size AM BSE image. (C) 30 μm pixel size simplified AM mineral map. (D) 30 μm voxel size XCT scan. The scheelite grains cause some streak artefacts. (E) 30 μm voxel size segmented XCT scan. Some of the silicates/carbonates in contact with scheelite are segmented as sulfides/oxides due to the streak artefacts but overall impact is minor. Abbreviations: Kfs – k-feldspar, Mol – molybdenite, Ms – muscovite, Pl – plagioclase, Py – pyrite, Sch – scheelite, Qz – quartz.

Only the disseminated sulfides can be segmented from the silicates. Yet, comparison between the mineral map and the top slice of the XCT data show a considerable under-segmentation of sulfides and over-segmentation of silicates (Figure 7, Table 4). Nevertheless, the veins and their orientation in the sample are clearly visible. XCT data confirms the separation of scheelite and molybdenite into two veins and shows the extent of the mineralization with depth in the sample (Figure 8). XCT also shows that the veins are dipping in the same direction. In addition, individual segmentation of both scheelite and molybdenite is possible and the quantitative difference between AM and XCT is similar for both minerals (Table 4). The main errors for scheelite come from streak artefacts and a pronounced partial volume effect at grain boundaries (despite the use of a Sobel filter) in the XCT data. For molybdenite, errors are especially miss-segmentation of fine grains as silicates and sulfides. This is because blurring is more pronounced in fine molybdenite grains (Hanna and Ketcham 2017), which therefore appear darker than larger molybdenite grains in the XCT images.

Table 4. Comparison of the area percentage of the classes segmented in the rock chip and mini core of samples 3 and 4 between the AM mineral maps and the corresponding slice of the different XCT scans. Additionally, the maximal difference in percent between the XCT slice used for comparison with AM and the following four slices is shown. Average and Max values refer to all slices within a specific segmented volume (rock chip, mini core) of an XCT data set. The XCT values are the mean values of six segmentations (n = 6) of the same volume, with the standard deviation in brackets

Rock Chip (Sample 3)	**Scheelite (vol%)**	**Molybdenite (vol%)**	**Sulfides/Oxides (vol%)**	**Silicates/Carbonates (vol%)**	
30 µm AM	0.26	0.37	0.88	98.1	
30 µm XCT	0.22 (0.004)	0.30 (0.041)	0.71 (0.11)	98.8 (0.11)	
Average	0.08 (0.0004)	0.33 (0.028)	0.68 (0.068)	98.9 (0.058)	
Max	0.38 (0.0004)	0.51 (0.034)	0.94 (0.090)	99.2 (0.047)	
Max. Difference in % (5 Slices)	+9.47 (1.38)	+17.1 (5.11)	+5.39 (1.40)	+0.10 (0.055)	
	−10.6 (1.74)	−4.88 (4.87)	−9.40 (5.11)	−0.087 (0.016)	
Relative Difference in % (30 µm AM)	−15.1 (1.73)	−17.3 (11.2)	−19.7 (11.9)	0.71 (0.12)	
Mini Core (Sample 3)					
1 µm AM	0.06	4.69	0.04	94.5	
30 µm AM	0.07	4.48	0.08	94.5	
Relative Difference in % (1 µm AM)	10.0	−4.47	93.4	0.02	
4 µm XCT	0.043 (0.009)	3.25 (0.027)	0.81 (0.30)	95.9 (0.31)	
Average	0.70 (0.0008)	1.71 (0.021)	0.56 (0.077)	97.5 (0.092)	
Max	2.25 (0.003)	8.25 (0.066)	2.09 (0.36)	100.0 (0.006)	
Max. Difference in % (5 Slices)	+37.1 (24.0)	+41.3 (0.27)	+0	+0	
	−2.70 (4.96)	−0	−52.6 (7.25)	−0.97 (0.20)	
Relative Difference in % (1 µm AM)	−33.1 (14.0)	−30.6 (0.58)	1983 (780)	1.48 (0.33)	
30 µm XCT	0	2.82 (0.37)	3.95 (0.29)	93.2 (0.27)	
Average	0.30 (0.004)	1.20 (0.11)	1.26 (0.11)	97.2 (0.08)	
Max	2.13 (0.013)	7.52 (0.59)	4.94 (0.60)	100.0 (0.005)	
Max. Difference in % (5 Slices)	+0	+9.77 (1.34)	+2.84 (3.47)	+3.33 (0.30)	
	−0	−45.7 (4.36)	−46.1 (5.32)	−0.40 (0.11)	
Relative Difference in % (1 µm AM)	−100	−39.8 (7.91)	10,011 (747)	−1.34 (0.29)	
Rock Chip (Sample 4)	**Scheelite (vol%)**	**Molybdenite (vol%)**	**Sulfides/Oxides (vol%)**	**Silicates/Carbonates (vol%)**	**Quartz/Albite (vol%)**
30 µm AM	<0.01	0.34	0.49	56.0	42.7
30 µm XCT	0	0.32 (0.021)	0.79 (0.079)	53.9 (1.92)	45.0 (1.96)
Average	0.017 (0.00009)	0.20 (0.014)	0.42 (0.044)	45.2 (1.49)	54.2 (1.50)
Max	0.11 (0)	0.43 (0.015)	1.04 (0.069)	55.7 (1.16)	77.7 (1.47)
Max. Difference in % (5 Slices)	+0	+7.26 (0.39)	+0	+1.81 (1.08)	+0.068 (0.18)
	−0	−6.45 (4.05)	−13.6 (2.94)	−0.024 (0.055)	−2.00 (1.14)
Relative Difference in % (30 µm AM)	−100	−5.99 (6.14)	61.0 (16.2)	−3.78 (3.43)	5.33 (4.59)

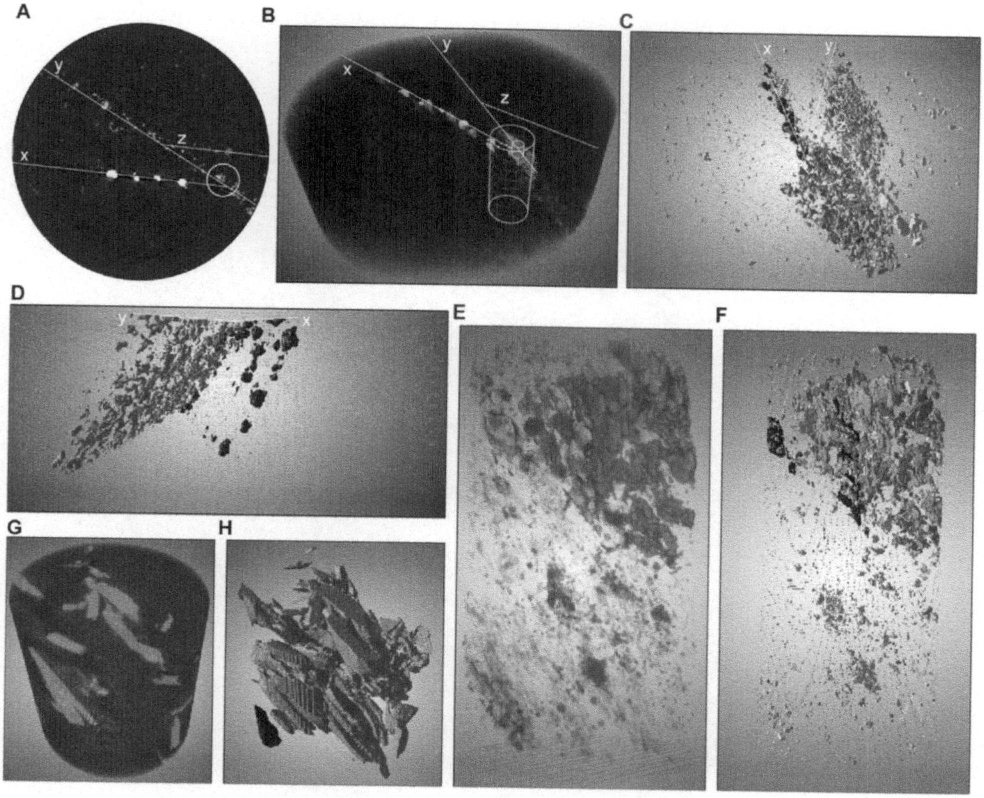

Figure 8. XCT images of the rock chip and mini core of sample 3. For color coding of the minerals, see Table 2. (A) Top slice of the sample showing quartz veins with mineralization of scheelite (x), molybdenite (y), and sulfides (z) in an aplite dyke. (B) 3D-rendered XCT image (30 μm voxel size) of the rock chip. The white cylinders indicate the position of the mini core and the subvolume within. (C) 3D segmented volumes of scheelite, molybdenite, and sulfides. All veins dip to the right and show mineralization extending with depth in the sample. (D) Different view of (C) without the sulfides. The control of the veins on mineralization and the contact between the veins are clearly displayed. (E) 3D-rendered XCT image (4 μm voxel size) of the mini core (stripped of silicates/carbonates). (F) 3D segmented volumes of scheelite, molybdenite, and sulfides in the mini core. Sulfides in the upper, molybdenite-dominated part are mostly wrongly segmented molybdenite grains. (G) 3D-rendered XCT subvolume (1 μm voxel size) within the mini core. (H) 3D segmented volumes of scheelite and molybdenite in the subvolume of the mini core. The molybdenite grains are flaky but aligned with the vein. Hence, grains appear as thin and elongated in the 2D sections.

The 1 μm pixel size scan by AM was focused on an area of molybdenite mineralization at the intersection of the molybdenite-bearing quartz vein with the scheelite-bearing quartz vein. In several molybdenite grains, bright inclusions (a few micron large) are visible in the BSE image (Figure 9). Some grains are identified as hessite associated with native Bi, others escape detection. The topography of many deformed and fractured molybdenite grains is particularly problematic. Uneven surfaces cause low and spurious x-ray counts in the EDS detector (Newbury and Ritchie 2015). To limit misidentification a cut-off value was set at 1,000 x-ray counts per pixel. As a result, many molybdenite grains contain unclassified areas (Figure 9).

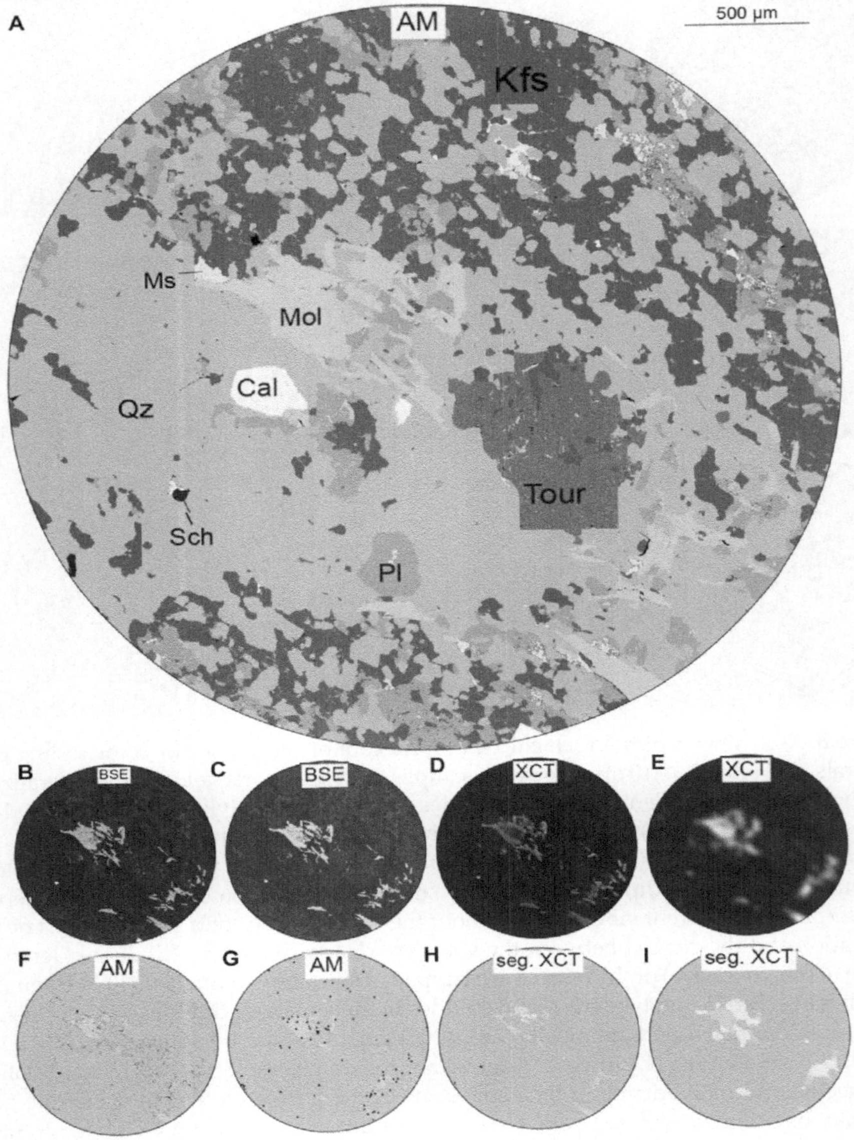

Figure 9. AM and XCT images of a region of interest in the polished rock chip of sample 3. For color coding of the minerals, see Table 2. (A) 1 μm pixel size AM mineral map. Uneven topography in some molybdenite grains caused issues with EDS analysis, rendering some areas in the molybdenite grains unclassified. (B) 250 nm pixel size AM BSE image. (C) 1.87 μm pixel size AM BSE image. (D) 4 μm voxel size XCT scan. (E) 30 μm voxel size XCT scan. (F) 1 μm pixel size simplified AM mineral map. (G) 30 μm pixel size simplified AM mineral map. (H) 4 μm voxel size segmented XCT scan. (I) 30 μm voxel size segmented XCT scan. Abbreviations: Cal – calcite, Kfs – k-feldspar, Mol – molybdenite, Ms – muscovite, Pl – plagioclase, Sch – scheelite, Tour – tourmaline, Qz – quartz.

At the intersection between the quartz veins, three (~40–80 μm) large scheelite grains were detected, but well separated from the molybdenite grains. The XCT data reveal the contact between scheelite and molybdenite mineralization at depth in the sample, but are unsuccessful in resolving heavy mineral inclusions in molybdenite (Figure 8). Streak artefacts and blurring of the scheelite and molybdenite grains affect segmentation of the XCT data. Particularly, grain edges and space between grains in clusters of molybdenite grains are segmented as sulfides due to blurring of bright molybdenite with dark quartz. This causes an extreme over-segmentation of sulfides compared to AM and shows why for example AM is needed to correctly interpret XCT data (Table 4).

The XCT scan (cylinder with 1 mm height and diameter) of 1 μm voxel size of a molybdenite cluster is still hardly of good enough quality to resolve heavy mineral inclusions. Yet their presence in molybdenite of the Liikavaara Östra Cu-(W-Au) deposit is known from AM and prior studies by synchrotron radiation x-ray fluorescence mapping (Warlo et al. submitted). The study of such small-scale features might be difficult to achieve in a lab and could require e.g., synchrotron-based XCT (Wang and Miller 2020). Nevertheless, the high-resolution XCT scan successfully visualizes the flaky texture of molybdenite well (Figure 8). All three XCT scans show molybdenite grains to be predominantly aligned with the vein (Figure 8). For that reason, many molybdenite grains in the surface maps appear as thin elongated grains (cut more or less perpendicular to the crystallographic c-axis) rather than tabular flakes.

Molybdenite-Rich Contact Between a Quartz Vein and Biotite Schist (Sample 4)

AM analysis shows gradually increasing biotitization of the K-feldspar and plagioclase matrix of the biotite schist towards the contact with the quartz vein (Figure 10). It also reveals ubiquitous epidote and tourmaline alteration within the biotite schist. The quartz vein contains disseminated calcite, which extends into the biotite schist. A thin straight calcite veinlet cuts through the quartz vein and continues as a zigzagging chlorite veinlet through the biotite schist. It also cuts disseminated pyrite associated with chalcopyrite (Figure 10). Mineralization of pyrite, chalcopyrite, pyrrhotite, and sphalerite within the quartz vein is strongly associated with calcite, which appears to replace the sulfides. Molybdenite mainly occurs along the contact of the quartz vein with the biotite schist and shows association with calcite but no signs of replacement like the other sulfides. Finally, a single rounded 100 μm large grain of scheelite is found in the quartz vein. In the XCT data, low or missing contrast in the grey values between many minerals results in a very simplified segmentation of five mineral classes (Figure 10, Table 4): 1) quartz and albite, 2) silicates and carbonates, 3) sulfides and oxides, 4) molybdenite, and 5) scheelite. This means that most of the features observed by AM concerning the biotite schist cannot be differentiated by XCT (Figure 10). In addition, calcite dissemination within the quartz vein is poorly resolved. However, it is possible to follow the molybdenite mineralization towards depth in the sample along the contact between the quartz vein and biotite schist (Figure 11). Additionally, larger scheelite grains are revealed at depth in the sample (Figure 11). Comparison of the top sample slice with AM shows a relatively good fit of segmented molybdenite, whereas sulfides/oxides are over-segmented (Table 4).

Under-segmentation of silicates/carbonates appears to compensate mainly for this difference (Table 4). The scheelite grain observed by AM is too fine-grained for proper resolution and segmented as quartz/albite.

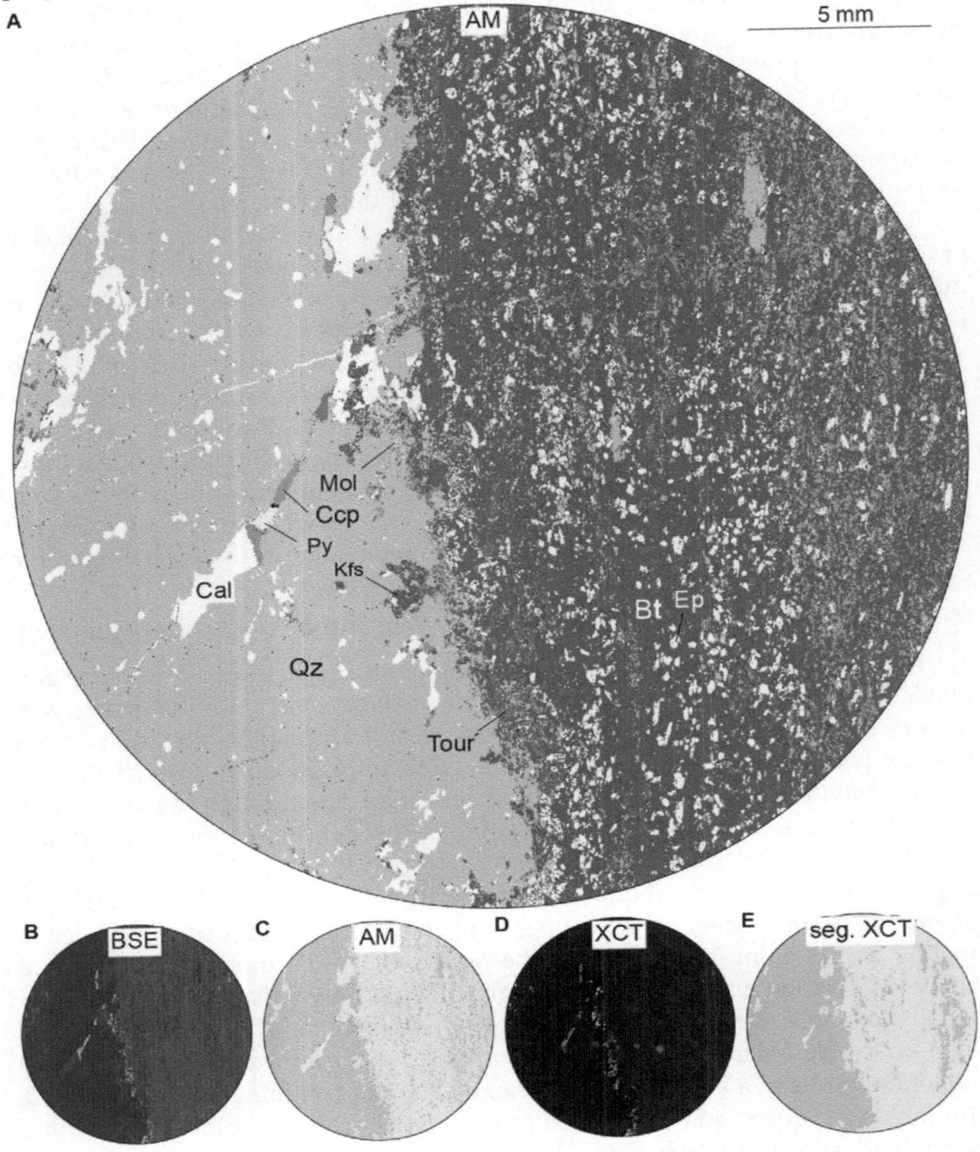

Figure 10. AM and XCT images of a polished rock chip of molybdenite mineralization along the contact between a quartz vein and biotite schist (sample 4). For color coding of the minerals, see Table 2. (A) 30 μm pixel size AM mineral map. (B) 1.87 μm pixel size AM BSE image. (C) 30 μm pixel size simplified AM mineral map. (D) 30 μm voxel size XCT scan. (E) 30 μm voxel size segmented XCT scan. Abbreviations: Bt – biotite, Cal – calcite, Ccp – chalcopyrite, Ep – epidote, Kfs – k-feldspar, Mol – molybdenite, Py – pyrite, Tour – tourmaline, Qz – quartz.

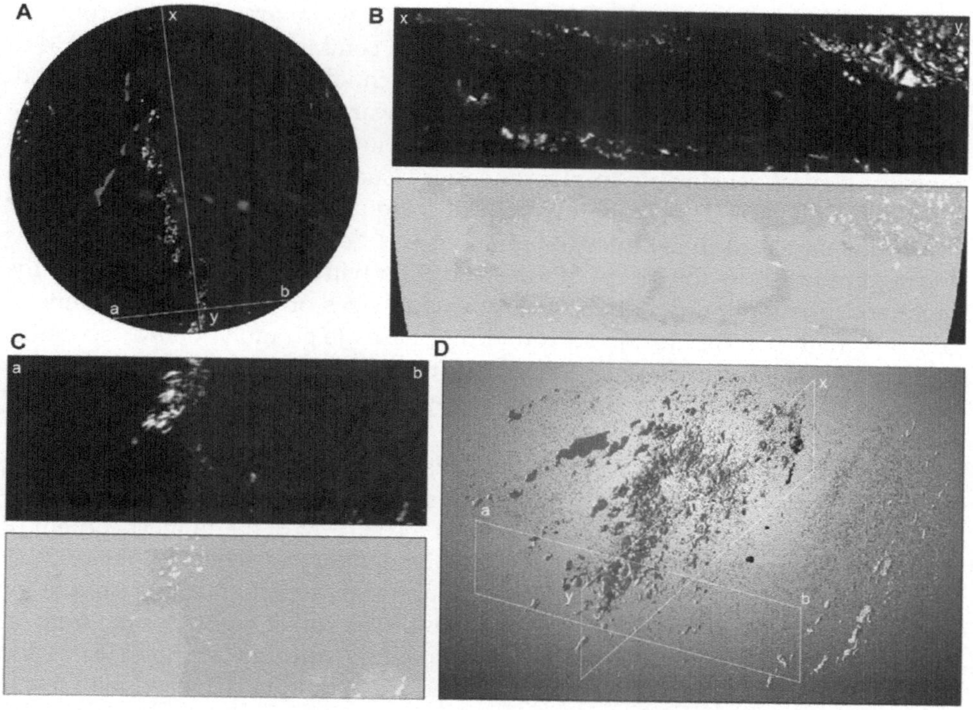

Figure 11. XCT images of the rock chip of sample 4. For color coding of the minerals, see Table 2. (A) Top slice showing the position of two profiles seen in (B,C). (B) Profile along the contact between the quartz vein and biotite schist. X-ray attenuation image (TOP) and segmented image (BOTTOM). (C) Profile across the contact between the quartz vein and biotite schist. X-ray attenuation image (TOP) and segmented image (BOTTOM). (D) 3D segmented volumes of molybdenite and sulfides. Volumes of silicates/carbonates and quartz/albite are indicated by dots.

DISCUSSION

The results reveal several challenges with the correlation and integration of AM and XCT data. Yet they also show how multi-scale analysis of AM and XCT may aid in ore characterization. Below both aspects are discussed.

Correlation and Integration of Automated Mineralogy and X-Ray Computed Tomography Data

Linking AM and XCT has a strong potential in (ore) geology to improve quantitative and textural information on a rock sample (Butcher 2020). XCT is already applied in combination with x-ray diffraction microtomography (e.g., Artioli et al., 2010; Valentini et al., 2011; Mürer et al., 2018; Takahashi and Sugiyama 2019) and x-ray fluorescence microtomography (e.g., Wildenschild and Sheppard 2013; Laforce et al., 2017; Suuronen and Sayab 2018), although mainly constrained to synchrotron facilities and very small sample sizes. In contrast, the recently developed Oreexplore GeoCore X10 merges XCT with x-ray fluorescence analysis for drill core pieces up to 4 × 1 m but limited to a maximal voxel size of 200 μm (Bergqvist et al., 2019). The combination of these analytical techniques in a single system is possible since all techniques utilize an x-ray beam. In contrast, AM is an electron beam-based technique. Thus, a combination of XCT and AM in a single system is impossible with the current technological knowledge. Instead, research and development have to focus on software solutions to improve cross-system integration. So far commercially available integration of XCT and AM only concerns geo-referencing of samples for easier visual correlation and presentation (e.g., Zeiss Atlas 5). This is mainly helpful for qualitative comparison, e.g., for the XCT data to aid in 3D interpretations of AM data. In this study, XCT scans of ore samples show the extent of mineralization with depth in the samples, the angle and orientation of veins, and shapes and sizes of well-separated mineral grains of high contrast to the surrounding phases (e.g., molybdenite, scheelite, heavy minerals). Yet, the segmentation and study of gangue minerals like silicates and carbonates in XCT datasets is challenging, due to too low grey value contrast between the minerals. While in this study AM analysis was mainly performed prior to XCT analysis, results suggest XCT also to be useful for targeted sampling for 2D analysis. In none of the XCT analyses, the topmost slice (the polished surface, which was also analyzed by AM) is the slice with the highest content of the minerals of interest (heavy minerals, molybdenite, and scheelite; Tables 3, 4). Hence, if the goal was to maximize the amount of these minerals in a sample for AM, the XCT scan data should be used to guide the sub-sampling of the sample. This would be especially helpful to better quantify trace metal minerals, as the required number of grains for statistical robustness could be achieved with fewer samples (Goodall and Scales 2007; Godel 2013).

On the other hand, integrating 2-dimensional AM data and 3-dimensional XCT data for quantitative analysis is not straightforward. The reliance on dedicated software to process 2-dimensional AM data is a challenge when merging with 3-dimensional XCT data. The AM software used in this study (Mineralogic Explorer) calculates valuable mineral parameters like mineral association, grain size distribution, metal deportment, etc. but only for the entire AM scan. Obtaining this information for a region of interest in the scan, e.g., for better fitting with the XCT data, is not possible. Further, the software is limited to the processing of AM data. AM and XCT images are easily overlaid in a graphical software but at the cost of losing the functionalities of the AM software. In addition, success in overlaying the images can depend on the quality of the stitching of the montaged AM mineral map. In this study, multiple steps were required to integrate the AM mineral maps into the XCT processing software (ORS Dragonfly) for geo-referencing and quantitative analysis. Currently, the best way to address data integration between AM and XCT seems to be through programming of dedicated algorithms in e.g., MatLab (Reyes et al., 2017; Guntoro et al., 2019b). However, this requires proficiency in programming. Furthermore, software is not the only challenge concerning quantitative correlation.

In this study, one issue of XCT analysis was x-ray attenuation artefacts at the top of the samples (cone-beam effect; Cnudde and Boone 2013). To prevent these artefacts, the epoxy mounted rock samples were stacked with their surfaces facing each other during XCT analysis. However, instead dense minerals caused artefacts into adjacent samples. Thus, to resolve this issue the topmost slices were excluded from segmentation. Consequently, comparison with AM of the same exact surface was hardly possible. To assess the impact of a potential slight vertical offset between the AM mineral maps and topmost segmented XCT slices on quantitative comparison, the relative variation in bulk mineralogy across the five topmost segmented XCT slices was calculated in all scans (Tables 3, 4). Results indicate that the variation in bulk mineralogy is generally smaller between XCT slices than between AM and XCT for major and minor phases. This is to say, segmentation plays a larger role in the quantitative differences between AM and XCT than a potential vertical offset. Yet, for trace phases, especially the heavy mineral phase in sample 2, variation between slices is larger than between AM and XCT. This is due to the small grain size of the heavy minerals, which are partly not detectable across more than one slice. To improve correlation a better approach may be to cut and polish a rock sample after XCT analysis, to perform AM on a surface corresponding to an inner slice in the XCT data, unaffected by the cone-beam effect. An even better alternative may be to try to remove the cone-beam effect altogether. This is achieved with a helical scanning trajectory (De Witte 2010; Cnudde and Boone 2013) or a multiscan with top and bottom scans centered on the edges of the sample.

A 30 μm step size was chosen for both AM and XCT analysis for better modal comparison. Yet, effective resolution differed significantly. For AM, a 30 μm pixel size was sufficiently large not to cause overlap of the beam interaction volume with more than one pixel. In contrast, in XCT analysis the impact of mineral grains on the calculated x-ray linear attenuation exceeded their own voxels due to instrument-related resolution limitations (Kyle and Ketcham 2015; Hanna and Ketcham 2017). This blurring resulted in a relatively worse resolution of XCT than AM. To achieve similar resolution, a better XCT instrument would need to be used and/or the step size of XCT analysis reduced. Yet, the latter would entail a reduction in scanning volume also (Kyle and Ketcham 2015). Moreover, other artefacts like beam hardening, particularly for high-Z mineral grains, would persist (Kyle and Ketcham 2015). Hence, a perfect correlation of XCT data with AM data is hardly possible.

The possibility to differentiate between individual minerals is much lower for XCT than AM due to the nature of the techniques (x-ray linear attenuation coefficients vs. energy dispersive x-ray spectra coupled with backscattered electron intensity). Thus, AM mineral maps require simplification for correlation with the segmented mineral classes of XCT. This simplification may be based on visual comparison of the data sets and calculations on the x-ray linear attenuation coefficients for the various minerals quantified by AM. The approach works well for coarse-grained minerals, easily identified in the data of both techniques, but is less reliable for fine-grained minerals, which are blurred and lack distinct grain boundaries in the XCT images. Furthermore, in this study, molybdenite grains are significantly darker than scheelite grains in all analyses of this study, regardless of size and shape, despite a higher calculated x-ray linear attenuation coefficient. This creates ambiguity regarding which segmented class some minerals belong to.

Reliable mineral segmentation of XCT data, particularly of complex geological samples, is challenging, as described in literature (e.g., Godel 2013; Guntoro et al., 2019a, 2019b; Wang and Miller 2020), but also demonstrated by the results of this study. The challenge starts with the generation of high-quality images used for segmentation: 1) the co-occurrence of minerals with largely different x-ray linear attenuation coefficients makes choosing a beam energy optimized to separation of all minerals impossible. The same applies to selecting of filters and beam hardening corrections post analysis (Herman 1979; Kyle and Ketcham 2015; Hanna and Ketcham 2017). Rocks abundant in heavy minerals require high beam energies for adequate x-ray penetration. This lowers the x-ray attenuation contrast between many minerals. In addition, analytical settings to minimize streak artefacts of highly attenuating minerals like scheelite lower overall beam intensity, thus requiring longer scan times (Gusenbauer et al., 2016). Post-analysis metal artefact reduction methods exist but are not yet common in industrial XCT analysis (Gusenbauer et al., 2016). Combining the results from multiple scans at varied beam energies may help to alleviate some of the issue concerning the x-ray linear attenuation coefficient but comes with a cost of increased scan time. 2) Choosing a voxel size to resolve features of interest adequately while minimizing scan time (and thereby cost of analysis) is difficult in rocks of heterogeneous grain size. Fine intergrowths of minerals with small differences in attenuation coefficient easily become visually homogenized through blurring. A small voxel size may not be cost effective due to long scan times, but too large a voxel size carries the risk of small grains to be unresolved. Additionally, scan volume limits the available scan resolution (Kyle and Ketcham 2015). 3) The edge contrast between mineral grains, especially for minerals of similar x-ray attenuation coefficient (e.g., silicates), is less pronounced than between e.g., grain and air, which makes segmentation more difficult in low-porosity rocks. 4) Minerals may exhibit a large range of grey scale values, even if chemical composition is uniform, based on grain size and associated minerals due to blurring (Kyle and Ketcham 2015). Other artificial grey scale variations (e.g., gradients, partial volumes, streaks, shadows, and rings) cause overlap in the grey scale value ranges of minerals of similar x-ray linear attenuation coefficient, which again constrains segmentation.

Recently, advancements have been made to improve segmentation of XCT data through matching of slices in the XCT data with corresponding AM quantitative maps via a MatLab algorithm, which are then used to determine global thresholds for segmentation (Reyes et al., 2017) or as training data for machine learning algorithms (Guntoro et al., 2019b; Guntoro et al., 2020). Results were shown to be superior to traditional segmentation techniques like grey scale thresholding or watershed segmentation (Reyes et al., 2017; Guntoro et al., 2019b). This integration of XCT and AM is promising and will perhaps help to routinize the use of XCT in ore geology and the mining industry in the future. However, the complexity of the method, the necessity of proficient programming skills, and a relatively long post-processing time per sample are still an obstacle to broad application. In this study, especially the segmentation of chalcopyrite and pyrite in the XCT scan of the mini core of sample 2 and molybdenite in the rock chips of samples 3 and 4 could benefit from the use of machine learning. Concerning chalcopyrite and pyrite, they are clearly differentiable with the naked eye in the 80 kV scan, yet histographic segmentation is unsatisfactory (Figure 6). The slight grey scale gradient from the center to the sidewalls of the cylinder causes overlap of grey scale values between the minerals (cupping effect; Cnudde and Boone 2013). Additionally, the noise in the images prevents the sobel filter from properly resolving grain boundaries between pyrite and chalcopyrite despite sharpening and denoising of the XCT images. Matching of the characteristic cubic shaped pyrite grains between the XCT data and AM may however overcome these hurdles. Similarly, molybdenite grains have rather distinct elongated shapes, which may support discrimination against other sulfides of similar x-ray linear attenuation coefficient. In this study, especially the decrease in brightness of molybdenite with grain size leads to false segmentation of small molybdenite grains as sulfides/oxides and silicates/ carbonates.

Multi-Scale Analysis

For every type of microanalysis method, time-to-result (and therefore the cost), has to be weighed up against the resolution needed to resolve the textures and mineral chemistry, which in turn may impact the quantity and quality of information gained. As a consequence, a common approach is sequential analysis – where the resolution is increased but at the same time the area scanned is made progressively smaller. This approach of 'zooming in' on a region of interest was also used in this study.

In the 30 μm pixel size AM mineral map of the rock chip of sample 2, several Au and Ag minerals are detected, but generally not in sizes larger than 1 or 2 pixels (Figure 5). The subsequent 1 μm pixel size scan on a much smaller area reveals numerous grains of Au, Ag, and Bi that are undetected in the 30 μm resolution scan (Figure 6). Thus, the 1 μm resolution provides much better information on parameters like grain size distribution, mineral association, etc. Obviously, a 1 μm pixel size scan of the entire polished surface of the rock chip would be ideal, yet it would consume an unreasonable amount of time (several days) to complete. In addition, if and to what extend identification of trace metal minerals could be expected would be purely based on observations of e.g., petrographic microscopy (often too low resolution) or manual SEM-EDS (mainly spot-checking, possibly time-consuming). Hence, the coarser but faster 30 μm scan serves as a good overview and allows targeted high-resolution analysis. Similar results are obtained by XCT, where with each scan at a smaller voxel size heavy mineral grains that the prior scans fail to resolve are detected (Figures 4, 6). Generally, coarser grains account for the majority of the bulk metal in an ore deposit and fine grains, even in high numbers, accumulate to only a small overall volume. In mining, loss of the finest-grained fractions is commonly accepted, partly due to missing tools and added costs for their recovery. However, for economic trace metals like Au, the fine-grained fractions can be substantial and loss to tailings a significant loss in potential profit (Vaughan 2004). Hence, these high-resolution analyses are needed to provide the necessary knowledge to develop production strategies.

Similarly, the study of molybdenite grains in the rock chip and mini core of sample 3 shows how high-resolution analyses are necessary to quantify micro- and nano-inclusions in molybdenite grains. Even the 1 μm resolution scans of AM and XCT only managed to resolve a small portion of the inclusions. Although the effect of these inclusions on processing of molybdenite requires more research, a study by Triffett et al. (2008) suggests foreign mineral grains trapped in molybdenite have potentially a negative effect on flotation.

CONCLUSION

XCT can aid AM in various ways: 1) The technique is well suited to provide three-dimensional visualization of textures and structures at centimeter-to micron-scale. The separation of molybdenite and scheelite mineralization onto two different quartz veins observed by AM in this study was confirmed to extend with depth in the sample. Furthermore, the XCT data revealed similar dipping angles and dip directions for the two quartz veins. In contrast, the seemingly straight contact between a quartz vein and biotite schist observed by AM was shown to be more rugged with depth by XCT. 2) XCT can help assess the stereological error of 2D analysis, although in this study it was limited to qualitative assessment of only well-separated high-contrast grains (e.g., Bi-minerals and scheelite). The stereological error was not quantified as no proper grain separation of the segmented XCT data was performed, e.g. molybdenite clusters were registered as a single grain rather than an accumulation of grains. Grain separation is possible but very time consuming and was not the focus of this study. 3) XCT can increase data volume on rare minerals to quickly achieve statistical robustness and guide (sub)-sample selection. In this study, heavy trace minerals in a single XCT slice constituted only a tiny fraction of the amount of heavy trace minerals within the entire volume of an XCT scan. Furthermore, the XCT slice corresponding to an AM mineral map was never the slice with the highest amount of heavy trace minerals. 4) XCT can be performed at various scales and allows zooming in onto an area of interest. In this study, successive sub-sampling (by e.g., mini-coring) allowed XCT analysis of a 1 × 1 mm cylindrical volume at 1 μm voxel size from within a ~20 cm long drill core piece, previously scanned at 45 μm voxel size. Further, despite the destruction caused by the sub-sampling, the spatial information of the rock samples was preserved in the XCT data to allow interpretation of the data in its original spatial context.

AM, on the other hand, can also guide subsampling for high-resolution XCT (e.g., drilling of mini cores) and help with the segmentation of the XCT data. Yet, quantitative integration of XCT and AM data remains challenging and is restricted by the quality of segmentation of the XCT data. This is particularly true for low-porosity rock samples with complex mineralogy like in this study. Despite this, recent studies look promising for improving segmentation and thereby gaining more quantitative mineral analysis of XCT data. Overall, linking XCT and AM provides a better characterization of an ore than applying only one technique, or using them separately (Warlo et al., Forthcoming 2021).

AUTHOR CONTRIBUTIONS

MW, AB, GB, and CW designed the study. MW collected the samples for analysis and performed the mini drilling. MW performed the Automated Mineralogy analysis including processing of the data. JK did the x-ray computed tomography analysis at the Geological Survey of Finland including corrections to the data post analysis. FF and HL performed the x-ray computed tomography analysis at Luleå University of Technology and also corrected the data post analysis. MW processed and segmented the x-ray computed tomography data in Dragonfly and did the correlation with the Automated Mineralogy data. MW compiled the results and wrote the manuscript. GB, CW, AB, JK, FF, and HL provided feedback to the manuscript.

ACKNOWLEDGMENTS

Ignacio Gonzalez-Alvarez is thanked for constructive editorial work and three reviewers for significantly improving the manuscript. The authors would also like to thank Boliden Mineral AB for permitting the study on rock samples of the Liikavaara Östra Cu-(W-Au) deposit. An earlier version of this manuscript has previously appeared in the doctoral thesis by Warlo (2021).

REFERENCES

1. Artioli, G., Cerulli, T., Cruciani, G., Dalconi, M. C., Ferrari, G., Parisatto, M., et al. (2010). X-ray Diffraction Microtomography (XRD-CT), a Novel Tool for Non-invasive Mapping of Phase Development in Cement Materials. *Anal. Bioanal. Chem.* 397 (6), 2131–2136. doi:10.1007/s00216-010-3649-0
2. Baker, D. R., Mancini, L., Polacci, M., Higgins, M. D., Gualda, G. A. R., Hill, R. J., et al. (2012). An Introduction to the Application of X-ray Microtomography to the Three-Dimensional Study of Igneous Rocks. *Lithos* 148, 262–276. doi:10.1016/j.lithos.2012.06.008
3. Bam, L., Miller, J., and Becker, M. (2020). A Mineral X-ray Linear Attenuation Coefficient Tool (MXLAC) to Assess Mineralogical Differentiation for X-ray Computed Tomography Scanning. *Minerals* 10 (5), 441. doi:10.3390/min10050441
4. Berger, M. J., Hubbell, J. H., Seltzer, S. M., Chang, J., Coursey, J. S., Sukumar, R., et al. (2010). XCOM: Photon Cross Sections Database. *NIST Stand. Reference Database* 8 (XGAM), NBSIR 87–3597. doi:10.18434/T48G6X
5. Bergqvist, M., Landström, E., Landström, E., and Luth, S. (2019). Access to Geological Structures, Density, Minerals and Textures through Novel Combination of 3D Tomography, XRF and Sample Weight. *ASEG Extended Abstr.* 2019 (1), 1–3. doi:10.1080/22020586.2019.12073146
6. Butcher, A. R. (2020). Upscaling of 2D Mineralogical Information to 3D Volumes for Geoscience Applications Using a Multi-Scale, Multi-Modal and Multi-Dimensional Approach. *IOP Conf. Ser. Mater. Sci. Eng.* 891 (1), 012006. doi:10.1088/1757-899X/891/1/012006
7. Cnudde, V., and Boone, M. N. (2013). High-resolution X-ray Computed Tomography in Geosciences: A Review of the Current Technology and Applications. *Earth-Science Rev.* 123, 1–17. doi:10.1016/j.earscirev.2013.04.003
8. De Man, B., Nuyts, J., Dupont, P., Marchal, G., and Suetens, P. (1998). Metal Streak Artifacts in X-ray Computed Tomography: a Simulation Study. *IEEE Nucl. Sci. Symp. Conf. Rec.* 3, 1860–1865. doi:10.1109/NSSMIC.1998.773898
9. De Witte, Y. (2010). *Improved and Practically Feasible Reconstruction Methods for High Resolution X-ray tomography*. Doctoral Thesis. Ghent, Belgium: Ghent University, 270.
10. European Commission (2020). *Critical Raw Materials Resilience: Charting a Path towards Greater Security and Sustainability*. Luxembourg: EU Publications, 23. COM 474 final.
11. Godel, B. (2013). High-resolution X-ray Computed Tomography and its Application to Ore Deposits: From Data Acquisition to Quantitative Three-Dimensional Measurements with Case Studies from Ni-Cu-PGE Deposits. *Econ. Geology.* 108 (8), 2005–2019. doi:10.2113/econgeo.108.8.2005

12. Goodall, W. R., and Scales, P. J. (2007). An Overview of the Advantages and Disadvantages of the Determination of Gold Mineralogy by Automated Mineralogy. *Minerals Eng.* 20 (5), 506–517. doi:10.1016/j.mineng.2007.01.010

13. Guntoro, P. I., Ghorbani, Y., Butcher, A. R., Kuva, J., and Rosenkranz, J. (2020). Textural Quantification and Classification of Drill Cores for Geometallurgy: Moving toward 3D with X-ray Microcomputed Tomography (MCT). *Nat. Resour. Res.* 29 (6), 3547–3565. doi:10.1007/s11053-020-09685-5

14. Guntoro, P. I., Ghorbani, Y., Koch, P.-H., and Rosenkranz, J. (2019a). X-ray Microcomputed Tomography (MCT) for Mineral Characterization: A Review of Data Analysis Methods. *Minerals* 9 (3), 183. doi:10.3390/min9030183

15. Guntoro, P. I., Tiu, G., Ghorbani, Y., Lund, C., and Rosenkranz, J. (2019b). Application of Machine Learning Techniques in mineral Phase Segmentation for X-ray Microcomputed Tomography (MCT) Data. *Minerals Eng.* 142, 105882. doi:10.1016/j.mineng.2019.105882

16. Gusenbauer, C., Reiter, M., Salaberger, D., and Kastner, J. (2016). "Comparison of Metal Artefact Reduction Algorithms from Medicine Applied to Industrial XCT Applications." Proceedings of 19th World Conference on Non-Destructive Testing (WCNDT 2016), Munich, June 2016. IEEE, 13–17.

17. Hanna, R. D., and Ketcham, R. A. (2017). X-ray Computed Tomography of Planetary Materials: A Primer and Review of Recent Studies. *Geochemistry* 77 (4), 547–572. doi:10.1016/j.chemer.2017.01.006

18. Herman, G. T. (1979). Correction for Beam Hardening in Computed Tomography. *Phys. Med. Biol.* 24 (1), 81–106. doi:10.1088/0031-9155/24/1/008

19. Ketcham, R. A., and Carlson, W. D. (2001). Acquisition, Optimization and Interpretation of X-ray Computed Tomographic Imagery: Applications to the Geosciences. *Comput. Geosciences* 27 (4), 381–400. doi:10.1016/S0098-3004(00)00116-3

20. Kyle, J. R., and Ketcham, R. A. (2015). Application of High Resolution X-ray Computed Tomography to mineral deposit Origin, Evaluation, and Processing. *Ore Geology. Rev.* 65, 821–839. doi:10.1016/j.oregeorev.2014.09.034

21. Laforce, B., Masschaele, B., Boone, M. N., Schaubroeck, D., Dierick, M., Vekemans, B., et al. (2017). Integrated Three-Dimensional Microanalysis Combining X-ray Microtomography and X-ray Fluorescence Methodologies. *Anal. Chem.* 89 (19), 10617–10624. doi:10.1021/acs.analchem.7b03205

22. Lätti, D., and Adair, B. J. I. (2001). An Assessment of Stereological Adjustment Procedures. *Minerals Eng.* 14 (12), 1579–1587. doi:10.1016/S0892-6875(01)00176-5

23. Mürer, F. K., Sanchez, S., Álvarez-Murga, M., Di Michiel, M., Pfeiffer, F., Bech, M., et al. (2018). 3D Maps of mineral Composition and Hydroxyapatite Orientation in Fossil Bone Samples Obtained by X-ray Diffraction Computed Tomography. *Sci. Rep.* 8 (1), 1–13. doi:10.1038/s41598-018-28269-1

24. Newbury, D. E., and Ritchie, N. W. M. (2015). Performing Elemental Microanalysis with High Accuracy and High Precision by Scanning Electron Microscopy/silicon Drift Detector Energy-Dispersive X-ray Spectrometry (SEM/SDD-EDS). *J. Mater. Sci.* 50 (2), 493–518. doi:10.1007/s10853-014-8685-2

25. Reich, M., Large, R., and Deditius, A. P. (2017). New Advances in Trace Element Geochemistry of Ore Minerals and Accessory Phases. *Ore Geology. Rev.* 81, 1215–1217. doi:10.1016/j.oregeorev.2016.10.020

26. Reyes, F., Lin, Q., Udoudo, O., Dodds, C., Lee, P. D., and Neethling, S. J. (2017). Calibrated X-ray Micro-tomography for mineral Ore Quantification. *Minerals*

Eng. 110, 122–130. doi:10.1016/j.mineng.2017.04.015Get

27. Schulz, B., Sandmann, D., and Gilbricht, S. (2020). SEM-based Automated Mineralogy and its Application in Geo- and Material Sciences. *Minerals* 10 (11), 1004. doi:10.3390/min10111004
28. Sutherland, D. (2007). Estimation of mineral Grain Size Using Automated Mineralogy. *Minerals Eng.* 20 (5), 452–460. doi:10.1016/j.mineng.2006.12.011
29. Suuronen, J.-P., and Sayab, M. (2018). 3D Nanopetrography and Chemical Imaging of Datable Zircons by Synchrotron Multimodal X-ray Tomography. *Sci. Rep.* 8 (1), 1–13. doi:10.1038/s41598-018-22891-9
30. Takahashi, H., and Sugiyama, T. (2019). Application of Non-destructive Integrated CT-XRD Method to Investigate Alteration of Cementitious Materials Subjected to High Temperature and Pure Water. *Construction building Mater.* 203, 579–588. doi:10.1016/j.conbuildmat.2019.01.128
31. Triffett, B., Veloo, C., Adair, B. J. I., and Bradshaw, D. (2008). An Investigation of the Factors Affecting the Recovery of Molybdenite in the Kennecott Utah Copper Bulk Flotation Circuit. *Minerals Eng.* 21 (12-14), 832–840. doi:10.1016/j.mineng.2008.03.003
32. Valentini, L., Dalconi, M. C., Parisatto, M., Cruciani, G., and Artioli, G. (2011). Towards Three-Dimensional Quantitative Reconstruction of Cement Microstructure by X-ray Diffraction Microtomography. *J. Appl. Cryst.* 44 (2), 272–280. doi:10.1107/S0021889810054701
33. Vaughan, J. P. (2004). The Process Mineralogy of Gold: the Classification of Ore Types. *JOM* 56 (7), 46–48. doi:10.1007/s11837-004-0092-8
34. Wang, Y., and Miller, J. D. (2020). Current Developments and Applications of Micro-CT for the 3D Analysis of Multiphase mineral Systems in Geometallurgy. *Earth-Science Rev.* 211, 103406. doi:10.1016/j.earscirev.2020.103406
35. Warlo, M. (2021). *Improving Trace Metal Characterisation of Ore Deposits Through Multi-Modal, Multi-Scale, and Multi-Dimensional Micro-Analysis*. Doctoral Thesis. Luleå, Sweden: Luleå University of Technology, 144.
36. Warlo, M., Bark, G., Wanhainen, C., McElroy, I., Björling, A., and Johansson, U. (Forthcoming 2021). *Extreme-Resolution Synchrotron X-Ray Fluorescence Mapping of Ore Samples*. Ore Geology Reviews.
37. Warlo, M., Wanhainen, C., Bark, G., Butcher, A. R., McElroy, I., Brising, D., et al. (2019). Automated Quantitative Mineralogy Optimized for Simultaneous Detection of (Precious/Critical) Rare Metals and Base Metals in a Production-Focused Environment. *Minerals* 9, 440. doi:10.3390/min9070440
38. Warlo, M., Wanhainen, C., Martinsson, O., and Karlsson, P. (2020). Mineralogy and Character of the Liikavaara Östra Cu-(w-Au) deposit, Northern Sweden. *GFF* 142 (3), 169–189. doi:10.1080/11035897.2020.1753807
39. Wildenschild, D., and Sheppard, A. P. (2013). X-ray Imaging and Analysis Techniques for Quantifying Pore-Scale Structure and Processes in Subsurface Porous Medium Systems. *Adv. Water Resour.* 51, 217–246. doi:10.1016/j.advwatres.2012.07.018
40. Withers, P. J., Bouman, C., Carmignato, S., Cnudde, V., Grimaldi, D., Hagen, C. K., et al. (2021). X-ray Computed Tomography. *Nat. Rev. Methods Primers* 1 (18), 1–21. doi:10.1038/s43586-021-00015-4

INDEX

R

S

T

U

V

W

X

Z